UNIT-1
General Properties of Plasma

Structure of the Unit

1.0 Objectives

This chapter of Plasma Physics is not simply aimed at people specializing in industrial plasmas or fusion plasmas but also provide a useful background for making understanding or for a broad range of topics in physics. This chapter includes simple understanding about the definition of plasma with fundamental parametrization. Further, the properties of Plasma have been explained followed by the Saha's ionization equation. As the varieties of plasma available in nature as well as generated artificially, thus production of plasma is then become necessary to understand for completeness.

1.1   Introduction

1.1.1 Preliminary

By the end of twentieth century, significance of plasma has been recognized and it is estimated that more than 99% of the matter in the universe exists in a plasma state. Plasma, known as the fourth state of matter, most of the known matter in the universe is in the ionized state. The states of solid, liquid, gas, and plasma represent correspondingly increasing freedom of particle motion. In a solid, the atoms are arranged in a periodic crystal lattice; they are not free to move, and the solid maintains its size and shape. In a liquid the atoms are free to move, but because of strong inter-atomic forces the volume of the liquid (but not its shape) remains unchanged. In a gas the atoms move freely, experiencing occasional collisions with one another. In plasma, the atoms are ionized and there are free electrons moving about – a plasma is an ionized gas of free particles. Here are some familiar examples of plasmas:

1. Lightning, Aurora Borealis, and electrical sparks. All these examples show that when an electric current is passed through plasma, the plasma emits light (electromagnetic radiation).

2. Neon and fluorescent lights, etc. Electric discharge in plasma provides a rather efficient means of converting electrical energy into light.

3. Flame. The burning gas is weakly ionized. The characteristic yellow color of a wood flame is produced by $579nm$ transitions (D lines) of sodium ions.

4. Nebulae, interstellar gases, the solar wind, the earth's ionosphere, the Van Allen belts. These provide examples of a diffuse, low temperature, ionized gas.

5. In Sun, the temperature of the core is about 17 million degrees. Its surface, the photosphere radiates at a temperature of 5700 K and the Corona has a temperature of more than one million degrees. Earth's atmosphere consists of plasma like the ionosphere, the plasmosphere and magnetosphere Terresric plasmas are formed in gas discharges as in lightinings, in sparks, in arcs, fluorescent lamps, arc lamps, plasma displays and plasma torches.

1.1.2 What is Plasma ?

Plasma is an ionized gas, consisting of free electrons, ions and atoms. We understand this with an example that when a solid is heated sufficiently that the thermal motion of the atoms break the crystal lattice structure and liquid is formed. When a liquid is heated enough that atoms vaporize, a gas is formed. When a gas is heated enough that the atoms collide with each other and electrons, ions produced in the process, a plasma is formed, i.e., fourth state of matter. The constituents of these states (solid, liquid, gas) are atoms and molecules. The transformation from one state to another is done by supplying energy, e.g. heat. Further transformation to fourth state (plasma) take place when the gas is further energized by very high temperature or subjected to energetic radiations (Fig. 1.1) results to electrons and ions and neutral atoms. Therefore, this fluid consisting of charged particles and neutral atoms or molecules is called Plasma. An ionized gas has unique properties. In plasma, charge separation between ions and electrons gives rise to electric fields and charged particle flows give rise to current and magnetic fields.

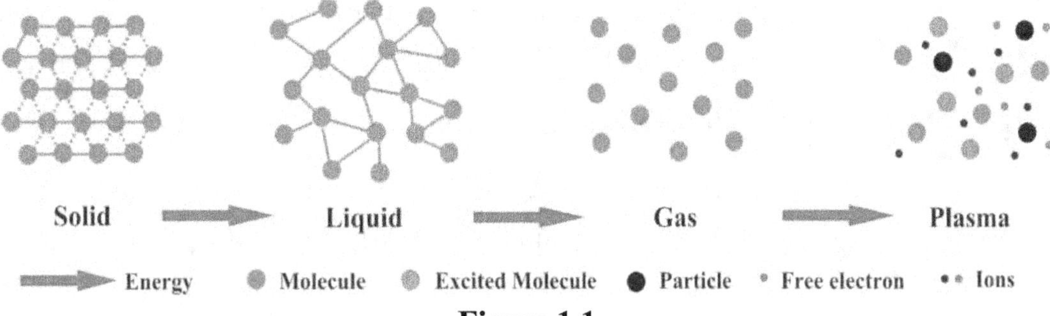

Figure 1.1

3

Plasma plays an essential role in many applications ranging from advanced lighting device and surface treatments for semiconductor applications or surface layer generation. Other applications are controlled fusion research, solar physics, astrophysics, plasma population, ionosphere physics and magnetospheric physics.

Some applications of plasma are followings:

1. Plasma may be kept in confinement and heated by magnetic and electric fields. These basic features of plasmas could be used to build "plasma guns" that eject ions at velocities up to 100 km/sec. Plasma guns could be used in ion rocket engines, as an example.

2. "Plasma motors"- It differs from ordinary motors by having plasma (not metals!) as the basic conductor of electricity. These motors could, in principle, be lighter and more efficient than ordinary motors. Similarly, one could develop "plasma generators" to convert mechanical energy into electrical energy. The whole subject of direct magnetohydrodynamic production of electrical energy is ripe for development. The current "thermodynamic energy converters" (steam plants, turbines, etc.) are notoriously inefficient in generating electricity.

3. Plasma could be used as a resonator or a waveguide, much like hollow metallic cavities, for electromagnetic radiation. Plasma experiences a whole range of electrostatic and electromagnetic oscillations, which one would be able to put to good use.

4. Communication through and with plasma. The earth's ionosphere reflects low frequency electromagnetic waves (below 1 MHz) and freely transmits high frequency (above 100MHz) waves. Plasma disturbances (such as solar flares or communication blackouts during re-entry of satellites) are notorious for producing interruptions in communications. Plasma devices have essentially the same uses in communications, in principle, as semi-conductor devices. In fact, the electrons and holes in a semi-conductor constitute a plasma, in a very real sense.

1.1.3 Definition of Plasma

In Greek ($\pi\lambda\alpha\sigma\mu\alpha$) plasma means 'moldable substance' to denote the clear fluid, which remains after the removal of all corpuscular material in blood. American scientist, Irving Langmuir proposed in 1922 that the electrons, ions and neutrals in an ionized gas could similarly be considered some kind of fluid medium and this medium is called plasma. Any ionized gas can not be called plasma, definition is as follows:

"Plasma is a quasineutral gas of charged and neutral particles which exhibits collective behavior where the meaning of quasineutral is the plasma is neutral enough, i.e. no density of electrons and ions are equal, called as plasma density. The meaning of collective behavior is motions that depend not only on local conditions but on the state of the plasma in remote regions as well".

**APPROXIMATE MAGNITUDES
IN SOME TYPICAL PLASMAS**

Plasma Type	n cm^{-3}	T eV	ω_{pe} sec^{-1}	λ_D cm	$n\lambda_D{}^3$	ν_{ei} sec^{-1}
Interstellar gas	1	1	6×10^4	7×10^2	4×10^8	7×10^{-5}
Gaseous nebula	10^3	1	2×10^6	20	8×10^6	6×10^{-2}
Solar Corona	10^9	10^2	2×10^9	2×10^{-1}	8×10^6	60
Diffuse hot plasma	10^{12}	10^2	6×10^{10}	7×10^{-3}	4×10^5	40
Solar atmosphere, gas discharge	10^{14}	1	6×10^{11}	7×10^{-5}	40	2×10^9
Warm plasma	10^{14}	10	6×10^{11}	2×10^{-4}	8×10^2	10^7
Hot plasma	10^{14}	10^2	6×10^{11}	7×10^{-4}	4×10^4	4×10^6
Thermonuclear plasma	10^{15}	10^4	2×10^{12}	2×10^{-3}	8×10^6	5×10^4
Theta pinch	10^{16}	10^2	6×10^{12}	7×10^{-5}	4×10^3	3×10^8
Dense hot plasma	10^{18}	10^2	6×10^{13}	7×10^{-6}	4×10^2	2×10^{10}
Laser Plasma	10^{20}	10^2	6×10^{14}	7×10^{-7}	40	2×10^{12}

Table1.2

The table and fig. (1.2) given below indicates typical values of the density and temperature for various types of plasmas. It is observed from the table that the density of plasma varies by a factor of 10^{16} and the temperature by 10^4 – this incredible variation is much greater than is possible for the solid, liquid, or gaseous states.

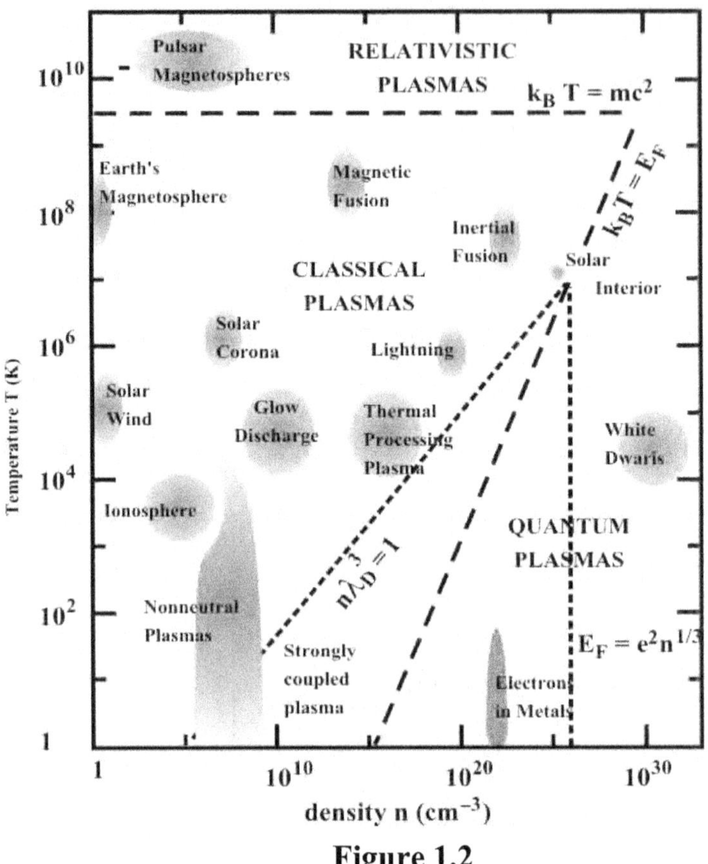

Figure 1.2

1.2 Basic Parameters: Speed, Energy and Temperature

Plasma is a collection of various charged particles which are free to move in response to the fields they generate or to the applied fields and on average it is almost electrically neutral. This implies that the densities of the charged particles (ions and electrons) are almost equal, i.e. $n_i \approx n_e = n_s$ and commonly termed as "quasi-neutrality". This condition exists uniformly throughout the volume of ionized gas except at near the boundaries and valid only when the spatial scale length of plasma is much greater than the characteristic length over which the charges or boundaries are electrically shielded, known as Debye length. Distribution in energy to ions and electrons are characterized by temperature T_i and T_e respectively, which are not found the same usually. In the plasma, the charged particle experiences a large number of collisions with each other or with other species (ions, electrons and/or neutrals) and thus it is not possible to analyze the motion of each particle. Therefore, dynamics of plasma is not then understood completely. Due to the collisions, there is a distribution of the velocities for each species and then each particle will move with a velocity, which is function of the macroscopic temperature and mass of that species.

For ensemble of N particles of mass m and velocity u, the average kinetic energy per particle is

$$\langle E \rangle = \frac{1}{2N} \sum_{i=1}^{N} m_i u_i^2 \,, \tag{1}$$

and the Maxwell – Boltzmann distribution function of speed at thermal equilibrium is given by

$$f(\upsilon) = N \sqrt{\frac{m}{2\pi k_B T}} \exp\left(-\frac{mu^2}{2k_B T} \right), \tag{2}$$

where k_B is the Boltzmann's constant and the width of the distribution is characterized by the temperature T.

The average kinetic energy of a particle in the Maxwellian distribution in one dimension is given by

$$\langle E \rangle = \frac{\int_{-\infty}^{\infty} \frac{mu^2}{2} f(u)\,du}{\int_{-\infty}^{\infty} f(u)\,du} \,. \tag{3}$$

By inserting eq. (2) in eq. (3), and integrating by parts, the average energy per particle in each dimension is

$$\langle E \rangle = \frac{1}{2} k_B T \,. \tag{4}$$

Generalizing the distribution to the three dimensions as

$$f(u,v,w) = N \sqrt{\frac{m}{2\pi k_B T}} \exp\left(-\frac{m(u^2 + v^2 + w^2)}{2k_B T} \right), \tag{5}$$

where u, v, w are the components of velocity in three coordinate axes.

Following the same procedure, the average energy per particle in three dimensions is

$$\langle E \rangle = \frac{3}{2} k_B T \,. \tag{6}$$

Using this, the average KE of a gas at 1K is $\sim 2.07 \times 10^{-23} J$. Typically, the energy corresponding to $k_B T$ used for plasma temperature, $k_B T = 1eV = 1.602 \times 10^{-19} C$.

Further, quasi-neutrality demands that $n_i \cong n_e \equiv n$, then the density of the particles is obtained as

$$n = \iint \int_{-\infty}^{+\infty} n\, f(u)\, du \,, \tag{7}$$

$$n = \iint \int_{-\infty}^{+\infty} n \sqrt{\frac{m}{2\pi k_B T}} \exp\left(-\frac{m(u^2 + v^2 + w^2)}{2k_B T}\right) du\,dv\,dw \,. \tag{8}$$

The kinetic temperature of species is essentially the average kinetic energy of the particles is

$$T = \frac{1}{3} m \left\langle u^2 \right\rangle , \tag{9}$$

where the average speed of a particle (in Maxwellian distribution) is then given by

$$\langle u \rangle = \left(\frac{8k_B T}{\pi m}\right)^{1/2} . \tag{10}$$

Thermal speed of species is then $u_{ts} = \sqrt{2k_B T / m}$ and ion thermal speed is usually smaller than the electron thermal speed: $u_{ti} \approx \sqrt{m_e / m_i}\, u_{te}$.

1.3 General Properties of Plasma

1.3.1 Plasma Oscillations and Frequency

Consider a one dimensional sitaution in which slab consisting entirely of one charge species in quasi neutral state, i.e. $n_i \cong n_e \equiv n$. Displacing the group of electrons by an infinitesimal distance Δx, due to which an electric field $\vec{E}$ sets up across the distance L (see fig. 1.3), which results to the charge density (charge per unit area) on each face of the slab is $\sigma = en\Delta x$ with equal and opposite polarity.

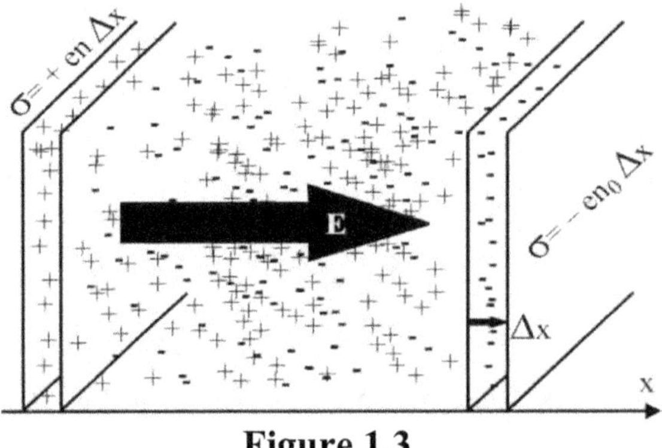

Figure 1.3

Along the x- direction, electric field generated inside the slab is

$$E_x = \frac{\sigma}{\varepsilon_0} = -\frac{ne\Delta x}{\varepsilon_0}. \tag{11}$$

Using Newton's law,

$$m_e \frac{d^2 \Delta x}{dt^2} = eE_x, \qquad \Rightarrow \frac{d^2 \Delta x}{dt^2} = -\frac{ne^2 \Delta x}{m_e \varepsilon_0} = -\omega_P^2 \Delta x, \tag{12}$$

where $\omega_P^2 = \dfrac{ne^2}{m_e \varepsilon_0}$ is known as Plasma Frequency.

Plasma frequency commonly written as

$$f_P = \frac{\omega_P}{2\pi} = 9000\sqrt{n_e}\ Hz. \tag{13}$$

Thus, if a group of electrons in two-component plasma are displaced slightly from their equilibrium position, they will experience a force that returns them to. When they arrive at the equilibrium position, they will have a kinetic energy equal to the potential energy of their initial displacement and will continue past until they reconvert their kinetic energy back to potential energy. The frequency of this simple period harmonic motion is known as plasma frequency. This phenomenon is known as plasma oscillation.

The plasma oscillations will only be observed if the plasma system is studied over a time period longer than plasma time period τ_p as

$$\tau_P = \frac{1}{\omega_P}. $$

1.3.2 Plasma Criteria

In a partially ionized gas where collisions are important, plasma oscillations can occur only if mean free time between the collisions (τ_c) is very much large compare to the oscillation time period (τ_P), i.e.

$$\frac{\tau_C}{\tau_P} \gg 1. \tag{14}$$

This is Plasma Criterion #I, for an ionized gas to be considered as plasma. It behaves like a neutral gas, if criterion does not hold. Plasma oscillation can be as certained by thermal motions of electrons, where the work done in displacement

Δx of electrons is

$$W = \int F dx \, , \tag{15}$$

Using eq. (11),

$$\Rightarrow \ W = \int_0^{\Delta x} eE(x) dx \ = \ \frac{e^2 n \Delta x^2}{2 \varepsilon_0} \, . \tag{16}$$

Equating work done by displacement with average thermal kinetic energy

$$\frac{e^2 n \Delta x^2}{2 \varepsilon_0} \simeq \frac{1}{2} k_B T \qquad \Rightarrow \ \Delta x^2 \simeq \frac{\varepsilon_0 k_B T}{e^2 n} \, . \tag{17}$$

The maximum distance an electron can travel, before it exit the system, is called *Debye Length*, $\Delta x_{max} = \lambda_D$ as

$$\lambda_D = \sqrt{\frac{\varepsilon_0 k_B T}{e^2 n}} \, . \tag{18}$$

Note that λ_D is independent of mass, and thus generally comparable for all different species. Clearly, the gas is considered a plasma of length scale of system is larger than Debye Length

$$\frac{\lambda_D}{L} >> 1 \, . \tag{19}$$

This is Plasma Criterion #2. Debye length is spatial scale over which charge neutrality is violated by spontaneous fluctuations. It has been emphasized that despite of requirement (eq. 19), plasma physics is capable of considering structures on the Debye scale. *Debye Sheath* is the most important example of this, i.e. the boundary layer which surrounds a plasma confined by a material surface.

1.3.3 Debye Shielding

Plasmas do not contain strong fields as they recognize to shield from them. One of the most important property of a plasma is the shielding of every charge in the plasma by a cloud of opposite charged particles is called the Debye Shielding.

A fundamental characteristic of the behavior of a plasma is its ability to shield out electric potential that are applied to it.

Suppose we tried to put an electric field inside the plasma by inserting two charged balls connected to a battery. The balls would attract particles of the opposite charge. A cloud of ions would surround the negative ball and a cloud of electrons would surround the positive ball. In cold plasma there is no thermal motion, there would be as same charges in the cloud as in the ball.

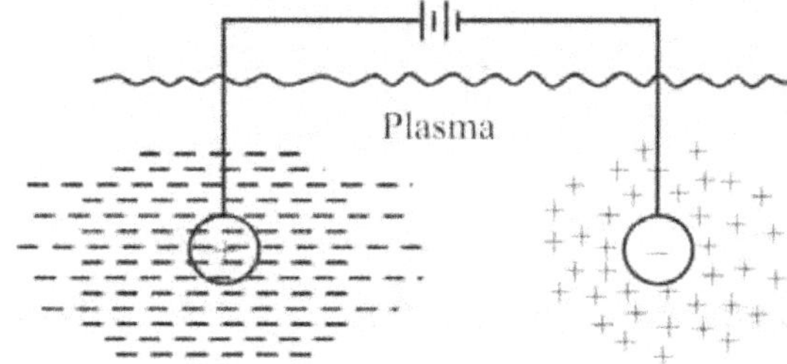

Figure: 1.4

The shielding would be perfect and no electric field present in the plasma outside of the clouds. On the other hand, if the temperature is finite, the particles which are at the edge of the cloud, where the electric field is weak. The particles have enough thermal energy to escape from the electrostatic potential well. The shielding is not complete, thus finite electric fields to exist there.

Now, we examine the mechanism by which the plasma strives to shield its interior from a disturbing electric field due to an external charged particle.

Consider a test particle to have a positive charge $+Q$ and choose a spherical co-ordinate system whose origin coincides with the position of the test particle. Then determining the electrostatic potential $\phi(\vec{r})$ that is established near the test charge $+Q$, due to the combined effects of the test charge and the distribution of charged particles surrounding it.

Since the positive test charge attracts the negatively charged particles and repels the positively charged ones the number densities of the electrons $n_e(r)$ and ions $n_i(r)$ will be slightly different near the test charge, whereas at large distances from it, the electrostatic potential vanishes, i.e. $n_i(\infty) = n_e(\infty) = n$.

At t = 0, electric scalar potential due to test charge is $\phi(\vec{r}) = \dfrac{Q}{4\pi\varepsilon_0 r}$ and as the time progresses, electron are attached while ions are repelled. As $m_i \gg m_e$, one may neglect the motion of ions. The number density for a system is usually written as,

$$n(r) = n_0 \exp\left(-\frac{U(r)}{k_B T}\right), \tag{20}$$

where the factor $\exp\left(-\dfrac{U(r)}{k_B T}\right)$ is known as the Boltzmann factor. In a plasma, this

is a steady-state problem under the action of a conservative electric field

$$\vec{E} = -\vec{\nabla}\phi(\vec{r}), \tag{21a}$$

$$U(r) = q\phi(\vec{r}). \tag{21b}$$

Therefore, the number density can be expressed as $n(\vec{r}) = n_0 \exp\left(-\dfrac{q\phi(\vec{r})}{k_B T}\right)$, which

is generalized for electron and ions as

$$n_e(\vec{r}) = n_0 \exp\left(\frac{e\phi(\vec{r})}{k_B T}\right), \tag{22a}$$

$$n_i(\vec{r}) = n_0 \exp\left(-\frac{e\phi(\vec{r})}{k_B T}\right). \tag{22b}$$

where we have assumed that the electrons and ions (of charge e) have the same temperature T.

At $t \gg 0$ and $n_e > n$, a new potential is set up with electric charge density $\rho(r)$, including the test charge Q, can be expressed as

$$\rho(r) = -e\{n_e(r) - n_i(r)\} + Q\delta(r), \tag{23}$$

where $\delta(r)$ denotes the Dirac delta function. Using eqs. (22) we get,

$$\rho(\vec{r}) = -en_0 \left\{ \exp\left(\frac{e\phi(\vec{r})}{k_B T}\right) - \exp\left(-\frac{e\phi(\vec{r})}{k_B T}\right) \right\} + Q\delta(\vec{r}). \tag{24}$$

Now, substituting $E(\vec{r}) = -\vec{\nabla}\phi(\vec{r})$ and eq. (24) into the following Maxwell's first equation,

$$\vec{\nabla}.\vec{E}(\vec{r}) = \frac{\rho(\vec{r})}{\varepsilon_0}, \tag{25}$$

This gives the Poisson's equation

$$\nabla^2\phi(\vec{r}) - en_0 \left\{ \exp\left(\frac{e\phi(\vec{r})}{k_B T}\right) - \exp\left(-\frac{e\phi(\vec{r})}{k_B T}\right) \right\} = -\frac{Q}{\varepsilon_0}\delta(\vec{r}). \tag{26}$$

For analytical calculation, it is assumed that the perturbing electrostatic potential is weak so that the electrostatic potential energy is much less than the mean thermal energy, i.e., $e\phi(\vec{r}) << k_B T$ and making use of Taylor's expansion ($e^x \approx 1 + x$), the eq. (26) simplifies to

$$\nabla^2 \phi(\vec{r}) - \frac{2}{\lambda_D^2}\phi(\vec{r}) = -\frac{Q}{\varepsilon_0}\delta(\vec{r} - 0),\tag{27}$$

where λ_D denotes the *Debye shielding length*,

$$\lambda_D = \left(\frac{\varepsilon_0 k_B T}{n_0 e^2}\right)^2 = \frac{1}{\omega_P}\left(\frac{k_B T_e}{m_e}\right)^{1/2}.\tag{28}$$

Since the electrostatic potential depends only on the radial distance r measured from the position of the test particle, thus using spherical coordinates, eq. (27) can be written (for $r \neq 0$) as

$$\frac{1}{r^2}\frac{d}{dr}\left\{r^2\frac{d}{dr}\phi(\vec{r})\right\} - \frac{2}{\lambda_D^2}\phi(\vec{r}) = 0,\tag{29}$$

because at $r = 0$, Q exists. In order to solve this equation we note initially that for a particle of charge $+Q$, in free space, the directed radially outward electric field is given by

$$\vec{E}(\vec{r}) = \frac{1}{4\pi\varepsilon_0}\frac{Q}{r^2}\hat{r},\tag{30}$$

and the corresponding electrostatic coulomb potential $\phi_C(\vec{r})$ due to this isolated charged particle in free space is

$$\phi_C(\vec{r}) = \frac{1}{4\pi\varepsilon_0}\frac{Q}{r}.\tag{31}$$

In the very close proximity of the test particle the electrostatic potential should be the same as that for an isolated particle in free space. Hence, it is appropriate to obtain the solution of eq. (29) in the from

$$\phi(r) = \frac{Q}{4\pi\varepsilon_0 r}\exp\left(-\frac{r}{\lambda_D}\right).\tag{32}$$

This result is commonly known as the *Debye Potential*. As $r \to 0$, potential is that of free charge in free space, but for $r >> \lambda_D$, potential falls off exponentially. It is understood that Coulomb potential of the point charge is shielded on a distance scale longer than the Debye length by shielding a cloud of approximate radius λ_D consisting of charge of the opposite sign. Therefore, it is guessed that a charged particle in a plasma interacts effectively only with particles situated at distances less than one Debye length away, and it has a negligible influence on particles lying at distance greater than one Debye length.

For numerical, expression for Debye length is

$$\lambda_D \cong 69\sqrt{T_e/n}\,,$$

where T_e is in K and n is in m^{-3}.

1.3.4 Plasma Parameter

The *Debye Number* N_D describes the no. of particles in the Debye sphere (volume) is defined as

$$N_D = n\left(\frac{4}{3}\pi\lambda_D^3\right). \tag{33}$$

The mean particle particle distance (Weigner-Seitz radius) between particles is defined as

$$r_D \cong n^{-1/3}, \tag{34}$$

and the distance of closest approach is given by

$$r_C \cong \frac{e^2}{4\pi\varepsilon_0 k_B T}\,. \tag{35}$$

Now we examine the significance of the ratio r_D/r_C: when this ratio is small, Debye sphere is saparsely populated means charged particles are dominated by one another's electrostatic influence and their kinetic energies are small compared to the interacting potential energies. Such plasmas are termed as *strongly coupled*. While if the ratio is large, then strong electrostatic interactions occur rarely between individual particles and a particle is influenced by all other particles within its Debye sphere. Such plasmas are termed a *weakly coupled*. But actually, strongly coupled plasma has more in common with a liquid than weakly coupled plasma.

Defining a dimensionless plasma parameter as

$$\Lambda = 4\pi n\lambda_D^3, \tag{36}$$

which is equal to the typical number of particles within the Debye sphere. Making use of eq. (18), eq. (34) & (35) with eq. (36), we may write-

$$\Lambda = \frac{1}{4\pi}\left(\frac{r_D}{r_C}\right)^{3/2} \propto \frac{k_B T^{3/2}}{n^{1/2}} .$$

(37)

It is noticed that for $\Lambda \ll 1$, Debye sphere is sparsely populated and thus corresponds to a strongly coupled plasma. Whereas, for $\Lambda \gg 1$, Debye sphere is densely populated corresponds to a weakly coupled plasma. The strongly coupled plasma tends to be a cold and dense while weakly coupled plasma tends to hot and diffuse.

Description	Plasma Parameter Magnitude	
	$\Lambda \ll 1$	$\Lambda \gg 1$
Coupling	Strongly coupled plasma	Weakly coupled plasma
Debye Sphere	Sparsely populated	Densely populated
Electrostatic Influence	Almost continuously	Occasional
Characteristics	Cold and dense	Hot and diffuse
Examples	Solid-density laser application plasmas; very cold and high pressure arc discharge; inertial fusion experiments; white dwarfs / neutron stars atmospheres	Ionospheric physics; Magnetic fusion devices; Space plasma physics; plasma balls

1.4   Saha's Ionization Formula

Saha equation gives a relationship between the particles and those bound in atoms. To derive the Saha's equation, choose a consistent set of energies and also choose *E=0*, when the electron veocity is zero, so *E= -I* for *n=1*. Ignore the energy of higher energy *n* levels, since if an electron has enough energy to reach *n=2*, then according to Bohr's equation it needs only ¼ more energy to ionize completely.

Derivation of Saha's equation:

At thermal equilibrium temperature T, the number of particles of a particular species in the energy range E to $E + dE$ is given by the distribution

$$dN = \frac{dg}{e^{-(\mu-E)/k_B T} \pm 1}. \tag{38}$$

Here the $+$ sign corresponds to fermions (proton, neutron, electron etc. all of which have half integral spin). The distribution is known as the Fermi-Dirac distribution. The $-$ sign corresponds to bosons which are particles (which can have integral spin). The distribution in this case is known as Bose-Einstein distribution.

In eq. (38), dg is the number of quantum states between energies E and $E + dE$ and μ is the chemical potential (the amount by which the energy of the system changes with the addition of a particle, keeping the volume and entropy of the system constant). In the limit $e^{-(\mu-E)/k_B T} >> 1$, the eq. (38) reduces to

$$dN = dg\, e^{\mu/k_B T}\, e^{-E/k_B T} \tag{39}$$

For free particles in momentum range p to $p + dp$, the number of states available are

$$dg = \frac{g_{int} V}{h^3} 4\pi p^2 dp. \tag{40}$$

where V is the volume in which the particles are confined and g_{int} is the internal states of the particle. For a particle of spin s, $g_{int} = 2s+1$. This formula does not apply to photons, which have integral spin. In this case, g_{int} is equal to two corresponding to the two states of polarizations (one spin up and other spin down).

Using eq. (39), we find the number of particles per unit volume in the momentum range p to $p + dp$ is given by

$$dn = g_{int} \frac{4\pi p^2 dp}{h^3} e^{\eta} e^{-E/k_B T} \tag{41}$$

where $\eta = \mu/k_B T$.

Now consider atoms in the Ith state of ionization. Defining

n_{Ij}: atoms in j^{th} state of excitation and i^{th} state of ionization.

n_{I1}: atoms in ground state of excitation and i^{th} state of ionization.

Using Boltzmann distribution, we find

$$\frac{n_{Ij}}{n_{I1}} = \frac{g_{Ij}}{g_{I1}} e^{-V_{Ij}/k_B T} \tag{42}$$

where V_{Ij} is the excitation potential, i.e. the energy required to excite an atom in ground state in the ith state of ionization to the jth excited state in the same state of ionization.

$$n_{Ij} = \sum n_{Ij} \frac{n_{I1}}{g_{I1}} Z_I(T) \tag{43}$$

where $Z_I(T)$ is the partition function. Next, consider the ionization reaction

$$V_I = V_{I+1} + e . \tag{44}$$

In equilibrium, we have $\mu_I = \mu_{I+1} + \mu_e$ or equivalently $n_I = n_{I+1} + n_e$. The kinetic energy of an electron is $E_e^2 = p^2/2m_e$ substituting this in eq. (41), we find

$$dn_e(p) = g_e \frac{4\pi p^2 dp}{h^3} \eta_e \, e^{-p^2/2m_e k_B T} , \tag{45}$$

which is just the Boltzmann distribution for a free electron gas. For atoms and ions, we also need to include the excitation and ionization energies. Hence,

$$E_{I,i} = \frac{p^2}{2m_I} + V_{I,i} + I_I \tag{46}$$

$$E_{I+1,j} = \frac{p^2}{2m_{I+1}} + V_{I+1,j} + I_{I+1} \tag{47}$$

Further, $m_I \approx m_{I+1} + m_e$. Substituting these relationships into eq. (41), and carrying out integrating over momentum p, we get

$$n_I = Z_I(T) e^{n_I} \left(\frac{2\pi m_I k_B T}{h^2} \right)^{3/2} \tag{48}$$

Here we have also summed over all the states of excitation of the atom using eq.43

$$n_{I+1} = Z_{I+1}(T) e^{n_{I+1}} \left(\frac{2\pi m_{I+1} k_B T}{h^2} \right)^{3/2} e^{-I_1/k_B T} . \tag{49}$$

We, therefore, find the ratio becomes

$$\frac{n_{I+1}}{n_I} = e^{-\eta_e} \frac{Z_{I+1}(T)}{Z_I(T)} e^{-I_1/k_B T}, \tag{50}$$

we also find

$$n_e = 2e^{\eta_e} \left(\frac{2\pi m_e k_B T}{h^2} \right)^{3/2} \tag{51}$$

where we have used $g_e = 2$ corresponding to the two spin states of electron. We can eliminate $e^{-\eta_e}$ in eq. (50) using this equation. This finally gives the Saha equation.

Since, 99% of the matter in the universe is in the plasma state, it would seem that we live in the 1% of the universe in which plasma do not occur naturally.

The reason for this can be seen from the Saha's equation, which tells us the amount of ionisation to be expected in a gas in thermal equilibrium

$$\frac{n_i}{n_n} \approx 2.4 \times 10^{21} \, e^{-I_1/k_B T} \tag{52}$$

where n_i and n_n the density of ionized atoms and of natural atoms. As the temperature is raised, then n_i / n_n rise abruptly and the gas is in a plasma state. This is the reason plasma exist in astronomical bodies with temperatures of millions of degrees, but not on the earth. The natural occurrence of plasma at high temperatures is the reason for the fourth state of matter.

1.5   Plasma Production

As there is wide range of plasma according to their uses, thus there are several means for its generation; however, one principle is common to all of them: there must be energy input to produce and sustain it.

Plasma is produced by supplying energy to neutral gas causing the formation of charge carriers. Electrons or photons with sufficient energy collide with the neutral atoms or molecules of the gas then electrons and ions are produced in the gas phase. For this case, plasma is generated when an electric current is applied across a dielectric gas or fluid (an electrically non-conducting material) as can be seen Fig. (1.5).

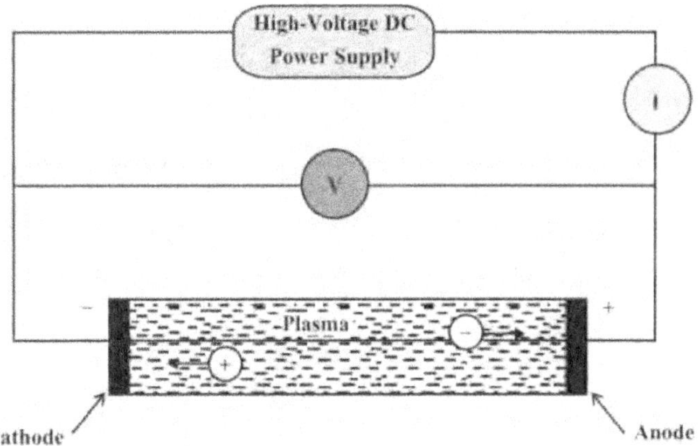

Figure 1.5

The potential difference and subsequent electric field pull the bound electrons (negative) toward the anode (positive electrode) while the cathode (negative electrode) pulls the nucleus. As the voltage increases, the current stresses the material (by electric polarization) beyond its dielectric limit (termed strength) into a stage of electrical breakdown, marked by an electric spark, where the material transforms from being an insulator into a conductor (as it becomes increasingly ionized). The underlying process is the Townsend avalanche, where collisions between electrons and neutral gas atoms create more ions and electrons, Fig. (1.6). The first impact of an electron on an atom results in one ion and two electrons. Therefore, the number of charged particles increases rapidly (in the millions) only "after about 20 successive sets of collisions", mainly due to a small mean free path.

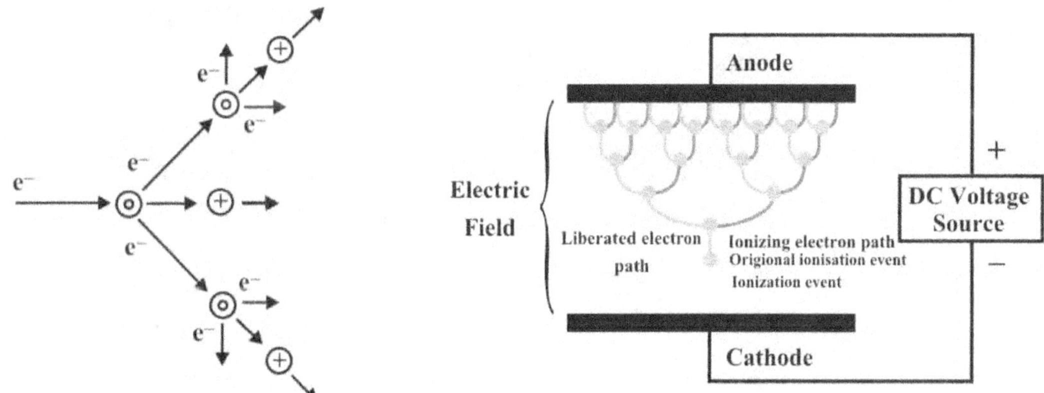

Figure 1.6

Electric arc is a cascade process of ionisation, where avalanche effect between two electrodes is produced. The original ionization event liberates one electron, and each subsequent collision liberates a further electron, so two electrons emerge from each collision: the ionizing electron and the liberated electron. With sufficient current density and ionization, a luminous electric arc (a continuous electric discharge similar to lightning) between the electrodes is produced. Electrical resistance along the continuous electric arc creates heat, which dissociates more gas molecules and ionizes the resulting atoms (where degree of ionization is determined by temperature).

There are various ways to supply the necessary energy for plasma production to a neutral gas; some of them are outlined below:

1.5.1 Gas Discharge Plasma (Low Pressure)

The most commonly and simplest method of providing a plasma is by applying an electric field to a neutral gas. A gas discharge plasma is a common form of plasma, which can have a variety of parameters. An electric field causing electrical breakdown of gas, which then generates different forms of plasma depending on the conditions of the process.

In the first stage of breakdown, a uniform current of electrons and ions arise. In the next stage of the breakdown process, the electric current establishes a distribution of charged particles in space. This is a gaseous discharge.

Glow Discharge Plasmas (GDP): Non-thermal plasmas generated by the application of DC or low frequency (<100 kHz) electric field to the gap between two metal electrodes. This is the most common plasma generated within fluorescent light tubes.

Capacitively Coupled Plasma (CCP): This is similar to glow discharge plasmas (GDP), but generated with high frequency RF electric fields, typically 13.56 MHz. These differ from glow discharges in that the sheaths are much less intense. These are widely used in the microfabrication and integrated circuit manufacturing industries for plasma etching and plasma enhanced chemical vapor deposition.

Inductively Coupled Plasma (ICP): This is same kind of plasmas as that of CCP except the electrode consists of a coil wrapped around the chamber where plasma is formed.

1.5.2 Gas Discharge Plasma (Atmospheric Pressure)

Arc Discharge: This is a high power thermal discharge of very high temperature

(~10,000 K). It can be generated using various power supplies. It is commonly used in metallurgical processes. For example, it is used to smelt minerals containing Al_2O_3 to produce aluminium.

Corona Discharge: This is a non-thermal discharge generated by the application of high voltage to sharp electrode tips. It is commonly used in ozone generators and particle precipitators.

Dielectric Barrier Discharge (DBD): This is a non-thermal discharge generated by the application of high voltages across small gaps wherein a non-conducting coating prevents the transition of the plasma discharge into an arc. It is often mislabeled 'Corona' discharge in industry and has similar application to corona discharges. It is also widely used in the web treatment of fabrics. The application of the discharge to synthetic fabrics and plastics functionalizes the surface and allows for paints, glues and similar materials to adhere.

Capacitive Discharge: This is non-thermal plasma generated by the application of RF power (e.g., 13.56 MHz) to one powered electrode, with a grounded electrode held at a small separation distance on the order of 1 cm. Such discharges are commonly stabilized using a noble gas such as helium or argon.

Piezoelectric direct discharge plasma: This is non-thermal plasma generated at the high-side of a piezoelectric transformer (PT). This generation variant is particularly suited for high efficient and compact devices where a separate high voltage power supply is not desired.

1.5.3 Photoresonant Plasma

The convenient way to produce plasma uses resonant radiation. The resonant radiation means whose wavelength corresponds to the energy of atomic transition in the atoms constituting the excited gas. As a result of the excitation of the gas, a high density of excited atoms is fermed, and collision of these leads to its ionisation and to plasma production. This plasma is called a photoresonant plasma.

1.5.4 Laser Plasma

A laser plasma is produced by laser irradiation of a surface and it is characteriszed by parameters such as the laser power and the time duration of the process. If a short nanosecond laser pulse is focused on to a surface, material evaporates from the surface in the form of a plasma.

If the number density of electrons in plasma increases above the critical density, the developind plasma screens the radiation, and subsequent laser radiation goes to heating the plasma. As a result, the temperature of the plasma increases tens of electron volts, and this plasma can be used as a source of X-ray laser.

1.5.5 Plasma Production by Electrons Beam

A widely used method of plasma production is based on the transition of electrons beams through a gas. Secondary electrons can be used for these processes. The electron beam as a source of ionisation is convenient for chemical lasers because the ionisation process finish in a short time.

1.5.6 Chemical Method of Plasma Production

A chemical method of plasma production is the use of plasmes. The chemical energy of reagents is spent on formation of excited particles. The transformation of chemical energy of ionized particles is not efficient, so the degree of ionisation in plames is small.

1.5.7 Plasma Production by Small Particles

Introduction of small particles and clusters into a weakly ionised gas can charge its electrical properties because these particles can absorb charged particles, i.e., electrons and ions or negative and positive ions can recambine on these particles. These process occurs in an aerosol plasma, i.e. an atmospheric plasma that contains aerosols.

1.5.8 Plasma Production by Fluxes of Ions

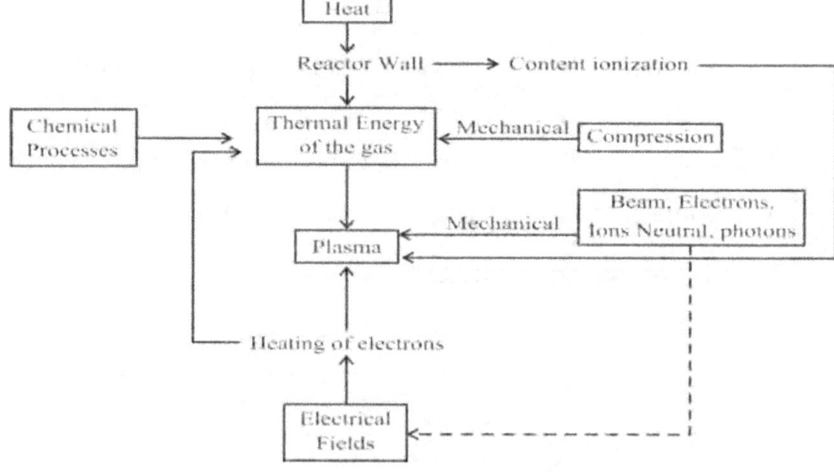

Figure 1.7

Plasma can be created under the action of fluxes of ions or neutrons when they pass through a gas. Ionisation near the earth's surface results from the decay of radioactive elements which are found in the earth's crust.

Thus, methods of plasma production are many and varied, and lead to the formation of different types of plasma.

1.6 Illustrative Examples

Example1. Compute the density (in units of m-3) of an ideal gas at 0^0C and 760 Torr pressure (I Torr =1 mm Hg).

Sol: Since a mole of an ideal gas contains Avogadro's number (6.022×10^{23} molecules) and occupies 22.4 liters. Therefore, number per m^3, i.e. $n=N/V$ is $6.022\times10^{23}/2.24\times10^{-2}=2.66\times10^{25}\,m^{-3}$.

Example 2. Compute the density (in units of m^{-3}) of an ideal gas in a vacuum of 10^{-3} Torr at room temperature 20^0C.

Sol. $\because PV=NRT$ and $n=N/V$,

Then $n=\dfrac{P}{RT}$. Taking n_0 at 0^0C from (a) and n_1 at 20^0C, we may then write

$$\frac{n_1}{n_0}=\frac{P_1T_0}{P_0T_1} \Rightarrow n_1=2.66\times10^{25}\frac{10^{-2}}{760}\frac{273}{293}=3.3\times10^{19}\,m^{-3}$$

Example 3. Compute λ_D and N_D for the earth's ionosphere with $n=10^{12}\,m^{-3}$, $k_BT=0.1eV$.

Sol. Using $\lambda_D=\sqrt{\dfrac{\varepsilon_0 k_BT}{e^2 n}} \Rightarrow 2.35\times10^{-3}\,m$

and $N_D=n\left(\dfrac{4}{3}\pi\lambda_D^3\right)\Rightarrow 54.45\times10^3$.

Example 4. Derive a constant A for a normalized one dimensional Maxwellian distribution $f(u)=A\exp\left(-mu^2/2k_BT\right)$, such that $\int\limits_{-\infty}^{\infty} f(u)du=1$.

Sol. Consider the integral in two dimensional space as

$$I^2=\int\limits_{-\infty}^{\infty} e^{-x^2}dx\int\limits_{-\infty}^{\infty} e^{-y^2}dy \simeq \int\limits_{-\infty}^{\infty}\int e^{-(x^2+y^2)}dxdy,$$

and transforming into cylindrical coordinates, we get

$$I^2 = \iint e^{-r^2} r\,dr\,d\phi = 2\pi \int_0^\infty e^{-r^2} r\,dr = \pi \int e^{-r^2} d(r^2) = -\pi \left(e^{-r^2} \right)_0^\infty = \pi$$

$$\Rightarrow I = \sqrt{\pi} \ .$$

Then, carrying out the integration

$$\int_{-\infty}^{\infty} f(u)\,du = 1 = A\left(\frac{2k_B T}{m}\right)^{1/2} \int e^{-mu^2/2k_B T} d\left[u\left(m/2k_B T\right)^{1/2} \right]$$

$$1 = AI\left(\frac{2k_B T}{m}\right)^{1/2} \Rightarrow A = \left(\frac{m}{2\pi k_B T}\right)^{1/2} .$$

1.7   Self Learning Exercise

Q.1 Calculate v_{rms} for proton and electrons at $10^6\,K$. {Hint: $v_{rms} = \sqrt{3k_B T/m}$ }

Q.2 Compute λ_D and N_D for a glow discharge with $n = 10^{16}\,m^{-3}$, $k_B T = 2eV$.

Q.3 What do you mean by Plasma oscillations and frequency?

1.8 Summary

This chapter starts with the development of the fourth state of matter, i.e. Plasma. To understand the dynamics, basic parameters are introduced, by which the interesting properties of plasma, namely plasma oscillations and frequency, Debye shielding parameters have been laid down. From these, the criteria of being plasma have been then fixed up. Further, since plasma is generated by continuous chain of ionization processes, is explained by Saha's ionization formula. Finally, chapter is ended up with the various type of plasma production, as these kind of plasma occur in lightning, fluorescent lamps, a variety of laboratory experiments, and a growing array of industrial processes and also a live quest to researcher, as it is known that 99% of the baryonic contents of the plasma, e.g. stars, nebulae, and even interstellar space and Solar System.

1.9 Glossary

Physical constants:

Boltzmann constant: $k_B = 1.38 \times 10^{-23}\,JK^{-1}$

Permittivity of vacuum: $\varepsilon_0 = 8.85 \times 10^{-12}\, Fm^{-1}$

Permeability of vacuum: $\mu_0 = 4\pi \times 10^{-7}\, Hm^{-1}$

Mass of electron: $m_e = 9.1 \times 10^{-31}\, Kg$

Mass of Hydrogen ion: $m_i = 1.67 \times 10^{-27}\, Kg$

Units of pressure: $1\,atm = 1.013 \times 10^{5}\, Pa = 760\, mmHg = 760\, torr$;

$1\,micron = 1\,m\,torr$

1.10 Exercise

Q.1   Compute λ_D and N_D for a θ-pinch with $n = 10^{23}\, m^{-3}$, $k_B T = 800\, eV$.

Q.2   For the radio-frequency discharge $k_B T = 3eV$, $n_e = 10^{17}\, m^{-3}$ and diameter about 100 mm, calculate λ_D, and N_D the number of electrons in a Debye sphere.

Q.3   Define the meaning of Plasma. What are the Plasma parameters? Explain.

Q.4   What do you mean by Debye Shielding? Establish a relationship between Plasma parameter and Debye shielding parameter.

Q.5   Discuss the Saha's ionization formula and derive it explicitly.

Q.6   Briefly discuss the production of Plasma by gas discharge method at both low and atmospheric pressure.

References and Suggested Readings

1.   A.R. Choudhuri: *The physics of fluids and plasmas* (Cambridge University Press, 1998)

2.   Francis F Chen: *Introduction to Plasma Physics and Controlled Fusion* (Plenum Press, New York, 1984)

3.   Richard Fitzpatrick: *Plasma Physics- An Introduction* (CRC Press, Taylor & Francis Group, 2014).

4.   L.D. Landau & E.M. Lifshitz: *Fluid mechanics* (Pergamon Press, 2nd edition, 1987).

5.   Referred useful and trusted internet resources.

UNIT- 2

Debye Shielding, Occurrence of Plasmas in Nature

Structure of the Unit

2.0 Objectives

2.1 Introduction

2.2 Definition of Plasma

2.3 Plasma Parameters

2.4 Coupling parameters

2.5 Quasi-Neutrality, Plasma Frequency and Debye length

2.6 Debye Screening

2.7 Alternative derivation of Debye Shielding

2.8 Occurrence of the plasma

    (A) Plasmas on Earth

    (B) Near-Earth Plasma

    (C)Solar, Stellar and Interstellar Plasma

2.9 Illustrative examples

2.10 Self Learning Exercise

2.11 Summary

2.12 Glossary

2.13 Answers to Self Learning Exercise

2.14 Exercise

    References and Suggested Readings

2.0 Objectives

To study

- Definition of a plasma

- Occurrence of the plasma

- Plasma parameter

- Characteristic values of plasma parameter

- Coupling parameters

- Occurrence of plasmas in nature

- The Sun and its atmosphere

- Solar Wind

- Ionosphere

2.1 Introduction

Nearly all the matter in the universe exists in the plasma state, occurring predominantly into this form in the Sun and stars and in interstellar space. Auroras, lightning, and welding arcs are also plasmas. Plasmas exist in neon and fluorescent tubes, in the sea of electrons that more freely within energy bands in the crystalline structure of metallic solids, and in many other objects.

On the earth, plasmas are found with dimension of microns to meters, that is, size spanning six orders of magnitude. Lightning is a natural plasma resulting from electrical discharges in the earth's lower troposphere. Such flashes are usually associated with clouds but also occur in snow and dust storms, active volcanoes, nuclear explosions. In this chapter we shall understand about the types of plasmas occurring in nature and their basic properties, plasma parameters etc.

2.2 Definition of a Plasma

1.   Plasma is an ionized gas in which all or a considerable part of the atoms have lost one or several of their electrons, thus becoming positive ions.

Today the plasma state as a fourth state of matter is well known, along with the concepts of gaseous, liquid, and solid states. This is due to the wide occurrence of the plasma state in nature, and also to the prospects regarding the practical utilization of plasmas in various branches of modern technology.

The term "plasma" was introduced by I. Langmuir in 1923, when he was studying phenomena in electrical discharges in gases. Thus, the first definition of a plasma was connected with the concept of an ionized gas. Let us preliminarily define a plasma as an ionized gas comprising a large number of positively and negatively charged particles and sometimes also neutral atoms and molecules. It is the presence of the large number of charged particles in the plasma that result in those specific properties which allow us to define the plasma as a fourth state of matter to distinguish it from ordinary gases.

The above definition of the plasma is neither exact nor complete. In fact, it is impossible to give a compressive definition of a plasma, since it could have to cover a great variety of phenomena under various conditions.

2.3 Plasma Parameters

As mentioned before, a plasma consists both of charged and neutral particles. Positively charged particles are ions (gaseous plasma) and holes (solid-state plasma) ; negatively charged particles are electrons and negative ions. Since the later usually are insignificant in plasma phenomena, they may be ignored here.

The composition of a neutral plasma may be rather complex since, besides atoms and molecules in the ground state, it will contain excited atoms and molecules in addition. Since a plasma is a gas, we may use the same characteristics for its description as for a common gas. Let us then introduce the main plasma parameters using simple molecular-kinetic notions.

First we consider the concentration (density) of particles of different type N_α the index α denoting the type of a particle. We shall mark all quantities for plasma electrons by the index e , for ions by i, and for neutrals by n. If several types of ions exist in the plasma, we define the concentration for ions of each type separately. The excitation of atoms and molecules will be of little interest to us. Therefore, N_n will mean total number of neutral particles per unit volume without regard to their state.

Alternatively, the composition of the plasma can be described as the ratio of the electron density to that of neutral particles, or the degree of ionization $r = \dfrac{N_e}{N_n}$.

According to this, a plasma is weakly ionized, if $r < 10^{-2}$ to 10^{-3}, and completely ionized, if $r \to \infty$ holds. (For the "degree of ionization" one frequently uses also.

$$r = \frac{N_e}{N_n + \sum_i N}$$

For this definition r = 1 stands for complete ionization, an r <1 for partial ionization).

Since plasmas consist of particles of different types, one must know their charge e_a and mass m_a. We have electrons with charge $e_e \equiv e$, where $|e| = 4.8 \times 10^{-10}$ esu, and a mass $m_e = m = 9.1 \times 10^{-28}$ g, and ions with a charge $e_i = -Ze$ (Z is multiplicity of ionization), and an approximate mass $m_i = M = A. 1.66 \times 10^{-24}$ g, where A is the atomic mass of the corresponding gas. For neutral particle $e_n = 0$, and $m_n \approx m_i = M$ holds.

In a solid state plasma the effective mass of charge carriers (electrons and holes) differs from that of free electrons. Thus to avoid misunderstandings we shall mark them as m_e^* and m_i^*, where it is necessary. In metals we have $m_e^* \approx m_e$, in semiconductors usually $m_e^* \approx (0.01$ to $0.1)$ m_e and $m_i^* \approx m_i$. The charge of a negative carrier is equal to that of an electron, and the charge of a positive carrier is $e_i = -e$.

Characteristics values of Plasma parameters :

	N_e	T
* Electron discharge plasma (glow discharge at a gas pressure of ~1 Torr)	$10^9 - 10^{11} cm^{-3}$	$10^4 K$
* Electron in metal (quantum mechanically Degenercatc plasma)	$\sim 10^{23} cm^{-3}$	$10^4 K$
* Plasma in semiconductors (electrons-hole plasma)	$\sim 10^{16}$ to $10^{18} cm^{-3}$	$< 10^2 K$
* F-layer of the ionosphere	$N_e \approx N_i \approx 10^6 cm^{-3}$ $N_n \leq 10^{-10} cm^{-3}$ $r \geq 10^{-4}$	$(3$ to $5) \times 10^3 K$

* Interplanetary plasma $\qquad 10^{-2}\, cm^{-3} \leq N_e \approx N_i \leq 10 cm^{-3}$

$10^4\, K$

* Solar corona $\qquad N_e \approx N_i \approx 10^4\, to\, 10^8\ cm^{-3}$

$10^6 - 10^8\, K$

* Fusion plasma $\qquad N_e \approx N_i = 10^{14}\ cm^{-3} \qquad >10^8\, K$

* White dwarf $\qquad N_e \approx 10^{25}\, to\, 10^{32}\ cm^{-3} \qquad 10^7\, to\, 10^8\ K$

2.4 Coupling Parameters

Plasmas are characterized by temperature and density parameters which vary over a wide range. It is therefore convenient to introduce dimensionless coupling constants, called coupling parameters, which characterize strengths of the particle interactions in those various charged-particle systems. We shall also consider parameters describing the extents to which the quantum-mechanical effects are involved in a plasma. This will help us in seceding whether a classical or quantum theoretical treatment of the plasma is valid.

Let us define a coupling constant η of a plasma as a ratio of an average Coulomb-interaction energy to an average kinetic energy :

$$\eta = \frac{e^2}{r_{av}T} \sim \frac{e^2 N^{\frac{1}{3}}}{T} \qquad (2.1)$$

Those plasmas with values of the coupling constant much smaller than unity may be called weakly coupled plasmas ; those with the coupling constant around or greater than unity, strongly coupled plasma.

When the charged-particle system under consideration obeys the classical statistics, the velocity distribution of individual particles is given by the Maxwellian

$$f_M(\vec{v}) = \left(\frac{m}{2\pi T}\right)^{1/2} \exp\left[-\frac{mv^2}{2T}\right]$$

where m denotes the mass of the particle. The average value of the kinetic energy particle is then calculated as $\left(\frac{3}{2}\right)T,$

which we take as T in an order – of – magnitude estimate.

For a high-density electron system such as conduction electrons in metal, one uses the Fermi energy

$$E_F = \frac{\hbar^2 (3\pi^2 N)^{2/3}}{2m} \tag{2.2}$$

Instead of the classical evaluation mentioned above. Here $\hbar = 1.0546 \times 10^{-27}$ erg-s is the Planck constant divided by 2π, m $= 9.1095 \times 10^{-28}$ g is the electron mass and N denotes the number density of electrons.

Since eq. (2.2) is an increasing function of N, $E_F >> T$ may be realized in a high density electron system. A degeneracy parameter of the electrons is thus defined by

$$(\varkappa) = \frac{T}{E_F} \tag{2.3}$$

When $(\varkappa) << 1$, the Fermi degeneracy brought about by the quantum statistics becomes more important than the effects of thermal motion represented by T. In these circumstances it is relevant to use the Fermi energy as an estimate of the kinetic energy.

For the degenerate plasma the coupling parameter η is the ration of the average potential energy to the Fermi energy

$$\eta = \frac{e^2 N^{\frac{1}{3}}}{E_F} \sim \left(\frac{\hbar\omega_p}{\varepsilon_{Fe}}\right)^2 \sim \frac{r_{av}^2}{r_{De}^2} \ll 1 \tag{2.4}$$

Comparing (2.4) and (2.1) we see that in the nondegenerate plasma an increase in the density disfavors, in the degenerate case if favors the applicability of the gas approximation. In metals this approximation is valid only for concentrations $N_e \gtrsim 10^{22}$ cm^{-3}, and in semiconductors with the effective mass.

$m^* = 10^{-2} m$ for concentrations $N_e \gtrsim 10^{16}$ $to 10^{17} cm^{-3}$. Thus, the condition (2.4) for real metals is only marginally fulfilled.

For a one-component plasma (OCP) containing n particles with electric charge Ze in a unit volume, the Coulomb energy per particle may be estimated as $(Ze)^2/a$, where a is the radius of a sphere with the characteristic volume $\frac{1}{N}$:

$$a = \left(\frac{4\pi N}{3}\right)^{-1/3} \tag{2.5}$$

This radius is usually referred to as the ion-sphere radius or the Wigner-Seitz radius.

The Coulomb coupling parameter for a plasma obeying the classical statistics, eq (2.1), can be expressed as

$$\eta = \frac{(Ze)^2}{r_{av}T} = 2.69 \times 10^{-5} \, Z^2 \left(\frac{N}{10^{12} \, cm^{-3}}\right)^{1/3} \left(\frac{T}{10^6 \, K}\right)^{-1} \tag{2.6}$$

This expression indicates that for $Z = 1$ and $T = 10^6$ K, the density N mush become as high as $\sim 10^{26}$ cm^{-3} to make $\eta = 1$.

Most of the classical plasmas that we encounter, however, are characterized by $\eta \ll 1$. For example :

(1) Gaseous-discharge plasma : $N = 10^{11}$ cm^{-3}, $T = 10^4$ K, $\eta = 10^{-3}$

(2) Thermonuclear fusion plasma : $N = 10^{16}$ cm^{-3}, $T = 10^8$ K, $\eta = 10^{-5}$

(3) Plasma in the solar corona: $N = 10^6$ cm^{-3}, $T = 10^6$ K, $\eta = 10^{-7}$

They are thus weakly coupled plasmas ; their thermodynamic properties, for instance, are analogous to those of an ideal gas.

A typical example of a strongly coupled classical plasma may be seen in the system of ions inside a highly evolved star. The interior of such a star is in a compressed, high-density state. The Fermi energy of the electron system takes on a value much greater than the binding energy of the electron around an atomic nucleus, all the atoms are thus in ionized states (pressure ionization). The electron system constitutes a weakly coupled degenerate plasma with an immensely large Fermi energy $(E_F \approx mc^2)$. It makes an ideal neutralizing background of negative charges to the ion system. Those atomic nuclei stripped of the electrons form an ion plasma obeying the classical statistics; their de-Broglie wavelengths are much smaller on the average than the inter particle spacing, i.e.

$$\frac{\hbar}{r_{av}\sqrt{MT}} \ll 1 \tag{2.7}$$

where M is the ion mass.

In the interior of a highly evolved star, the coupling parameter η for such an ion plasma is usually greater than unity ; in a white dwarf one estimates that $\eta = 10$ to 200.

As an example of a strongly coupled plasma in the laboratory, we can think of plasmas produced by shock compression. Other examples of strongly coupled laboratories plasmas include liquid and/or ultrahigh pressure metals, super ionic conductors, and cryogenic non-neutral plasmas, contained in electromagnetic traps, such as the laser-cooled pure ion plasmas and the strongly magnetized pure electron plasmas.

2.5 Quasi-Neutrality, Plasma Frequency and Debye Length

One should note that not every ionized gas is a plasma. It must also possess the property of quasi-neutrality, i.e on the average it must remain neutral for sufficiently long time and space intervals. The assumption of quasi-neutrality implies :

$$\sum_{\alpha} e_\alpha N_\alpha = 0 \tag{2.8}$$

Where $e_\alpha N_\alpha$ are, respectively, the charge and density of particles of type α .

For a plasma containing single charged ions of only one type this condition becomes :

$$N_e = N_i, \qquad \text{Since an electron carries the charge } e = - e_i.$$

We can estimate the time scale of charge separation imagine a plasma electron to deviate from the initial equilibrium position. A restoring force appears, defined in order of magnitude by the average interparticle force

$$F \approx \frac{e^2}{r_{av}^2}, \text{ where } r_{av} \text{ is the average distance between the particles :}$$

$$\frac{4}{3}\pi r_{av}^3 N_e \simeq 1. \qquad \Rightarrow r_{av} = \left[\frac{3}{(4\pi N_e)}\right]^{\frac{1}{3}}$$

As a result the electron oscillates with the frequency given by

$$m\ddot{r}_{av} = F = \frac{e^2}{r^2} \qquad \Rightarrow m\omega^2 r_{av} = F$$

$$\therefore \omega \approx \left[\frac{F}{mr_{av}}\right]^{\frac{1}{2}} \approx \left[\frac{4\pi e_2 N_e}{3m}\right]^{\frac{1}{2}} \sim \omega_p$$

where the quantity

$$\omega_p = \left[\frac{4\pi e^2 N_e}{m}\right]^{\frac{1}{2}} \tag{2.9}$$

It is called the electron Langmuir frequency, electron plasma frequency, or simply plasma frequency, and is a very important characteristic parameter.

Consequences of the Coulomb Interaction :

A plasma is a collection of charged particles. The Coulomb force with which the charged particles interact is known to be long-ranged. Consequently the physical properties of a plasma

exhibit remarkable differences from those of an ordinary neutral gas.

Most of the salient features in plasmas can be understood by investing the behavior of the one component plasma (OCP). Here we will adopt the OCP model to study the basic consequences of the Coulomb interaction in the static properties of the plasma; this can easily, be extended also to cases with various charged species.

We begin with a consideration of the Coulomb cross sections for transfer of momentum and energy. This example is intended to illustrate how a naive substitution of Coulomb potential in the calculation of plasma properties would lead to a false prediction. It will at the same time point to the necessity of taking into account the organized or collective behavior of many charged particles brought about by the long-ranged coulombs forces; the plasma thus exhibits a medium like character.

A.   Cross Sections for the Coulomb Scattering :

Suppose that charged particles with masses m_1 and m_2 and electronic charges $Z_1 e$ and $Z_2 e$ scatter each other by the Coulomb force with the relative velocity $v = |\vec{v}_1 - \vec{v}_2|$, and impact parameter b. In the frame of reference comoving at the centre-of-mass.

Velocity,
$$\vec{V} = \frac{m_1 \vec{v}_1 + m_2 \vec{v}_2}{m_1 + m_2} \tag{2.10}$$

The angle χ of scattering is related to the impact parameter

$$\cot\left(\frac{\chi}{2}\right) = \frac{b\mu v^2}{Z_1 Z_2 e^2} \tag{2.11}$$

where $\mu = \dfrac{m_1 m_2}{m_1 + m_2}$ is the reduced mass.

The differential cross section of $d\sigma$ for scattering into an infinitesimal solid angle $d\Omega$ around a scattering angle is give by Rutherford formula

$$\frac{d\sigma}{d\Omega} = \left(\frac{Z_1 Z_2 e^2}{2\mu v^2 \sin^2 \dfrac{\chi}{2}}\right)^2 \tag{2.12}$$

The cross section σ_m for momentum transfer can then be calculated by integrating (2.12) over solid angles with a weighting function $(1-\cos\chi)$ which represents the fractional change of momentum on scattering. If for the moment a finite angle χ_{min} is chosen for the lower limit of the χ integration, we calculate.

$$\sigma_m = \int_{\chi_{min}}^{\pi} d\chi\, 2\pi \sin\chi\,(1-\cos\chi)\left[\frac{Z_1 Z_2 e^2}{2\mu v^2 \sin^2 \dfrac{\chi}{2}}\right]^2 \tag{2.13}$$

$$= 4\pi\left(\frac{Z_1 Z_2 e^2}{\mu v^2}\right)^2 \ln\left[\frac{1}{\sin \dfrac{\chi_{min}}{2}}\right]$$

The logarithmic factor, whose appearance is typical of the Coulomb interaction, is called a Coulomb logarithm. The cross section is seen to diverge logarithmically as χ_{min} approaches zero.

The cross section for the momentum transfer is proportional to the electric resistivity of the plasma experiments tell us that ordinary plasmas are characterized by finite values of resistivity, non infinite ones. This points to the inadequacy of the foregoing treatment when it is applied directly to the charged particles in the plasmas.

The origin of the logarithmic divergence mentioned above can be traced to those scattering acts which takes place scattering angles. Charged particles in the plasma located at long distances from a scattering centre will undoubtedly be influenced by many other scattering centres with similar strengths of interaction. Hence a simple picture of binary Coulomb scattering cannot correctly describe the behavior of charged particles interaction at large distances in the plasma.

2.6 Debye Screening

It may now be recognized that as simple picture of binary collisions is inadequate to describe the interaction between the charged particles in a plasma. It leads to the cross-section for momentum transfer that diverges logarithmically.

We now take up the problem of determining an effective interaction between charged particles in a plasma.

We shall thus calculate the potential field around a Coulomb scattering centre in an OCP, by taking explicit account of the statistical distribution of other charged particles.

Consider a point charge $Z_0 e$ located at the organ $(\vec{r} = \vec{o})$;

In vacuum it produces a potential field

$$\phi_0(\vec{r}) = \frac{Z_0 e}{r} \tag{2.14}$$

In the plasma such a potential field acts to disturb the spatial distribution of charged particles. The space charge so induced around the point charge in turn produces an extra potential field, which should be added to the original potential $\phi_0(\vec{r})$; a new effective potential $\Phi(\vec{r})$ is thus obtained as a summations of the two. The space-charge distribution in the plasma is determined, not from the bare potential $\Phi_0(\vec{r})$, but from the total potential field in a self-consistent fashion.

The charge $Z_0 e$ produces an electric field polarizing the plasma. As in result, besides the density of the external charge density $Ze\delta(\vec{r})$, the induced charge density $\rho(\phi)$ appears in the plasma. Φ is the unknown potential and obeys the Poisson equation :

$$\Delta\phi = -4\pi\rho(\phi) - 4\pi Ze\delta(\vec{r}) \tag{2.15}$$

The density of the induced charge is

$$\rho = \sum_{\alpha} e_{\alpha} \, \tilde{N}_{\alpha}(\phi) \tag{2.16}$$

where $\tilde{N}_{\alpha}(\phi)$ is the density of particles of type α, when the field Φ is present in the plasma.

In equilibrium the distribution of particles in potential field is governed by the usual Boltzmann formula.

$$\tilde{N}_{\alpha}(\phi) = N_{\alpha} \exp\left(\frac{-e_{\alpha}\phi}{KT_{\alpha}}\right) \tag{2.17}$$

where N_{α} is the unperturbed particle density in the absence of the charge Ze. Then the Poisson equation results in

$$\Delta\phi = -4\pi Z_{0}e\,\delta(\vec{r}) - 4\pi\sum_{\alpha} e_{\alpha}N_{\alpha}\exp\left(\frac{-e_{\alpha}\phi}{KT_{\alpha}}\right) \tag{2.18}$$

$$\approx -4\pi Z_{0}e\,\delta(\vec{r}) + \sum_{\alpha}\frac{4\pi e_{\alpha}^{2}N_{\alpha}}{KT_{\alpha}}\phi \tag{2.19}$$

where we have expanded the exponentials in the range where $e_{\alpha}\phi \ll KT_{\alpha}$, and the quasineutrality condition $\sum_{\alpha} e_{\alpha}N_{\alpha} = 0$

On developing the potential $\phi(\vec{r})$ into the Fourier series

$$\phi(\vec{r}) = \int d\vec{k}\, e^{i\vec{k}.\vec{r}}\,\phi(\vec{k}) \tag{2.20}$$

Applying the Laplace operator to both sides of (2.20)

$$\Delta\phi = -\int k^{2} e^{i\vec{k}.\vec{r}}\,\phi_{k}\, d\vec{k}$$

$$\therefore\ (\Delta\phi)_{k} = -k^{2}\phi_{k} \tag{2.21}$$

Fourier analyzing equation (2.19) and making use of the result (2.21)

$$\int(\Delta\phi)_{k}e^{i\vec{k}.\vec{r}}d\vec{k} = -\frac{4\pi Z_{0}e}{(2\pi)^{3}}\int d\vec{k}\,e^{i\vec{k}.\vec{r}} + \sum_{\alpha}\frac{4\pi e_{\alpha}^{2}N_{\alpha}}{KT_{\alpha}}\int\phi_{k}e^{i\vec{k}.\vec{r}}d\vec{k}$$

where we have made use of the relation

$$\delta(\vec{r}) = \frac{1}{(2\pi)^{3}}\int e^{i\vec{k}.\vec{r}}d\vec{k}$$

Denoting

$$\sum_{\alpha} \frac{4\pi N_\alpha e_\alpha^2}{KT_\alpha} = \frac{1}{\lambda_D^2} , \tag{2.22}$$

and using $(\Delta\phi)_k = -k^2 \phi_k$ we get $\left(k^2 + \frac{1}{\lambda_D^2}\right)\phi_k = \mp\frac{4\pi Z_0 e}{(2\pi)^3}$

$$\therefore \ \phi_k = \mp\frac{4\pi Z_0 e}{(2\pi)^3}\left(k^2 + \frac{1}{\lambda_D^2}\right)^{-1}$$

$$\therefore \ \phi(\vec{r}) = \frac{4\pi Z_0 e}{(2\pi)^3}\int_{-\infty}^{\infty} d\vec{k}.e^{i\vec{k}.\vec{r}}\left(k^2 + \frac{1}{\lambda_D^2}\right)^{-1} \tag{2.23}$$

If we integrate the expression (2.23) using the theory of analytic functions by closing the contour along a large circle in the upper half plane of the complex $\vec{k}$ and evaluating the residue at $k = \dfrac{i}{\lambda_D}$, we obtain

$$\phi(\vec{r}) = \frac{Z_0 e}{r}\exp\left[\frac{-r}{\lambda_D}\right] \tag{2.24}$$

$$\frac{1}{\lambda_D^2} = \sum_{\alpha} \frac{4\pi e^2 N_\alpha}{KT_\alpha} = \frac{1}{\lambda_{De}^2} + \frac{1}{\lambda_{Di}^2} \tag{2.25}$$

where λ_{De}, λ_{Di} are the Debye lengths of electrons and ions, respectively.

From the expression (2.24) we note that the field of a test charge in a plasma differs from the field in vacuum since it is screened at larger distances $r > \lambda_D$. The screening is a consequence of the displacement of charged particles around the test charge. At distances $r < \lambda_D$ the field of the test charge in the plasma practically does not differ from that in vacuum.

Debye length is an important characteristic length for the weakly coupled plasmas. Putting numbers for the known physical parameters in eq. (2.22), we find the following convenient numerical formula for the Debye length :

$$\lambda_D = \frac{6.9\sqrt{T/\eta}}{Z} \tag{2.26}$$

where T and η are to be measured in units, of K and cm^{-3}. As an example for an OCP, Z = 1 n =10^{10} ,T=10^4; λ_D is computed to be approximately 7×10^{-3}

cm; the Debye length usually takes on a value smaller than a macroscopic scale of distance.

The meaning of Eq. (2.24) is clear by comparison with coulomb potential Eq (2.14). For $r \ll \lambda_D$, the effective potential is virtually identical to the bare Coulomb potential, Eq (2.14). For $r \gg \lambda_D$, the potential field decreases exponentially; one can take $\phi(\vec{r}) \approx 0$.

In otherwords, the potential field around a point charge is effectively screened out by the induced space charge field in the OCP for distances greater than the Debye length.

Let us consider the number of plasma particles involved in such an act of Debye screening. For such a purpose. We define the Debye number, N_D, the average number of particles contained in a shire with a radius λ_D :

$$N_D = \left(\frac{4\pi n}{3}\right)\lambda_D^3 = 1.38 \times 10^6 \, Z^{-3} \left[\frac{n}{10^{12}\,cm^{-3}}\right]^{-1/2} \left[\frac{T}{10^6\,k}\right]^{3/2}$$

2.7 Alternative Derivation of Debye Shielding

In the following we show how a plasma modifies or shields the electric field of a discrete charge.

We place a charge q at rest in a plasma with an initially uniform electron density n_0 and treat the ions as a fixed neutralizing background. The electric potential $(\vec{E} = -\nabla\phi)$ is determined by Poisson equation.

$$\nabla^2\phi = -4\pi q\delta(\vec{x}) + 4\pi e(n_e - n_o) \tag{1}$$

where the charge is located at r = 0 for convenience.

In the static limit, the force equation for electron fluid is reduced to

$$nm\frac{d\vec{u}_e}{dt} = -n_e e\vec{E} - \theta_e \nabla n_e, \quad \text{(Condition of equilibrium)}$$

$$0 = n_e e\vec{E} + \theta_e \nabla n_e$$

$$n_e e\vec{E} = -\theta_e \nabla n_e$$

where an isothermal equation of state has been used.

$$\therefore \qquad \frac{\nabla n_e}{n_e} = \frac{+e}{\theta_e}\nabla\phi \qquad\qquad (\because \vec{E} = -\nabla\phi)$$

Integrating $\qquad \ln n_e = \frac{e}{\theta_e}\phi + \ln n_0$

$$\ln n_e = n_0 \exp\left(\frac{e\phi}{\theta_e}\right)$$

Noting that $\quad \dfrac{e\phi}{\theta_e} \ll 1$, we expand the exponential

$$\nabla^2\phi - 4\pi e[n_o e^{\frac{e\phi}{\theta_e}} - n_o] = 4\pi q\delta(\vec{x})$$

$$\nabla^2\phi - 4\pi n_0 e[1 + \frac{e\phi}{\theta_e} + \dots - 1] = 4\pi q\delta(\vec{x})$$

or $\qquad \nabla^2\phi - \dfrac{4\pi n_o e^2\phi}{\theta_e} = -4\pi q\delta(\vec{x})$

or $\qquad \nabla^2\phi - \dfrac{\phi}{\lambda_{De}^2} = -4\pi q\delta(\vec{x})$,

where $\lambda_{De} = \sqrt{\dfrac{\theta_e}{4\pi n_o e^2}}$

If θ_e is expressed in units of eV and n_o in units of cm^{-3},

$$\lambda_{De} = 743\left(\frac{\theta_e}{n_o}\right)^{1/2}$$

The equation (2) is easily solved by Fourier transforming and then inverting, which gives

$$\phi = \frac{q}{r}\exp\left(\frac{-r}{\lambda_{De}}\right)$$

The plasma electrons shield out the field of a discrete charge in a characteristic distance which is λ_{De}. In general, the ions also contribute to the shielding.

Note : $\quad \lambda_{De} = \sqrt{\dfrac{\theta_e}{4\pi n_o e^2}} \qquad$ Or θ_e is in eV, $\qquad n_o$ in $\dfrac{1}{cm^3}$

$$\lambda_{De} = 743\left(\frac{\theta_e}{n_o}\right)^{\frac{1}{2}} \text{ cm}$$

2.8 Occurrence of the Plasma

Plasma is the most widespread state of matter in nature. The Sun and stars can be regarded as enormous lumps of hot plasma. The outer layer of Earth's atmosphere- ionosphere-consists of plasma.

The radiation belts found outside the ionosphere around the earth are plasma formations of very low density. We encounter plasma in various natural and gas discharges, since any gas discharge (lightning, spark, arc, etc) involves plasma formation. Finally, we mention the solid-State plasmas, i.e. the electron plasmas of metals and electron-hole plasmas of semiconductors.

Studies in plasma physics were always stimulated mainly by the prospects of practical applications. At first plasma attracted attention as a peculiar conductor of electric current and a light source. Nowadays new techniques in plasma studies stem from important current technological problems for which plasma physics serves as a scientific foundation .The most important among these problems are controlled thermonuclear fusion and magneto hydrodynamic conversion of thermal energy into electrical energy. In the near future plasma physics will, possibly, make a significant contribution to accelerator technology.

In the scheme of "controlled thermonuclear research", energy is released in fusion processes between light nuclear of hydrogen isotopes such as deuterium and tritium. Only two reactions are of practical interest :

$$D + D \begin{cases} \nearrow\ {}^3He + n + 3.25\ \text{MeV} \\ \searrow\ T + p + 4.0\ \text{MeV} \end{cases}$$

$$D + T \longrightarrow {}^4He + n + 17.6\,\text{MeV}$$

Since atomic nuclei are positively charged, they repel each other by the Coulomb forces. To induce nuclear fusion reactions effectively by overcoming such repulsive forces, those nuclei have to collide vigorously with each other. The minimum conditions for net production of energy by a magnetic-confinement scheme are estimated to be dense, high temperature plasma of 10^{14} to 10^{15} cm^{-3} and $\sim 10^{8}$ K held for more than a second. Stable confinement and heating of a plasma are essential problems in realization of the controlled release of such nuclear fusion energy.

It is a well known fact that most of the matter in the nature, with a few exceptions such as the surface of cold planets (the earth, for example) exists as a plasma. The physical sizes and characteristics of plasmas in the universe are described in the followings :

(A) Plasmas on Earth :

On the earth, plasmas are found with dimensions of microns to meters. The magnetic fields associated with these plasmas range from about 0.5 gauss (the earth's ambient field) to mega gauss field strengths. Plasma lifetimes on earth span 12 to 19 orders of magnitude:

Laser produced plasmas have properties measurable in picoseconds. Pulsed power plasmas have nanosecond to microsecond lifetimes   and magnetically confined fusion oriented plasmas persist for appreciable fractions of a second. Quiescent plasma sources, including fluorescent light sources, continuously produce plasmas whose lifetimes may be measured in hours, weeks, or years, depending on the cleanliness of the ionization system.

Lightning is a natural plasma resulting from electrical discharges in the earth's lower troposphere. The maximum time duration of a lightning flash is about 2s in which peak currents as high as 200 kA can occur. The conversion from air molecules to a singly ionized plasma occurs in few microseconds, with hundreds of Mega Joules of energy dissipated and plasma temperatures reaching ~ 3eV. The discharge channel avalanches at about one-tenth speed of light, and the high current carrying core expands to a diameter of a few centimeters. The total length of the discharge is typically 2–3 km.

Lightning has been observed on Jupiter, Saturn, Uranus, and Venus. The energy released in a single flash on the earth, Venus, and Jupiter is typically 6×10^8 J, 6×10^{10} J, and 2.5×10^{12} J, respectively.

Nuclear driven atmospheric plasmas were a notable exception to the generally short-lived energetic plasmas on earth. For example, the 1.4 megaton (5.9×10^{15} J) starfish detonation, 400 km above Johnston Island, on July 9, 1962, generated plasma from which artificial "Van Allen belts" of electrons circulating the earth were created.

(B) Near-Earth Plasma :

The earth's "ionosphere" and magnetosphere" constitute a cosmic plasma system that is readily available for extensive and detailed experimentation. It contains a rich variety of plasma a populations ranging from more than 10^6 cm^{-3} to less than 10^{-2} cm^{-3}, and temperatures from 0.1 eV to more than 10 keV.

The earth's magnetosphere is that region of space defined by the interaction of the solar wind with the earth's dipole-like magnetic field. It extends from approximately 100 km above the earth's surface, where the proton neutral atom collision frequency is equal to the proton gyrofrequency, to about ten earth radii (~63,800 km) in the and sunward direction and to several hundred earth radii in the anti-sunward direction. It is shown in figure 1.

Ionosphere :

The ionosphere is a layered plasma region closest to the surface of the earth whose properties change continuously during a full day. The ionosphere was first detected by radio waves and then bye radar.

First to be identified was a layer of molecular ionization, called the E-layer. This region extends over a height range of 90-140 km and may have a nominal density of 10^5 cm^{-3} during the periods of low solar activity.

A D-region underlies this with a nominal day time density of 10^3 cm^{-3}.Overlying the E region is the F layer of ionization, the major part of the ionosphere, starting at about 140 km. In the height range 100–150 km, strong electric currents are generated by a process analogous to that of a conventional electric generator, or dynamo. The region and in consequence, is often termed the dynamo region and may have densities 10^6 cm^{-3}. The F layer may extend.

1000 km in altitude where it eventually merges with the plasmas of the magnetopause and solar wind. The interaction of the supersonic solar wind with the intrinsic dipole magnetic field of the earth forms the magnetosphere whose boundary, called the magnetopause, separates interplanetary and geophysical magnetic fields and plasma environments.

Upstream of the magnetopause a collisionless bow shock is formed in the solar wind –magnetosphere interaction process. At the bow shock the solar wind becomes thermalized and subsonic and continues its flow around the magnetosphere as magneto sheath plasma, ultimately rejoining the undisturbed solar wind.

In the anti-solar direction, observations show that the earth's magnetic field is stretched out in an elongated geomagnetic tail to distances of several hundred earth radii. The field lines of the geomagnetic tail intersect the earth at high latitudes ($\sim 60°–75°$) in both the northern and southern hemisphere (polar horns), never the geomagnetic poles.

Deep within the major magnetosphere is the plasmosphere, a population of cold ($\leq 1eV$) iono-spheric ions and electrons corotating with the earth. Table 1 lists some typical values of parameters in the earth's magnetospheres.

(C) Solar, Stellar and Interstellar Plasma

Plasmas in the Solar System

The space environment around the various planetary satellites and rings in the solar system is filled with plasmas such as the solar wind, solar and galactic cosmic rays (high energy charged particles) and particles trapped in the planetary magnetospheres.

The nuclear core of the sun is a plasma at about a temperature of 1.5 keV

(1 eV = 11600 K). Beyond this, our knowledge about the Sun's interior is highly uncertain.

We do have information about the Sun's surface atmospheres which are known as follows : the photosphere the chromosphere, and the inner corona. These plasma layers are superposed on the Sun like onion skins.

The photosphere (T ~ 0.5 eV) is only a very weakly ionized atmosphere the degree of ionization being 10^{-4}–10^{-5} in the quiet regions and perhaps 10^{-6}–10^{-7} in the vicinity of sunspots.

The chromosphere (T ~ 4eV) extends 5000 km above the photosphere and is a transitions region to the inner corona.

The highly ionized inner corona extends some 10^{5} km above the photosphere. From a plasma physics point of new the corona is perhaps the most interesting region of the Sun. The corona is the sight of unstable magnetic field configurations, X-Ray emission, an plasma temperatures in the range 70–263 eV.

Table 1:-

Parameters associated with the Sun

Density:

At centre	$(20^{26}\,cm^{-3})$	$160g\,/\,cm^3$
At surface	$(10^{15}\,cm^{-3})$	$10^{-9}g\,/\,cm^3$
At corona	$(10^{7}\,cm^{-3})$	$10^{-16}g\,/\,cm^3$

Temperature

At centre	$1.5\,keV$
At surface	$0.5\,eV$
At sunspots	$0.37\,eV$
At chromosphere	$0.38 - 4.5\,eV$
In corona	$70eV - 263eV$
Emission	$3.826 \times 10^{26}\,W$
Magnetic Field Strengths :	
In sunspots	3.5 kG
Elsewhere on sun	1 to 100 G

Plasmas Beyond the Solar System:

Beyond the solar system we find a great-variety of natural plasmas in starts, in the stellar space, galaxies, intergalactic space, and far beyond. The galactic plasma has an extent equal to the dimensions of over galaxy itself $\sim$ 35kpc or 10^{21}m. The most salient feature of the galactic plasma are 10^{-3}G poloidal- toroidal plasma filaments extending nearly 250 light year (60 pc, 1.5×10^{18} m) at

the galactic center.

The vast regions of nearly neutral hydrogen found in the galaxy are weakly ionized plasmas. These regions extend across the entire widths of the galaxy and are sometimes found between interacting galaxies. They are deflected by the 21 cm radiation they emit, Galaxies may have bulk plasma densities of 10^{-1} cm^{-3} ; groups of galaxies, 3×10^{-2} cm^{-3}, and rich clusters of galaxies, 3×10^{-3} cm^{-3}.

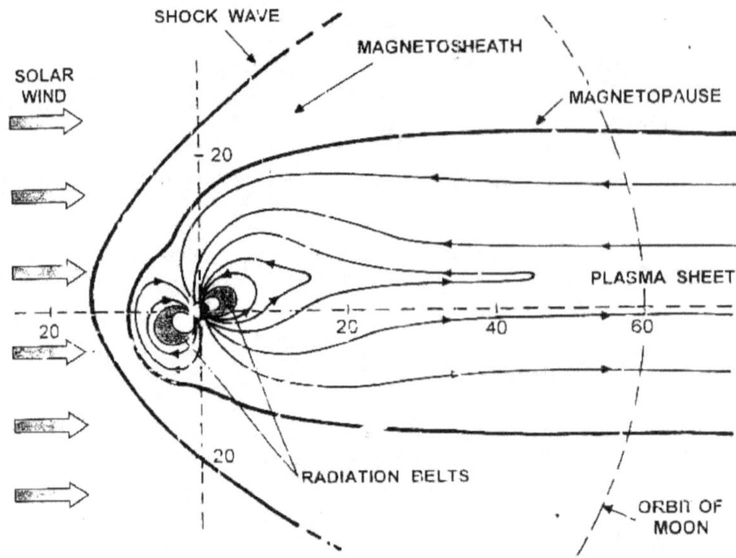

Figure 1. Schematic configuration of the magnetosphere.The dark descents represent the regions of trapped energetic particles (Van Allen radiation belts). The turbulent region between the shock wave (bow wave) and the magnetopause is known as magneto sheath. Geocentric distances are indicated in units of Earth radii.

2.9 Illustrative Examples

Example 1 Complete the following table, which lists typical parameters for the electrons in four different plasmas :

	n_0 (in m^{-3})	T (K)	ω_{pe} (rad s^{-1})	λ_D (m)	N_D
Solar atmosphere	10^{18}	10^4			
Solar corona	10^{13}	10^6			
Ionosphere	10^{10}	10^3			
Tokomak	10^{19}	10^8			

Sol. $\omega_{pe} = \left(\dfrac{n_0 e^2}{m\varepsilon_0} \right)^{\frac{1}{2}}$ is the plasma frequency

To obtain quantitative information, we make two further steps. First, we replace the universal constants e^2, m, and ε_0 by their numerical values. Secondly , we write

n_o as $\left(\dfrac{n_o}{Characteristic\ density}\right)\times characteristic\ density$. We choose the characteristic density to be that of a plasma in a present-day medium-size fusion experiment, 10^{19} m^{-3}. This enables us to write ω_{pe} given by equation (1) in the from

$$\omega_{pe} \approx 1.8\times10^{11}\left(\frac{n_o}{10^{19}\,m^{-3}}\right)^{\frac{1}{2}} rad\ s^{-1}.$$

Thus, when $n_0 = 10^{19}$ m^{-3}, $\omega_{pe} = 1.8 \times 10^{11}$ rad s^{-1}.

When $n_0 = 10^{18}$ m^{-3}, $\omega_{pe} = 1.8 \times 10^{11} \times (10^{-1})^{1/2} rad\ s^{-1} = 5.6 \times 10^{10}$ rad s^{-1}.

For solar corona $n_0 = 10^{13}$ m^{-3},

$$\text{Therefore } \omega_{pe} = 1.8 \times 10^{11}\left(\frac{10^{13}}{10^{19}}\right)^{\frac{1}{2}} \text{rad s}^{-1}$$

$$= 1.8 \times 10^{11} \times 10^{-3} \text{ rad s}^{-1}$$

$$= 1.8 \times 10^{8} \text{ rad s}^{-1}$$

$$\text{For ionosphere}: \omega_{pe} = 1.8 \times 10^{11}\left(\frac{10^{10}}{10^{19}}\right)^{\frac{1}{2}} \text{rad s}^{-1} = \frac{1.8\times10^{11}}{10^4}\times\left(\frac{1}{10}\right)^{\frac{1}{2}} \text{rad s}^{-1}$$

$$= 1.8 \times 10^{7} \times 0.3 = 5.6 \times 10^{6} \text{ rad s}^{-1}$$

We define the Debye length

$$\lambda_D = \left(\frac{\varepsilon_o K_B T}{n_0 e^2}\right)^{\frac{1}{2}} = \left(\frac{K_B T}{m}\right)^{\frac{1}{2}} \frac{1}{\omega_{pe}} \frac{1}{\omega_{pe}}$$

K_B is Boltzmann constant. We again choose typical values appropriate to medium-size fusion experiments, with characteristic thermal energy 1 keV and number density 10^{19} m^{-3}. Thus we can write λ_D

$$\lambda_D = 7.4\times10^{-5}\left\{\left(\frac{K_B T}{1keV}\right)\middle/\left(\frac{n_o}{10^{19}\,m^{-3}}\right)\right\}^{\frac{1}{2}} m$$

Another convenient formula for λ_D is

$$\lambda_D = 6.9\frac{\sqrt{\dfrac{T}{n}}}{Z} \text{ (cm)}$$

Where T and n are to be measured in units of K and cm^{-3}. Z = 1 for one-component-plasma.

We now calculate λ_D for solar atmosphere :

$$\lambda_D = 6.9\sqrt{\frac{T}{n(cm^{-3})}}\,cm$$

$$T = 10^4 \text{ K, } n = 10^{18}\,\frac{1}{m^3} = 10^{18} \times \frac{1}{10^6\,cm^3} = 10^{12}\,cm^{-3}$$

$$\therefore \qquad \lambda_D = 6.9\left[\frac{10^4}{10^{12}}\right]^{\frac{1}{2}}cm \quad = 6.9 \times 10^{-4}\text{ cm}$$

For solar corona : $\qquad T = 10^6 \text{ K,} \qquad n = 10^{13}\,\frac{1}{m^3} = 10^{13}\,\frac{1}{10^6}cm^{-3} = 10^7\text{ cm}^{-3}$

$$\therefore \qquad \lambda_D = 6.9\sqrt{\frac{10^6}{10^7}}cm = 2.18\text{ cm}$$

For Ionosphere :

$$T = 10^3 \text{ K, } n_0 = \frac{10^{10}}{10^7}\,cm^{-3} = 10^4\text{ cm}^{-3}$$

$$\therefore \qquad \lambda_D = 6.9\sqrt{\frac{10^3}{10^4}}cm = 2.18\text{ cm}$$

For Tokomak $\lambda_D = 6.9\sqrt{\dfrac{T}{n(cm^{-3})}} \qquad ,T = 10^8 \text{ K, } n = 10^{19}\text{ m}^{-3} = 10^{13}\text{ cm}^{-3}$

Thus $\lambda_D = 6.9\sqrt{\dfrac{10^8}{10^{13}}}cm = 6.9 \times 10^{-2} \times \sqrt{\dfrac{1}{10}}cm = 2.18 \times 10^{-2}$ cm.

In order to calculate the number of particles N_D which lie within a Debye length of the charge is

$$N_D = \frac{4\pi}{3}\lambda_D^3 n_0 = 1.38 \times 10^6\,Z^{-3}\left(\frac{n_0}{10^{12}\,cm^{-3}}\right)^{\frac{1}{2}}\left(\frac{T}{10^6\,K}\right)^{\frac{3}{2}}$$

n_0 is number density of electrons in $\dfrac{1}{cm^3}$.

For example in solar atmosphere

$$n_0 = 10^{18}\,\frac{1}{m^3} = 10^{12}\,\frac{1}{cm^3} \qquad ,T = 10^4 \text{ k, } Z = 1$$

$$\therefore N_D = 1.38 \times 10^6 \left(\frac{10^{12}\,cm^{-3}}{10^{12}\,cm^{-3}}\right)^{-\frac{1}{2}} \times \left(\frac{10^4\,K}{10^6\,K}\right)^{\frac{3}{2}} = 1.38 \times 10^6 \times 1 \times (10^{-2})^{3/2}$$

$$= 1.38 \times 10^6 \times 10^{-3} \quad = 1.38 \times 10^3 \text{ Ans.}$$

Similarly for other plasma.

2.10 Self Learning Exercise

Q.1 Estimate the plasma frequency, the Debye length of electrons and the plasma parameter for the ionospheric plasma.

$$(Ne \approx N_i = 10^7\,cm^{-3}, T_e \approx T_i \approx 1eV)$$

Q.2 Estimate the plasma frequency, the Debye length of electrons and the plasma parameter for the thermonuclear plasma $(N_e \approx N_i \simeq 10^{15}\,cm^{-3}, T_e \approx T_i \approx 10keV)$

2.11 Summary

In this chapter, we have introduced the fundamental parameters that characterize the collective, self-consistent behavior of plasmas. First we introduced the natural frequency of oscillation of the number density of electrons, which is the electron plasma frequency ω_{pe}. We have described various types of plasmas occurring in space, sun, an earth. We have introduced their characteristic value of the densities, temperature etc.

2.12 Glossary

Plasma: Plasma is an ionized gas in which all or a considerable part of the atoms have lost one or several of their electrons, thus becoming positive ions. Plasma consists of charged particles that respond collectively to electromagnetic forces. The charged particles are usually clouds or beams of electrons or ions or a mixture of electrons and ions, but also can be charged grains or dust particles.

Magnetopause: The interaction of the supersonic solar wind with the intrinsic dipole magnetic field of the earth forms the magnetosphere whose boundary, called the magnetopause, separates interplanetary and geophysical magnetic fields and plasma environments.

2.13 Answers to Self Learning Exercise

Ans.1: $\omega_{pe} = 1.7 \times 10^8 s^{-1}$, $r_{De} \approx 0.25cm$,Plasma parameter $\eta \approx 3 \times 10^{-5}$.

Ans.2: (Plasma frequency) $\omega_{pe} \approx 1.7 \times 10^{14} s^{-1}$, Debye length $\approx 1.5 \times 10^{-7}$ cm and plasma-parameter $\approx 1.5 \times 10^{-6}$.

2.14 Exercise

Q.1 Electron plasma frequency $\omega_{pe}\left(in\dfrac{rad}{s}\right)$ is defined as $\omega_{pe} = \left(\dfrac{n_0 e^2}{m \, \epsilon_0}\right)^{1/2}$ where n_0 is in the unit of $\dfrac{1}{m^3}$. Show that the plasma frequency defined by $f = \dfrac{1}{2\pi}\omega_{pe}$ (in the unit of Hertz) can be written as

$$f_p = a\sqrt{n(m^{-3})}\ \text{Hz.}$$

$$\omega_{pe} = 56.6\ n_e^{\frac{1}{2}}\left(in\dfrac{rad}{s}\right)$$

Q.2 Debye length $\lambda_D = \left(\dfrac{\epsilon_0 \, K_B T}{ne^2}\right)^{\frac{1}{2}}$. Show that it can be written as

$$\lambda_D = 6.9.0\left(\dfrac{T}{n}\right)^{\frac{1}{2}} \text{(in m)where T is in Kelvin, n is m } \dfrac{1}{m^3}$$

Q.3 Write approximate magnitudes of n_0, Temperature T, ω_{pe}, λ_D, $n_0\,\lambda_D^3$ of the following plasmas

(i) Solar corona (ii) Ionosphere (iii) Thermonuclear plasma.

References and Suggested Readings

1. Plasma physics   -F.F Chen.

2. Statistical Plasma physics Vol. I -S.Ichemann

3. Fundamentals of Plasma physics -J.A. Bittencourt

UNIT-3
Applications of Plasma Physics :
Controlled Thermonuclear Physics

Structure of the Unit

3.0 Objectives

In this chapter, we try to understand the problem of obtaining controlled nuclear fusion in the laboratory, and configure a device for the containment of high temperature plasma at a density and for a period of sufficient time for nuclear fusion reactions to take place, generating more energy required to create and confine the plasma.

The dynamics of conducting plasma fluid under the application of electric and magnetic field is important and thus working of Magnetohydrodynamic (MHD) generator has been given for understanding the production of electrical power. The importance of MHD mechanism is used in propulsion system.

3.1 Controlled Thermonuclear Reactions

Thermonuclear fusion is the process by which nuclei of low atomic weight such as hydrogen combine to form nuclei of higher atomic weight such as helium by using extremely high temperatures. Two isotopes of hydrogen, deuterium (composed of a hydrogen nucleus containing one neutrons and one proton) and tritium (a hydrogen nucleus containing two neutrons and one proton), provide the most energetically favorable fusion reactants. In the fusion process, some of the mass of the original nuclei is lost and transformed to energy in the form of high energy particles. Energy from fusion reactions is the most basic form of energy in the universe; our sun and all other stars produce energy through thermonuclear fusion reactions.

We shall first discuss the physics of fusion reactions rather generally, and then talk about the promising reactions for practical energy product ion.

Considering the fusion reaction

$$a + b \rightarrow fusion\ products + Kinetic\ Energy\ (few\ MeV)$$

where a and b are the initial nuclei. This fusion reaction cannot occur, however, unless nuclei can overlap, or get with a distance of about 10^{-13} cm from one another. In particular, they must overcome the Coulombic repulsion by having sufficient kinetic energy when they are separated. Sometimes, larger energy is required to overcome the centrifugal barrier effects for particular fusion with substantial probability. As the temperature of the plasma is increased, the contents of the plasma gain more kinetic energy and thus fusion may start to occur at the ignition temperature.

Beside the fusion energy, plasma has its own property of cooling off or radiating energy. Plasma does not radiate or lose energy rapidly as the black body does (proportional to T^4), and thus becomes prominent mechanism for radiation of energy in plasma, is known as Bremsstrahlung. If n is number density of ions, Z is their average charge, and T is the ion temperature, then the power radiated is proportional to $Z^2 n^2 \sqrt{k_B T}$. Further, the additional loss of energy accounts for escaping the particles. The worst situation comes when the escaping particles can damage the walls of the cavity, destroy the superconducting magnets or contaminate the plasma with impurities and release harmful radiations.

Therefore, Lawson criterion is necessary to produce more energy by fusion than is lost by Bremsstrahlung and escape. It is defined as the product of the plasma density n and the time τ for which the plasma remains confined above the ignition temperature. The most favorable Lawson criterion is $n\tau \geq 10^{14}$ for deuterium-tritium reaction.

The most commonly fusion reaction is the deuterium –tritium reaction:

$$D + T \rightarrow He^4 (3.5\,MeV) + n(14.4\,MeV)$$

The D-T reaction has the lowest ignition temperature of any fusion reaction. Most of the fusion energy is released in the form of 14 MeV neutrons, which finally converts into thermal energy. The kinetic energy of charged fusion products may be directly converted into electrical energy. Finally, the energetic neutrons may cause loss of radiation for this D-T reaction to materials in confinement system.

Another reaction commonly considered is D-D reaction: two sets of products occur with equal probability:

$$D + D \rightarrow T(1MeV) + p(3\,MeV)$$

$$\rightarrow He^3(.8MeV) + n(2.5\,MeV)$$

In this reaction only one third of the fusion energy is given to neutrons instead of breeding the tritium. Drawback of this reaction is higher ignition temperature (35 KeV) and Lawson number (10^{16}). Further, the following three reactions which have the desirable feature of producing no neutrons at all, but it still requires higher ignition temperature and Lawson number for fusion:

$$D + He^3 \rightarrow He^4 + p(18.4\,MeV)$$

$$p + Li^6 \rightarrow He^4 + He^3 (4\,MeV)$$

$$p + B^{11} \rightarrow 3He^4 (8.7\,MeV)$$

Plasma is dependent on the two parameters n and $k_B T$, i.e. Plasma density and thermal energy respectively. Plasma applications cover a wide range of n and $k_B T$.

Plasma density varies from 10^6 to 10^{34} /m^3 and thermal energy varies from 0.1 to 10^6 eV.

3.1.1 Temperature Requirements

Temperature is a measure of the average kinetic energy of particles, so by heating the material it will gain energy. After reaching sufficient temperature, given by the Lawson criterion, the energy of accidental collisions within the plasma is high enough to overcome the coulomb barrier and the particles may fuse together. For example, in a deuterium–tritium fusion reaction, the minimum energy required to overcome the coulomb barrier is 0.1 MeV. Converting between energy and temperature shows that the 0.1 MeV barriers would be overcome at a temperature in excess of 1.2 billion Kelvin.

There are two effects that lower the actual temperature needed. One is the fact that temperature is the average kinetic energy, which implies that some nuclei at this temperature would actually have much higher energy than 0.1 MeV, while others would be much lower. The other effect is quantum tunnelling. The nuclei do not actually have to have enough energy to overcome the coulomb barrier completely. If they have sufficient amount of energy, they can tunnel through the remaining barrier. For these reasons fuel at lower temperatures will still undergo fusion events, at a lower rate.

3.1.2 Confinement

In thermonuclear fusion, the major problem is how to confine the hot plasma? Actually, the plasma cannot be in direct contact with any solid material because of high temperature, thus it has to be located in a vacuum. But it is known that high temperatures also imply high pressures, and as the temperature raises the plasma tends to expand immediately and therefore some force is necessary to act against this thermal pressure. This force can be either gravitational (in stars), magnetic forces (magnetic confinement fusion reactors), or the fusion reaction which may occur before the plasma starts to expand, so in fact the plasma's inertia is keeping the material together.

Gravitational confinement:

The force, which is capable of confining the fuel well enough to satisfy the Lawson criterion, is gravity. The gravitational confinement is only found in stars however mass needed is so great. The least massive stars capable of sustained fusion are red dwarfs, while brown dwarfs are able to fuse deuterium and lithium if they are of sufficient mass. In heavy stars, after the supply of hydrogen is exhausted in their cores, their cores start fusing helium to carbon. In the most massive stars, the process is continued until some of their energy is produced by fusing lighter elements to iron. As iron has one of the highest binding energies, reactions producing heavier elements are generally endothermic.

Magnetic confinement:

Without the mass required to obtain a high gravitational field, fusion on earth must be controlled by means other than gravity. It is more feasible for controlled fusion purposes to work at low gas densities and increase the temperature to values considerably higher than that in the center of the sun. At these high temperatures, all matter is in the plasma state. Fortunately, plasma consists of a gas of charged particles that experience electromagnetic interactions and can be confined by a magnetic field of appropriate geometry. The magnetic field acts as a container that is not affected by heat, like ordinary solid containers, and cannot be a source of impurities which would prevent the fusion reaction. The motion of electrically charged particles is constrained by a magnetic field. In the absence of the magnetic field, heated particles will move in straight lines in random directions, quickly striking the walls of the container. When a uniform magnetic field is applied the charged particles will follow spiral paths encircling the magnetic lines of force. The motion of the particles across the magnetic field lines is restricted and so is the access to the walls of the container.

Inertial confinement:

In inertial fusion a pulse of radiation from a *driver* is focused on a small fuel capsule, rapidly heating its surface. An inward shock wave produced by the outward expansion of hot surface material compresses the pellet core. When the deuterium-tritium fuel in the core is compressed to a density of more than 10^{30} particles/m^3, ignition occurs at a temperature of 10^8 K. Inertia holds the pellet material together long enough for considerable thermonuclear burn to occur, releasing more energy than deposited by the driver source. The pulse of radiation provided by the driver may be light from a high energy laser source focused directly on the target, or, more effectively, X-rays created by laser light striking the internal walls of an hollow metallic cylinder which contains the fusion target. Other inertial fusion driver concepts involve heavy or light-ion accelerators. Very large laser facilities are presently approaching the conditions of ignition by inertial confinement.

Electrostatic confinement:

There are also electrostatic confinement fusion devices. These devices confine ions using electrostatic fields. The best known is the Fusor: This device has a cathode inside an anode wire cage. Positive ions fly towards the negative inner cage, and are heated by the electric field in the process. If they miss the inner cage they can collide and fuse. Ions typically hit the cathode, creating prohibitory high conduction losses. Also, fusion rates in fusors are very low due to competing physical effects, such as energy loss in the form of light radiation.

3.2   Magnetohydrodynamic (MHD) Generator

Background: When an electrical conductor is moved so as to cut lines of magnetic induction, the charged particles in the conductor experience a force (Lorentz) in a direction mutually perpendicular to the field $\vec{B}$ and to the velocity of the conductor. The negative charges tend to move in one direction and the positive charges in the opposite direction. This induced electric field, or motional emf, provides the basis for converting mechanical energy into electrical energy. At the present time nearly all electrical power generators utilize a solid conductor which is caused to rotate between the poles of a magnet. In hydroelectric generators, the energy required to maintain the rotation is supplied by the gravitational motion of river water. Turbo generators generally operate using a high-speed flow of steam or other gas. The heat source required to produce the high-speed gas flow may be supplied by the combustion of a fossil fuel or by a nuclear reactor (either fission or possibly fusion). Later on, it was recognized by Faraday that one could use fluid conductor as the working substance in a power generator. To verify this, Faraday immersed electrodes into the Thames river at either end of the Waterloo Bridge in London and connected the electrodes at mid span on the bridge through a galvanometer. Faraday reasoned that the electrically conducting river water moving through the earth's magnetic field should produce a transverse emf. Small irregular deflections of the galvanometer were observed. The production of electrical power through the use of a conducting fluid moving through a magnetic field is referred to as magneto-hydrodynamic power generation.

Thus, the interaction of moving conducting fluids with electric and magnetic fields provides for a rich variety of phenomena associated with electro-fluid-mechanical energy conversion. Effects from such interactions can be observed in liquids, gases, two-phase mixtures, or plasmas. An MHD generator is an energy conservation device which can be used with high temperature heat source like nuclear reactor. The maximum efficiency of such a system is determined by the temperature of the heat source. The MHD generator develops DC power and the conversion to AC is done using an inverter.

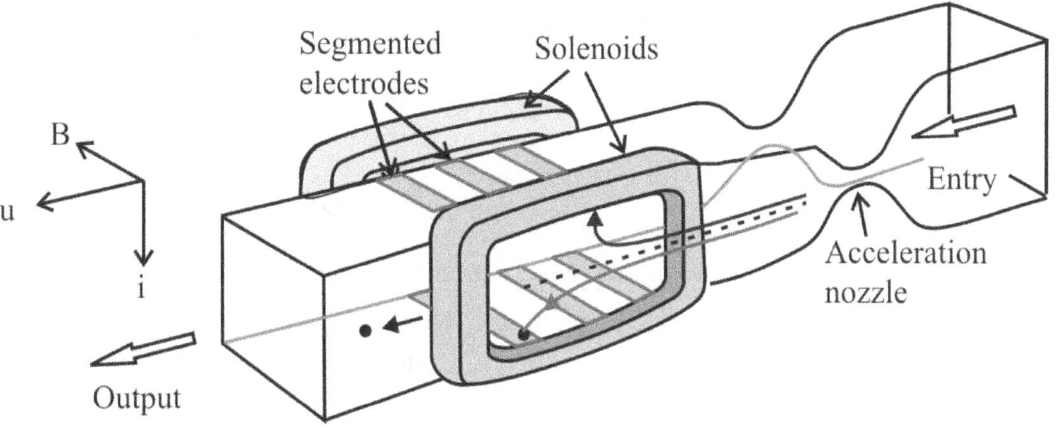

Figure 1

Principle: The principal of MHD power generation is based on Faraday's law of electromagnetic induction, which states that when a conductor and a magnetic field moves relative to each other, then voltage is induced in the conductor, which results in flow of current across the terminals. As the name implies, the magneto hydro dynamics generator shown in the figure1, is concerned with the flow of a conducting fluid in the presence of magnetic and electric fields.

Theory: In conventional generator or alternator, the conductor consists of copper strips (windings) while in an MHD generator the hot ionized gas or conducting fluid replaces the solid conductor. A field of magnetic induction B is applied transverse to the motion of an electrically conducting gas flowing in an insulated duct with the velocity $\vec{u}$. Charged particle will experience an induced electric field $\vec{u} \times \vec{B}$, which further develop an electric current in perpendicular direction to both

$\vec{u}$ and $\vec{B}$. This current is measured then by the pair of electrodes on opposite side of the duct in contact with the gas.

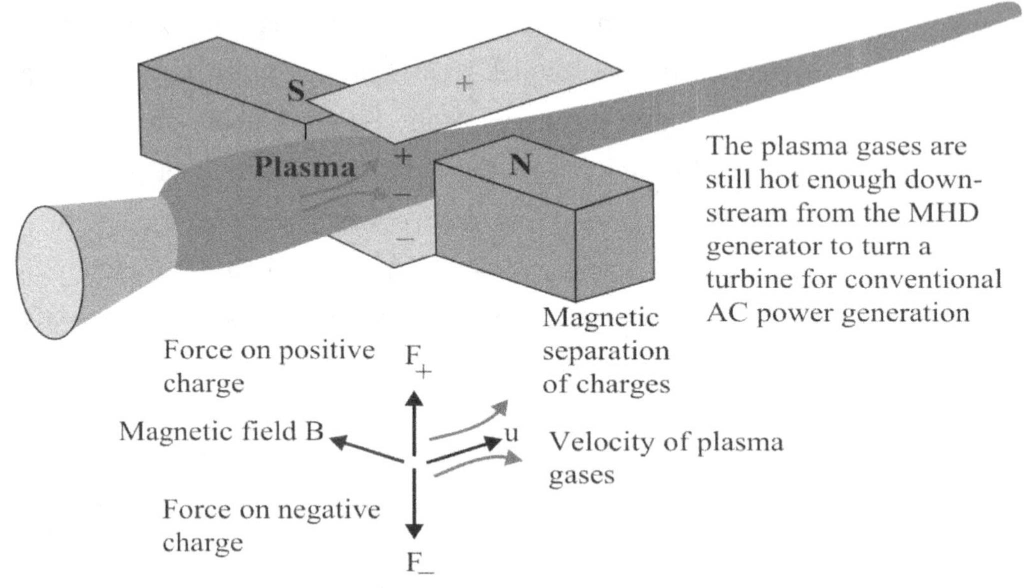

Figure 2

Electrodes in the MHD generator perform the same function as brushes in a conventional DC generator. Neglecting the magnitude of the current density for a weakly ionized gas is given by the generalized Ohm's law as

$$\vec{j} = \sigma(\vec{E} + \vec{u} \times \vec{B}) \tag{1}$$

Where $\vec{E}$ applied electric field is added to the induced field. In terms of coordinate system (according to figure)

$$j_y = \sigma(E_y - uB) \tag{2}$$

For open circuit $j_y = 0$ then electric field $E_y = uB$. For short circuit, $E_y = 0$ and $j_y = -\sigma uB$. For general load conditions, introducing a *loading parameter* $\left(0 \leq K \leq 1\right)$ as

$$K \equiv \frac{E_y}{uB} \tag{3}$$

$$\Rightarrow j_y = -\sigma uB(1 - K) \tag{4}$$

The negative sign indicates that the conventional current flows in the negative y-direction. Since the electrons flow in the opposite direction, the bottom electrode must serve as an electron emitter, or cathode, and the upper electrode is an anode.

The electric power delivered to the load per unit volume of a MHD generator gas is

$$P = \vec{j}_y . \vec{E}_y \tag{7}$$

$$P = \sigma u^2 B^2 K (1-K) \tag{8}$$

This power density has a maximum value, for $K = 1/2$, is

$$P_{max} = \frac{\sigma u^2 B^2}{4} \tag{9}$$

The rate of which energy is extracted from the gas by the electromagnetic field per unit volume is $-\vec{u}.\left(\vec{j} \times \vec{B}\right)$. Therefore, we define the electrical efficiency of a MHD generator as

$$\eta_e = \frac{\vec{j}.\vec{E}}{\vec{u}.(\vec{j} \times \vec{B})} \tag{10}$$

For the generator being discussed, efficiency $\eta_e = K$.

3.3 MHD System

The MHD generator needs a high temperature ionized gas source (either coolant from a nuclear reactor or more likely high temperature combustion gases generated by burning fossil fuels, coal) in a combustion chamber. The possible components of MHD system are shown in diagram below:

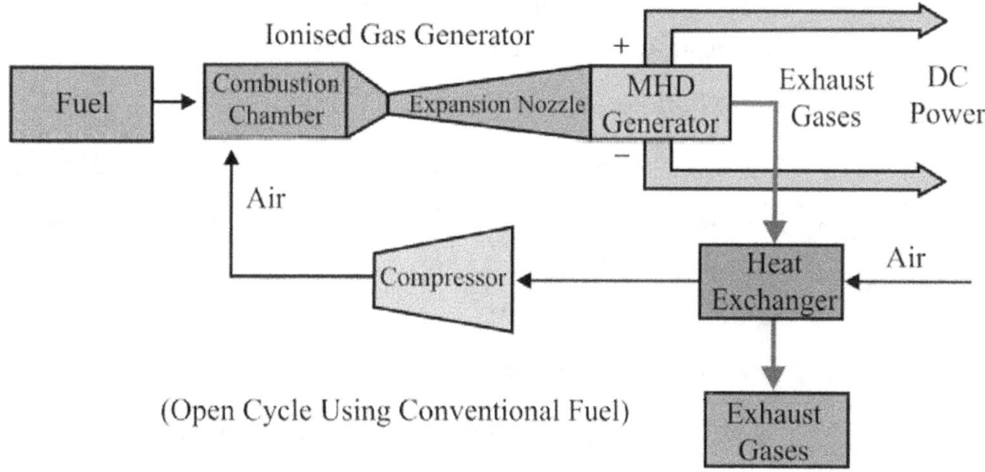

Figure 3

The expansion nozzle reduces the gas pressure and consequently increases the plasma speed through the generator duct to increase the output power. But unluckily, at the same time, the pressure drop causes the plasma temperature to fall, which also increases the plasma resistance.

The Plasma:

The main requirement of MHD system is creating and managing the conducting gas plasma since the system depends on the plasma having a high electrical conductivity. Therefore, to achieve high conductivity, the gas must be ionized by detaching the electrons from the atoms or molecules leaving the positively charged plasma. The plasma flows through the magnetic field at high speed, the flow of the positively charged particles providing the moving electrical conductor necessary for inducing a current in the external electrical circuit. Suitable working gases are obtained from combustion, noble gases, and alkali metal vapours.

Methods of Ionizing the Gas:

As we know that there are various methods for ionizing the gas, all of which depend on imparting sufficient energy to the gas. Mainly it is achieved by heating or irradiating the gas with X rays or Gamma rays. In addition to this, it has been suggested to use the coolant gases such as helium and carbon dioxide employed in some nuclear reactors as the plasma fuel for direct MHD electricity generation rather than extracting the heat energy of the gas through heat exchangers to raise steam to drive turbine generators. Seed materials such as Potassium carbonate or Cesium are often added in small amounts, typically about 1% of the total mass flow to increase the ionization and improve the conductivity, particularly of combustion gas plasmas.

Containment:

Since the plasma temperature is typically over 1000 °C, the duct containing the plasma must be constructed from non-conducting materials capable of withstanding these high temperatures. The electrodes must of course be conducting as well as heat resistant.

The Faraday Current:

A powerful electromagnet provides the magnetic field through which the plasma flows, and two electrodes on opposite sides of the plasma are employed across which the electrical output voltage is generated. The current flowing across the plasma between these electrodes is called the Faraday current. This provides the main electrical output of the MHD generator.

The Hall Effect Current:

The very high Faraday output current which flows across the plasma duct itself interacts with the applied magnetic field creating a Hall Effect current perpendicular to the Faraday current. The total current generated will be the vector sum of the transverse (Faraday) and axial (Hall effect) current components. The Hall Effect current, along the axis of the plasma, will constitute an energy loss unless it is captured in some way. Various configurations of electrodes have been devised to capture both the Faraday and Hall effect components of the current in order to improve the efficiency.

One of the configurations is to split the electrode pair into a series of segments physically side by side (parallel) but insulated from each other, with the segmented electrode pairs connected in series to achieve a higher voltage but with a lower current. The electrodes are skewed at a slight angle from perpendicular to be in line with the vector sum of the Faraday and Hall effect currents, as shown in the diagram below, thus allowing the maximum energy to be extracted from the plasma.

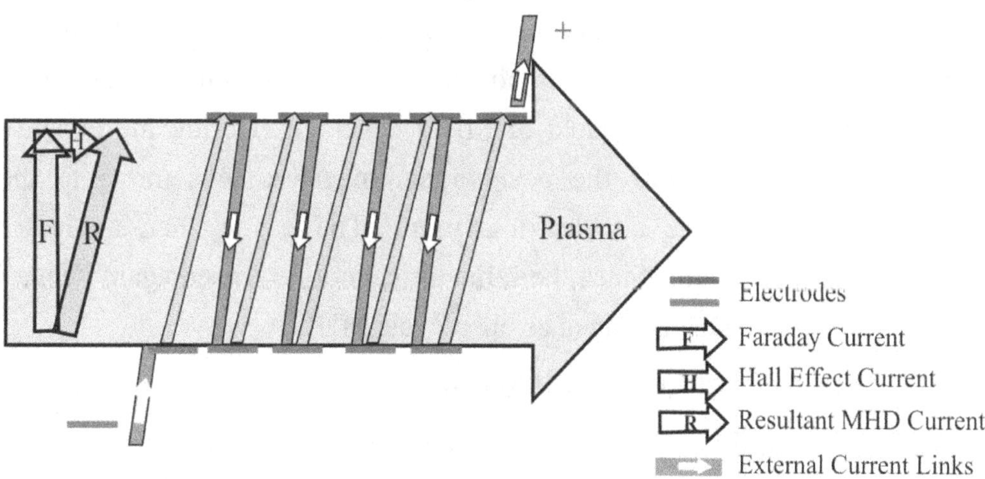

Figure 4

Power Output:

The output power is proportional to the cross sectional area and the flow rate of the ionized plasma. The conductive substance is also cooled and slowed in this process. MHD generators typically reduce the temperature of the conductive substance from plasma temperatures to just over 1000 °C.

Efficiency:

Typical efficiencies of MHD generators are around 10 to 20 percent mainly due to the heat lost through the high temperature exhaust. This limits the MHD's potential applications as a standalone device but they were originally designed to be used in combination with other energy converters in hybrid applications where the output gases (flames) are used as the energy source to raise steam in a steam turbine plant. Total plant efficiencies of 65% could be possible in such arrangements.

3.4 MHD Cycles and Working Fluids

Majorly there are two types of MHD cycles, namely Open Cycle MHD and Closed Cycle MHD. A brief account of these and the working fluids used are given below.

Open Cycle MHD System:

In open cycle MHD system, atmospheric air at very high temperature and pressure is passed through the strong magnetic field. Coal is first processed and burnet in the combustor at a high temperature of about 2700°C and pressure about 12 atp with pre-heated air from the plasma. Then a seeding material such as potassium carbonate is injected to the plasma to increase the electrical conductivity. The resulting mixture having an electrical conductivity of about 10 Siemens/m is expanded through a nozzle, so as to have a high velocity and then passed through the magnetic field of MHD generator. During the expansion of the gas at high temperature, the positive and negative ions move to the electrodes and thus constitute an electric current. The gas is then made to exhaust through the generator. Since the same air cannot be reused again hence it forms an open cycle and thus is named as open cycle MHD.

Closed Cycle MHD System:

In this system, the working fluid is circulated in a closed loop. Thus, in this case inert gas or liquid metal is used as the working fluid to transfer the heat. The liquid metal has typically the advantage of high electrical conductivity, hence the heat provided by the combustion material need not be too high. Contrary to the open loop system there is no inlet and outlet for the atmospheric air. Hence, the process is simplified to a great extent, as the same fluid is circulated time and again for effective heat transfer.

Advantages of MHD Generation:

(i)   Here only working fluid is circulated, and there are no moving mechanical parts. This reduces the mechanical losses almost to nil and makes the operation more dependable.

(ii) The temperature of working fluid is maintained the walls of MHD.

(iii) It has the ability to reach full power level almost directly.

(iii) The price of MHD generators is much lower than conventional generators.

(iv) MHD has very high efficiency, which is higher than most of the other conventional or non-conventional method of generation.

3.5 Plasma Propulsion

Humans have been developing propulsion methods for years, and there is a constant drive to develop faster, more powerful drive systems. Recent technological developments allow us to propel massive machines with an elaborate selection of thruster types. Some of these burn high energy fuel to generate thrust, whereas other devices use electricity and magnetic fields.

A magnetohydrodynamic drive or MHD propulsor is a method for propelling vessels using only electric and magnetic fields with no moving parts, using magnetohydrodynamics. The working principle involves electrification of the propellant (gas or water) which can then be directed by a magnetic field, pushing the vehicle in the opposite direction.

The major problem with MHD is that with current technologies it is more expensive than a propeller driven by an engine. The extra expense is from the large generator that must be driven by an engine. Such a large generator is not required when an engine directly drives a propeller. MHD is attractive to engineers because it has no moving parts, which means that a good design might be silent, reliable, efficient and inexpensive. If fuel cells become common, MHD propulsors may have lower costs in some applications than electric motors driving propellers. In magnetohydrodynamic thrusters magnetic field is generated by passing an electric

Current through a liquid conductor, such as sea water. Using another magnetic field, the liquid can be pushed in a chosen direction, therefore generating thrust.

A number of experimental methods of spacecraft propulsion are based on magnetohydrodynamic principles. In these the working fluid is usually plasma or a thin cloud of ions. Some of the techniques include various kinds of ion thruster, the magnetoplasmadynamic thruster, and the variable specific impulse magnetoplasma rocket. Plasma rockets open up new and exciting possibilities for fast space transportation. The same MHD generator principle in reverse has been used to develop engines for interplanetary missions. A current is driven through a plasma by applying a voltage to the two electrodes. The $\vec{j} \times \vec{B}$ force shoots the plasma out of the rocket and the ensuing reaction force accelerates the rocket. The plasma ejected must always be neutral, otherwise, the spaceship will charge to a high potential.

Plasma rockets are dependent upon the availability of electric power, which is still limited in space. In space, electricity is generated mainly by solar arrays, and major changes in solar technology have increased the availability electric power. Another problem with plasma rockets is the confinement of plasma. The temperature of plasma is comparable to temperature in the interior of sun. No known material could survive direct contact with such plasma. A magnetic field can be constructed to both heat and guide a hot plasma. So that plasma never touches material walls.

In future, plasma propulsion could be very valuable in number of ways. For example, low power rockets could play an important role in space missions. Fusion driven plasma rockets could be instrumental in carrying us in space.

3.6   Self Consistent Formulation

The interaction of charged particles with electromagnetic fields is defined by the Lorentz force. For a particle of charge q and mass m, moving with v velocity in the presence of electric field (E) and magnetic field $\vec{B}$, the equation of motion is

$$\frac{d\vec{p}}{dt} = q\left(\vec{E} + \vec{u} \times \vec{B}\right), \tag{11}$$

It is possible to describe the plasma dynamics by solving the equation of motion for each particle in the plasma under the influence of externally applied fields and internally applied fields by the other plasma particles.

A self consistent formulation can be used since the internal fields associated with the presence and motion of the plasma particles. The electromagnetic fields obey Maxwell's equations. The plasma charge and current density can be given as

$$\text{Plasma charge density: } \rho_p = \frac{1}{\Delta V}\sum_i q_i \tag{12}$$

$$\text{Plasma current density: } \vec{j}_p = \frac{1}{\Delta V}\sum_i q_i \vec{v}_i \tag{13}$$

where the summation is over all charged particles contained in a small volume element ΔV.

Although this self consistent approach is possible only in principle, it cannot be carried out in practice without average scheme, as a large number of variables are involved. According to the classical mechanics, in order to determine position and velocity of each particle in the plasma it is necessary to know initial position and velocity of each particle. For a system like plasma, these initial conditions are unknown because the observable macroscopic properties of plasma are due to the average collective behavior of a large number of particles and thus it is not possible to know the detailed individual motion of each particle.

3.7 Self Learning Exercise

Q.1 What do you mean by Controlled thermonuclear reactions? How the fusion energy of plasma is increased?

Q.2 What do you mean by "confinement of plasma"? Discuss the various types of confinement mechanism.

3.8 Summary

In this chapter, we have discussed thermonuclear fusion process, with need of constraining the temperature and confinement for plasma production. Lawson criterion is used to produce more energy by fusion than is lost by Bremsstrahlung and escape. On the basis of law of electromagnetic induction, the Magnetohydrodynmic (MHD) generator has been discussed in detail. The working of MHD system (Open and closed cycle) has been described for complete understanding. Finally, this chapter ends with the discussion of propulsion system followed by the self consistent formulation of plasma.

3.9 Glossary

ATP: stands for Ambient Temperature and Pressure.

MHD: Magnetohydrodynmic

Siemens: Unit of electrical conductivity.

Thermonuclear: Reactions that occur only at very high temperatures.

Pellet: A small, rounded, compressed mass of a substance.

Fusor: A fusor is a device that uses an electric field to heat ions to conditions suitable for nuclear fusion.

3.10 Exercise

Q.1 Discuss the principle and working of Magnetohydrodynamic generator and obtain the expression of efficiency. To improve the efficiency of MHD generator, how the Hall effect currents is managed?

Q.2 Give a brief account of Open cycle and closed cycle MHD system.

Q.3 Write a short note on Propulsion system.

References and Suggested Readings

1. A.R. Choudhuri: The physics of fluids and plasmas (Cambridge University Press, 1998)

2. S. Glasstone & R. H. Lovberg: Controlled Thermonuclear Reactions (R E Krieger Publishing Company, Huntington, NewYork 1975)

3. Francis F Chen: Introduction to Plasma Physics and Controlled Fusion (Plenum Press, NewYork, 1984)

4. P.H. Roberts: An introduction to magnetohydrodynamics (American Elsevier, 1967)

5. L.D. Landau & E.M. Lifshitz: Fluid mechanics (Pergamon Press, 2nd edition, 1987).

5. Referred some useful and trusted internet resources.

UNIT-4
Charged Particle Motion in Electromagnetic Field and Drifts

Structure of the Unit

4.0 Objectives

To learn

- Charged particle motion in a constant uniform electric field.
- Charged particle motion in uniform magnetostatic fields.
- Charged particle motion in nonuniform magnetostatic fields.
- Invariance of the orbital magnetic moment and of magnetic flux.
- Magnetic mirror effect
- The Longitudinal adiabatic invariants
- Drift due to an external force ; $\vec{E} \times \vec{B}$ drift.
- Gradient drift
- Curvature drift
- Combined gradient -curvature drift

4.1 Introduction

Every Plasma has a natural tendency to disperse unless there is some restraining force, the energetic particles that compose the plasma will travel away from their initial positions at high velocity and the plasma will cease to exist. Plasmas are prevented from dispersing by magnetic fields, which act on the charged particles through the Lorentz force. The dynamics of charged particles therefore form an important area of plasma physics, which we shall examine in this chapter. We start by considering the simplest possible example, which consists of a charged particle moving in a spatially uniform magnetic field $\vec{B}$ which does not vary in time.

4.2   The Equation Motion for a Particle of Charge Q under the Action of the Lorentz Force

The interaction of charged particles with electromagnetic fields is governed by the " Lorentz force" . For a particle of charge q and mass m, moving with velocity $\vec{v}$, in the presence of electric field $\vec{E}$ and magnetic field $\vec{B}$, the equation of motion is

$$\frac{d\vec{p}}{dt} = q(\vec{E} + \vec{v} \times \vec{B})$$

where $\vec{p} = m\vec{v}$ denotes the particle momentum.

The electromagnetic fields obey Maxwell equations :

$$\vec{\nabla} \times \vec{E} = -\frac{\partial \vec{B}}{\partial t} \quad , \quad \vec{\nabla} \times \vec{B} = \mu_o\left(\vec{J} + \epsilon_o \frac{\partial \vec{E}}{\partial t}\right)$$

$$\vec{\nabla}.\vec{E} = \frac{\rho}{\epsilon_o} \quad , \quad \vec{\nabla}.\vec{B} = 0$$

where ρ, $\vec{J}$, ϵ_o, and μ_o denote respectively, the total charge density, total electric current density, the electric permittivity, and magnetic permeability of free space.

Charged particle in an electric field obeys the differential equation:

$$\frac{d\vec{p}}{dt} = q\vec{E} \tag{1}$$

For a constant electric field, it can be integrated, giving $\vec{p}(t) = q\vec{E}t + \vec{p}_0$ (2)

where $\vec{p}_0 = \vec{p}(0)$ denotes the initial momentum.

Using the nonrelativistic expression

$$\vec{p} = m\vec{v} = m\frac{d\vec{r}}{dt} \tag{3}$$

and integrating, we obtain the following expression for the particle position $\vec{r}(t)$;

$$\vec{r}(t) = \frac{1}{2}\left(\frac{q\vec{E}}{m}t^2\right) + \vec{V}_0 t + \vec{r}_0 \tag{4}$$

where $\vec{r}_0$ is the initial position and $\vec{V}_0$ the initial velocity of the particle.

If $q > 0$, the particle moves with constant acceleration $\frac{q\vec{E}}{m}$ in the direction of $\vec{E}$. If $q < 0$ then the particle moves in the opposite direction of $\vec{E}$

4.3   Motion of a Charged Particle in Constant, Uniform Magnetic Field

* The equation of motion is (in the non-relativistic time.

$$\frac{d}{dt}(m\vec{v}) = q(\vec{v} \times \vec{B}) \tag{1}$$

$$\left(\frac{v}{c} << 1\right)$$

*Let the magnetic field $\vec{B}$ be in z-direction, i.e. $\vec{B} = (0, 0, B)$. From eq.(1), we obtain

$$\dot{v}_x = \frac{q}{m}B\ v_y \tag{2}$$

$$\dot{v}_y = -\frac{q}{m}B\ v_x \tag{3}$$

$$\dot{v}_z = 0 \tag{4}$$

From eq. (4), it follows

$$v_z = \text{constant} = v_0 \text{ (say)} \qquad \text{or} \qquad \frac{dz}{dt} = v_0$$

Integrating

$$z = v_0 t + z_0 \tag{5}$$

That is the particle moves in the direction of magnet field with constant velocity. The trajectory of the particle in the x-y plane i.e. in a plane perpendicular to $\vec{B}$ field is obtained from equations (2) and (3)

Substituting v_y from eq. (2) into eq. (3), we obtain

$$\ddot{v}_x = -\left(\frac{qB}{m}\right)^2 v_x \tag{6}$$

or $\ddot{v}_x + \Omega_c^2 v_x = 0$, where $\Omega_c = \frac{qB}{m}$ is called the cyclotron frequency of the charged particle. Eq. (6) can be rewritten as

$$\frac{d^2 v_x}{dt^2} + \Omega_c^2 v_x = 0 \tag{7}$$

This is the homogeneous differential equation for a harmonic oscillator of frequency Ω_c, whose solution is

$$v_x(t) = v_\perp(\Omega_c t + \theta_0) \tag{8}$$

Where $v_\perp$ is the constant speed of the particle in the (x,y) plane (normal to $\vec{B}$), θ_0 is a constant of integration that depends on the relation between the initial velocities $v_x(0)$ and $v_y(0)$, according to

$$\tan\theta_0 = \frac{v_y(0)}{v_x(0)} \tag{9}$$

To determine $v_y(t)$, we substitute (3) in (2) to obtain

$$\ddot{v}_y + \Omega_c^2 v_y = 0 \text{ , whose solution is}$$

$$v_y(t) = -v_\perp \sin(\Omega_c t + \theta_0) \tag{10}$$

We note that $\quad v_x^2 + v_y^2 = v_\perp^2$

The equation (8) can be integrated to give $x(t) = +\dfrac{v_\perp}{\Omega_c}\sin(\Omega_c t + \theta_0) + X_0$ $\qquad$ (11)

Similarly, integrating (10), we obtain

$$y(t) = \dfrac{v_\perp}{\Omega_c}\cos(\Omega_c t + \theta_0) + Y_0 \qquad (12)$$

from (11) and (12) we find that

$$(x-X_0)^2 + (y-Y_0)^2 = \left(\dfrac{v_\perp}{\Omega_c}\right)^2 = r_c^2 \qquad (13)$$

The particle trajectory in the plane normal to $\vec{B}$ is therefore a circle with center at (X_0,Y_0) and radius equal to $\left(\dfrac{v_\perp}{\Omega_c}\right)$ and the particle moves along helical path (Fig.1)

The constants of integration (X_0, Y_0) are the coordinates of the centre of a circular motion. The amplitude $\dfrac{v_\perp}{\Omega_c}$ that occurs in (11) and (12) is known as the gyro-radius of the charged particle (also known as Larmor radius of the electrons).

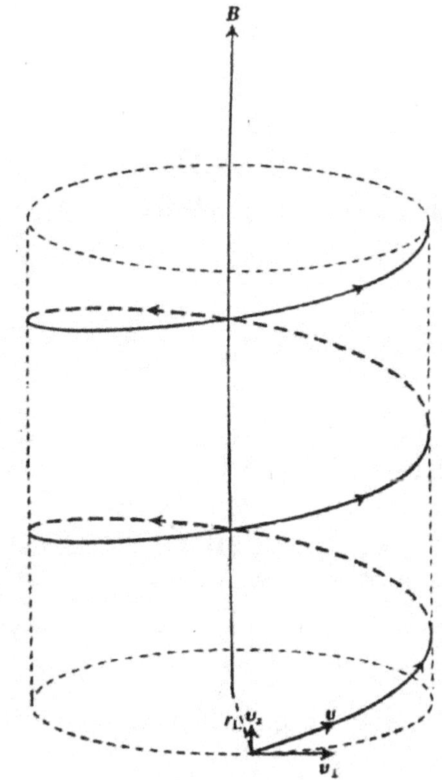

Figure1: *Helical path of an electron in a uniform magnetic field.*

4.4 Illustrative Examples

Example 1.   Find r_c (Larmor radius of electrons)

Sol:
$$r_c = \frac{v_\perp}{\Omega_c} = \frac{v_\perp}{\dfrac{eB}{m}} \tag{14}$$

The Larmor radius can be expressed as

$$r_c = 7.6 \times 10^{-5} \left(\frac{v_\perp}{1.3 \times 10^7 m\ s^{-1}} \right) \left(\frac{B}{1Telsa} \right)^{-1} m \tag{15}$$

r_c is the distance from the centre of the circular motion to the electron itself. The normalizing velocity in eq. (14) is that of an electron of energy 1kev.

Example 2   Find the magnetic moment associated with the circulating electron orbit in the presence of magnetic field.

Sol. We note that the circulating electron constitute a current. If we project the motion of the electron onto a plane which is perpendicular to the magnetic field, a charge $-e$ passes a given point on its orbit every $\dfrac{2\pi r_c}{v_\perp}$ seconds, therefore the

$$\text{current } I = \frac{ev_\perp}{2\pi r_c} \tag{16}$$

The associated magnetic moment μ is the product of this current with area enclosed by the orbit. Hence

$$\mu = \pi r_c^2\, I$$

$$= \frac{\frac{1}{2}mv_\perp^2}{B} \tag{17}$$

The magnetic field to which μ gives rise opposes the applied field $\bar{B}$: This diamagnetic effect is by eq. (17), proportional to the perpendicular energy.

4.5 Guiding Centre Drift due to Non-Magnetic Forces

From Eqs. (11), (12), and (5) that the path of an electron in a constant homogeneous magnetic field is helical, The helix is produced by uniform circular motion about a point that moves with constant velocity parallel to the magnetic field.

We write the electron position as

$$\vec{x} = \vec{x}_c + \vec{r}_c \qquad (\vec{r}_c \text{ is Larmor radius})$$

$$\text{or} \qquad \vec{x} = \vec{x}_c + \frac{\vec{v} \times \vec{\Omega}_{ce}}{\Omega_{ce}^2} \qquad (18)$$

which implicitly defines the guiding centre position: (See Fig. 2)

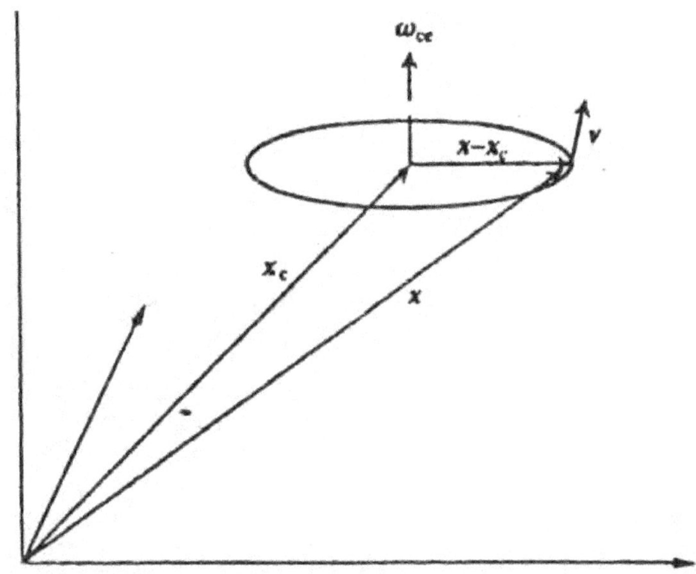

Figure 2: *Guiding center position x_c*

• Differentiating eq. (18) with respect to time, and using

$$\dot{\vec{v}} = \vec{\Omega}_{ce} \times \vec{v}, \text{ we find}$$

$$\dot{\vec{x}}_c = v_z \hat{e}_z \qquad (\hat{e}_z \text{ is unit vector in z-direction}) \qquad (19)$$

• From the Fig. (2) we also note that $\left| \vec{x} - \vec{x}_c \right| = r_c$

Now $\vec{x}_c$ is the mean position of the electron if the rapid rotation with frequency Ω_{ce} is averaged out. This mean position often contains all the information.

• By calculating $\vec{x}_0$, we can follow the path of the electron over a timescale which is long compared to $\dfrac{1}{\Omega_{ce}}$

•Consider, for example, and electron subjected to an impulsive collision, which by definition leaves its position unchanged but instantaneously changes $\vec{v}$ to $\vec{v} + \Delta\vec{v}$

$$\text{Then } \vec{x} = \vec{x}_{c_1} + \frac{\vec{v} \times \vec{\Omega}_{ce}}{\Omega_{ce}^2} = \vec{x}_{c_2} + \frac{(\vec{v} + \Delta\vec{v}) \times \vec{\Omega}_{ce}}{\Omega_{ce}^2} \qquad (20)$$

The instantaneous step $\Delta\vec{x}_c$ in guiding centre position perpendicular to the magnetic field is a measure of the effect of the instantaneous momentum transfer, (Figure 3)

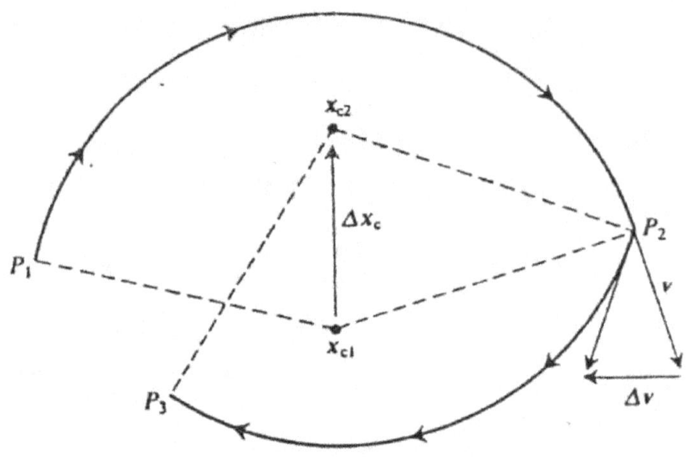

Magnetic field direction $\otimes$

Figure 3: *Instantaneous guiding centre step $\Delta\bar{x}$ due to impulsive collision.*

Consider now a continuous non-magnetic force $\bar{F}$ per unit mass. The force equation becomes

$$m\dot{\bar{v}} = -e\bar{v}\times\bar{B} + \bar{F}' \tag{21}$$

$$\text{or } \dot{\bar{v}} = -\bar{v}\times\bar{\Omega}_c + \frac{\bar{F}'}{m}$$

$$\text{or } \dot{\bar{v}} = \bar{\Omega} + \bar{\Omega}_c + \frac{\bar{F}'}{m}$$

$$\text{or } \dot{\bar{v}} = \bar{\Omega}_{ce}\times\bar{v} + \bar{F} \quad (\because \frac{\bar{F}'}{m} = \bar{F}) \tag{22}$$

Differentiating (20) with respect to time and using eqn. (22), we find

$$\dot{\bar{x}}_c = v_z\hat{e}_z + \frac{\bar{\Omega}_{ce}\times\bar{F}}{\Omega_{ce}^2} \tag{23}$$

The second term on the right-hand side is the guiding centre drift velocity, $\bar{v}_d$
It is perpendicular both to the applied force and to the magnetic field.

• As an example, let us take an electric field, for which $\bar{F} = \dfrac{-e\bar{E}}{m}$ .The

perpendicular drift of the guiding centre is $\quad \bar{v}_d = -\dfrac{e}{m}\dfrac{\bar{\Omega}_{ce}\times\bar{E}}{\Omega_{ce}^2} = \dfrac{\bar{E}\times\bar{B}}{B^2}$ $\tag{24}$

This is a typical example of the three components of electron motion : rapid perpendicular rotation with frequency Ω_{ce} at a radius r_c from $\bar{x}_c$; particle motion with constant velocity v_z ; and a slow perpendicular drift $\bar{v}_d$.

4.6 Guiding Centre Drift due to Magnetic Forces

In most of the plasma devices, we find that the magnetic field is not uniform in space or time. The division of particle motion into three distinct components ceases to be exact.

This division, however remains a useful approximation so long as the variation of the magnetic field remains small on the scale of the cyclotron motion. That is

$$\left(\frac{2\pi}{\Omega_{ce}}\right)\frac{\partial B}{\partial t} << B \tag{25a}$$

$$r_L \left|(\nabla B)_\perp\right| << B \tag{25b}$$

$$v_z\left(\frac{2\pi}{\Omega_{ce}}\right)\left|(\nabla B)_{||}\right| << B \tag{25c}$$

We shall assume eq. (25 a–c) to hold in the discussion which follows.

We consider first a magnetic field whose strength increases in a direction perpendicular to the direction of the field itself. The field strength experienced by particle changes periodically as it circulates with frequency Ω_{ce}.

• The periodic shortening and lengthening of r_L causes the guiding centre to drift perpendicular to the magnetic field (See Fig. 5). For clarity, the gradient in magnetic field strength has been replaced by a discontinuity, so that this diagram actually violates the condition (25b) (Fig.4)

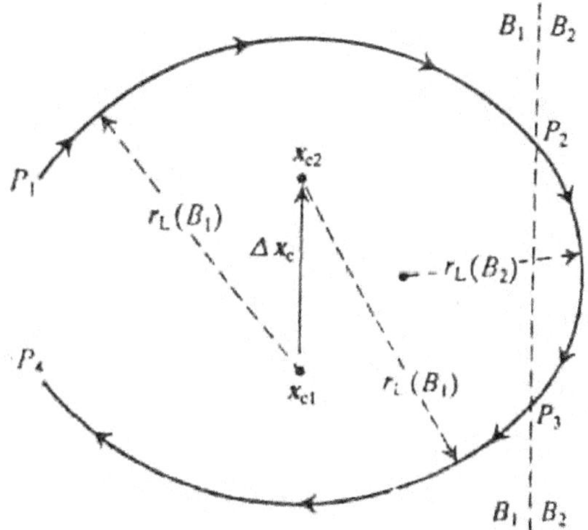

Magnetic field direction $\otimes$;Field Strength $B_2 > B_1$

Figure 4

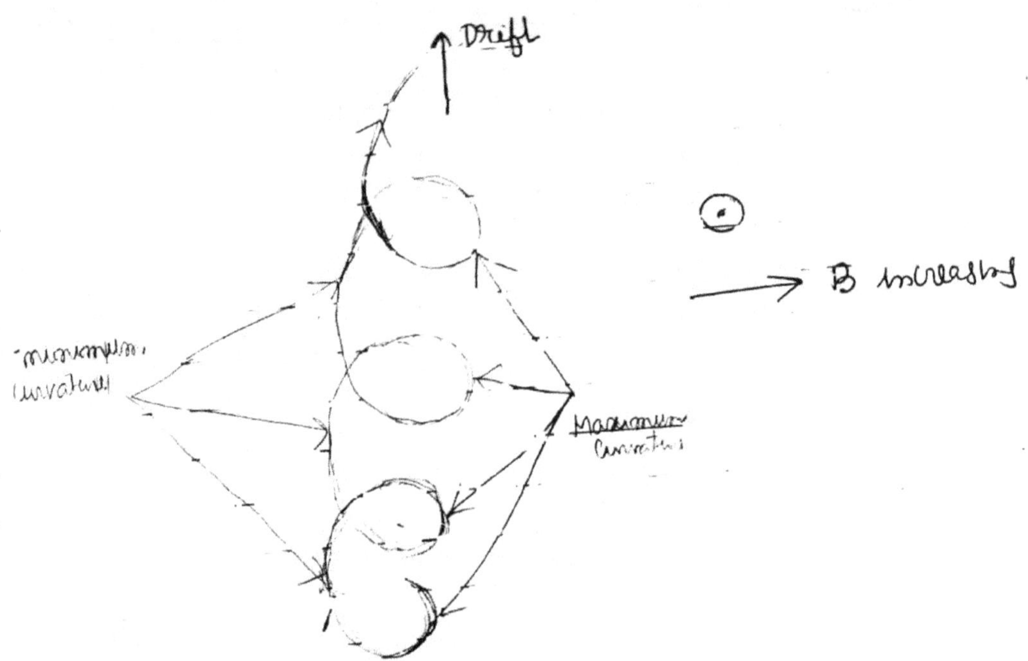

Figure 5 : Indicating *the origin of the drift of the guiding centre in a magnetostatic field $\vec{B}$ that a gradient variation of B in a direction perpendicular to $\vec{B}$.*

• The direction of this drift depends on the sense in which the particle is rotating, and is therefore opposite for electrons and ions.

$$\text{Let} \qquad \vec{B} = (B_0 + y\frac{\partial B_z}{\partial y})\hat{e}_z \qquad\qquad (26)$$

and define a scale length $\qquad l_c = \left\{ \frac{1}{B_0}\frac{\partial B_z}{\partial y} \right\}^{-1}.$

• Using eqn. (26) in (1), and denoting the cyclotron frequency $\frac{eB_0}{m}$ of the plasma at $y=0$ by Ω_{ce}, the electron motion is governed by:

$$\frac{d}{dt}(m\vec{v}) = -e\left(\vec{v} \times (B_0 + y\frac{\partial B_z}{\partial y})\hat{e}_z \right)$$

$$\text{or} \qquad \dot{v}_x = -\frac{e}{m}\left[V_y\,B_0 + V_y\,y\,\frac{\partial B_z}{\partial y} \right]$$

$$\text{or} \qquad \dot{v}_x = -\Omega_{ce}V_y - \frac{\Omega_{ce}}{l_c}y\,v_y \qquad\qquad (27a)$$

$$\text{and} \qquad \dot{v}_y = \Omega_{ce}v_x + \frac{\Omega_{ce}}{l_c}y\,v_x \qquad\qquad (27b)$$

In the uniform field limit $l_c \longrightarrow \infty$ and eqns. (27a-b) reduce to eqs. (2,3).

We now use the uniform field orbit eqs. (8,10) and calculate the effect of the additional terms in eqs. (27a, b); setting $x_0=y_0=0$ for convenience:

$$-\frac{\Omega_{ce}}{l_c}\,y\,v_y = -\frac{\Omega_{ce}}{l_c}\,\frac{v_\perp}{\Omega_{ce}}\cos(\Omega_{ce}t+\phi).(-v_\perp\sin(\Omega_{ce}t)$$

or

$$-\frac{\Omega_{ce}}{l_c}\,y\,v_y = \left(\frac{\Omega_{ce}}{2l_c}\right)\left(\frac{v_\perp}{\Omega_{ce}}\right)v_\perp\sin 2(\Omega_{ce}t+\phi)$$

or

$$-\frac{\Omega_{ce}}{l_c}\,y\,v_y = \frac{\Omega_{ce}}{2l_c}\,r_L\,v_\perp\sin 2\,(\Omega_{ce}t+\phi) \tag{28a}$$

where $\quad r_L = \dfrac{v_\perp}{\Omega_{ce}} \qquad$ (Larmor radius)

Similarly the additional term in the expression for $\dot{v}_y$ (viz. eqn. 27b) is

$\dfrac{\Omega_{ce}}{l_c}\,y\,v_x$ becomes after substitution of y and v_x from eq. 12 and eq.(8):

$$\frac{\Omega_{ce}}{l_c}\,y\,v_x = \frac{\Omega_{ce}}{l_c}\,\frac{v_\perp}{\Omega_{ce}}\cos(\Omega_{ce}\,t+\phi).\,v_\perp\cos(\Omega_{ce}t+\phi)$$

or

$$\frac{\Omega_{ce}}{l_c}\,y\,v_x = \frac{\Omega_{ce}r_L\,v_\perp}{2l_c}\left(1+\cos 2(\Omega_{ce}t+\phi)\right) \tag{28b}$$

From eq. (28b) we note that there is an extra non-oscillatory contribution to $\dot{v}_y$ only; all other terms arising from the spatial non-uniform of the magnetic field strength average to zero over a single orbit.

• The average additional acceleration of the electron is $\bar{a}=(0,\dfrac{v_\perp^2}{2l_c},0)$. Substituting $\bar{a}$ for the force per unit mass $\vec{F}$ in the drift velocity formula eqn. (23):

$$\vec{v}_d = \frac{\Omega_{ce}\times\vec{F}}{\Omega_{ce}^2} = \frac{\Omega_{ce}}{\Omega_{ce}^2}\frac{v_\perp^2}{2l_c}(\hat{e}_z\times\hat{e}_y)$$

$$\vec{v}_d = -\frac{r_L\,v_\perp}{2l_c}\hat{e}_x \tag{29}$$

Note that v_d is slower $v_\perp$ by the factor $\left(\dfrac{2l_c}{r_L}\right)$ which is assumed large.

The additional drift velocity $\bar{v}_d$ is in a direction perpendicular both to $\vec{B}_0$ and to ∇B we can express the drift velocity $\bar{v}_d$ in a general way as

$$\bar{v}_d = \frac{r_L\,v_\perp}{2}\,\frac{\vec{B}\times\vec{\nabla}B}{B^2} \tag{30}$$

Fig.3 depicts this drift.

4.7 Drift in the Case of Gradient in the Magnetic Field Strength Parallel to the Direction of the Magnetic Field

We now consider the case of a gradient in the magnetic field strength parallel to the direction of the magnetic field : $\dfrac{\partial B_z}{\partial z} \neq 0,$

From the equation $\vec{\nabla}.\vec{B} = 0$ and expressing it in the cylindrical coordinates (R, θ, z),

$$\frac{1}{R}\,\frac{\partial}{\partial R}(R\,B_R) + \frac{\partial B_z}{\partial z} = 0, \qquad\qquad \text{we find}$$

or $\qquad \dfrac{B_R}{R} + \dfrac{\partial B_R}{\partial R} = -\dfrac{\partial B_z}{\partial Z}$

or $\qquad \dfrac{2B_R}{R} \approx -\dfrac{2B_z}{\partial z}$

$$\Rightarrow B_R = -\frac{R}{2}\left(\frac{\partial B_z}{\partial z}\right) \tag{31}$$

This small radial field is crucial to the particle dynamics of the system B_R is zero on the axis of symmetry, however the electron senses a finite value of magnetic field B_R as it orbits at a distance $R=r_L$ from the magnetic field line on which its guiding centre lies. The electron therefore experiences an additional acceleration:

$$\vec{a} = \frac{-e}{m}\,v_\perp \hat{e}_\theta \times \left(-\frac{1}{2}r_L\,\frac{\partial B_z}{\partial z}\hat{e}_R\right)$$

$$= \frac{-e}{m}\,v_\perp\left(-\frac{1}{2}r_L\right)\frac{\partial B_z}{\partial z}(\hat{e}_\theta \times \hat{e}_R)$$

$$= \frac{e}{2m}\,v_\perp\frac{v_\perp}{\Omega_c}\,\frac{\partial B_z}{\partial z}\,(-\hat{e}_z)$$

$$= \frac{e}{2m}\,\frac{v_\perp^2\,m}{e\,B_0}\,\frac{\partial B_z}{\partial z}\,(-\hat{e}_z)$$

$$\vec{a} = -\frac{v_\perp^2}{2l_c}\,\hat{e}_z \tag{32}$$

where we have defined a scale length $l_c = \left(\dfrac{1}{B_0}\,\dfrac{\partial B_z}{\partial z}\right)$

It may be noted that the acceleration a is independent of the charge: the effects of $v_\perp \hat{e}_\theta \to -v_\perp \hat{e}_\theta$ and $-e \to e$ cancel. The effect of the acceleration is to reduce the parallel velocity of a particle as it moves into a region of higher field strength .

From eqn. (32) we note that the electron experiences a force as it mores into a region of higher field strength :

$$F_z = ma = -\frac{m\, v_\perp^2}{2l_c}\, \hat{e}_z$$

Substituting the value of l_c, we find

$$F_z = -\frac{m\, v_\perp^2}{2\, B_0}\frac{\partial B_z}{\partial z}$$

$$= -\mu\frac{\partial B}{\partial z}\left(\mu = \frac{1}{2}\frac{m\, v_\perp^2}{B_0}\text{ is the magnetic moment}\right) \tag{33}$$

This force is purely magnetic in origin. Therefore it cannot change the total electron energy. When v_z decreases owing to F_z, $v\perp$ must increase:

$$F_z\, V_z = -v_z\frac{\partial}{\partial z}\left(\frac{1}{2}m\, v_\perp^2\right) \tag{34}$$

where $v_z\dfrac{\partial}{\partial z}$ gives the rate of change with time following the guiding centre motion.

4.8 Electron Motion in a Magnetic field that varies with time

Consider a magnetic field that varies with time. A time varying magnetic field induces an electric field described by Maxwell equation

$$\vec{\nabla}\times\vec{E} = -\frac{\partial\vec{B}}{\partial t} \tag{35}$$

Any parallel component E_z will accelerate particles along the magnetic field lines. In addition, if $\vec{E}$ includes a perpendicular component, the particle perpendicular energy may be changed as follows :

In the time $T = \dfrac{2\pi}{\Omega_{ce}}$ taken for one cyclotron gyration, the energy acquired by an

electron is $\quad \Delta\left(\dfrac{1}{2}m\, v_\perp^2\right) = -e\oint_{\text{Orbit}}\vec{E}.d\vec{l} = -e\int(\vec{\nabla}\times\vec{E}).d\vec{S}$

$$= -e\int\frac{\partial\vec{B}}{\partial t}.d\vec{S} = e\,\pi r_L^2\frac{\partial B}{\partial t}$$

$$= \left(\frac{e}{T_0}\pi r_L^2\right)T_0\frac{\partial B}{\partial t}$$

$$= \mu\, T_0\frac{\partial B}{\partial t} \tag{36}$$

It follows that

$$\frac{\partial}{\partial t}\left(\frac{1}{2}m\,v_\perp^2\right) = \mu\frac{\partial B}{\partial t} \tag{37}$$

We can now consider how magnetic fields, for which both $\frac{\partial B}{\partial t}$ and $\frac{\partial B}{\partial z}$ are non-zero affect the perpendicular motion of the electron.

Combining (37) ,(34) and (33), we find

$$\frac{d}{dt}\left(\frac{1}{2}m\,v_\perp^2\right) = \mu\frac{dB}{dt} \tag{38}$$

where $\frac{d}{dt} = \frac{\partial}{\partial t} + v_z\frac{\partial}{\partial z}$ is the total rate of change with time.

Since $\frac{m\,v_\perp^2}{2} = \mu B$ by eqn. (19) it followsfrom eqn. (38) that μ and hence $\frac{v_\perp^2}{B}$, is a constant of the particle motion; it is sometimes known as the first " adiabatic invariant".

• The magnetic flux passing through the orbit of the electron is

$$\int \vec{B}.d\vec{S} = \pi r_L^2\, B_z = \pi\left(\frac{v_\perp}{\Omega_c}\right)^2 B_z$$

$$= \pi\frac{v_\perp^2.m^2}{e^2\,B^2}\,B$$

$$= \pi\left(\frac{m}{e}\right)^2\frac{v_\perp^2}{B} \tag{39}$$

which is proportional to μ and therefore constant.

4.9 Curvature Drift

We will now consider the effects on charged particle motion associated with the curvature of the magnetic field lines. We investigate now the effect of curvature terms $\frac{\partial B_x}{\partial z}$ and $\frac{\partial B_y}{\partial z}$ on the motion of a charged particle. We will assume that these terms are so small that the radius of curvature of the magnetic field lines is very large compared to the particle cyclotron radius.

Let us introduce a local coordinate system gliding along the magnetic field line with the particle longitudinal velocity $\vec{v}_\parallel$.

• Since this is not an inertial system because of the curvature of the field lines, a centrifugal force will be present. This local coordinate system can be specified by the orthogonal set of unit vectors $\hat{B}, \hat{n}, \hat{n}_2$, where $\hat{B}$ is a unit vector along the field line, $\hat{n}_1$ is along the principal normal to the field line, and $\hat{n}_2$ is along the binormal to the curved magnetic field line, as indicated in Fig. 6

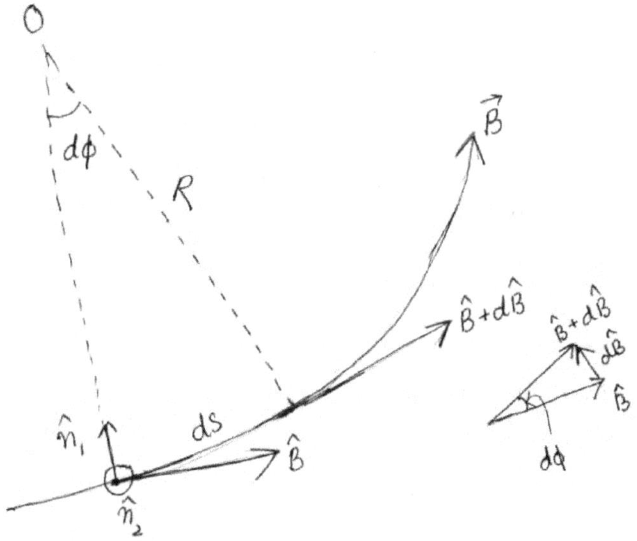

Figure 6 : *Curved magnetic field line showing the unit vector $\hat{B}$ along the field line, the principal normal $\hat{n}_1$, and the binomial $\hat{n}_2$, at an arbitrary point. Note that $\hat{n}_2 \times \hat{n}_1 = \hat{B}$. The local radius of curvature is R.*

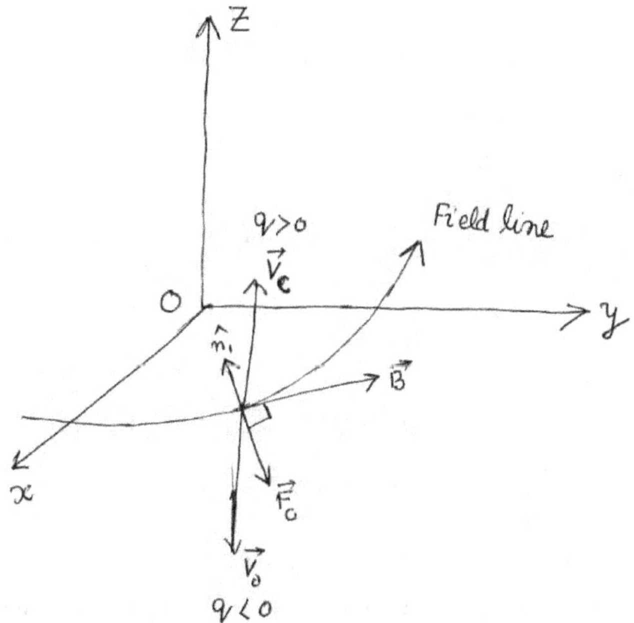

Figure 7: *Relative direction of the particle guiding center drift velocity $\vec{v}_0$, due to the curvature of the magnetic field line.*

The centrifugal force $\vec{F}_C$ acting on the particle, as seen from this non inertial systems, is given by

$$\vec{F}_C = -\frac{mv_{\parallel}^2}{R}\hat{n}_1 \tag{40}$$

where R denotes the local radius of curvature of the magnetic field line and $\vec{V}_{\parallel}$ is the particle instantaneous longitudinal speed. From Equation (24)

The curvature drift associated with this force is

$$\vec{V}_C = \frac{\vec{F}_C \times \vec{B}}{q\,B^2} = -\frac{mv_{\parallel}^2}{R\,q\,B^2}(\hat{n}_1 \times \vec{B}) \tag{41}$$

To express the unit vector $\hat{n}_1$, in terms of the unit vector $\hat{B}$ along the magnetic field line, we let ds represent an element of arc along the field line subtending an angle $d\phi$,

$$ds = Rd\phi \tag{42}$$

If $d\hat{B}$ denotes the change in $\vec{B}$ due to the displacement ds (See Fig.6) then $d\hat{B}$ is in the direction of $\hat{n}_1$ and its magnitude is

$$|d\hat{B}| = |\hat{B}|\,d\phi = d\phi \tag{43}$$

consequently, $d\hat{B} = \hat{n}_1 d\phi$ $\tag{44}$

Dividing this equation by (42) gives,

$$\frac{d\hat{B}}{ds} = \frac{\hat{n}_1\,d\phi}{R\,d\phi} = \frac{\hat{n}_1}{R} \tag{45}$$

The derivative $\dfrac{d}{ds}$ along $\vec{B}$ may be written as $(\vec{B}.\vec{\nabla})$, so that (45) becomes

$$\frac{\hat{n}_1}{R} = (\hat{B}.\vec{\nabla})\hat{B} \tag{46}$$

using this result into equation (40), we find

$$\vec{F}_0 = -mv_{\parallel}^2(\vec{B}.\vec{\nabla})\hat{B} \tag{47}$$

This force is obviously perpendicular to the magnet field $\vec{B}$, since it is in the $-\hat{n}_1$, direction , and gives rise to a curvature drift whose velocity is

$$\vec{v}_0 = -\frac{mv_{\parallel}^2}{q\,B^2}\left[(\hat{B}.\vec{\nabla})\hat{B}\right] \times \vec{B} \tag{48}$$

Since $\vec{B} = B\,\hat{B}$ and writing $w_\parallel = \dfrac{mV_\parallel^2}{2}$ for the particle longitudinal kinetic energy, equation (47) and (48) can be written, respectively, as

$$\vec{F}_0 = -\frac{2W_\parallel}{q\mathrm{B}^2}\Big[(\vec{B}.\vec{\nabla})\vec{B}\Big]_\perp \tag{49}$$

$$\vec{V}_0 = -\frac{2W_\parallel}{q\mathrm{B}^4}\Big[(\vec{B}.\vec{\nabla})\vec{B}\Big]\times\vec{B} \tag{50}$$

This, at each point, the curvature drift is perpendicular to the plane of the magnetic field line, as shown in Figure 7.

4.10 Combined Gradient- Curvature Drift

The curvature drift and the gradient drift appear together and both point in the same direction, since the term $\nabla\mathrm{B}$ points in the direction opposite to $\vec{F}_C$ (See Fig.7). These two drifts can be added to form the combined gradient-curvature drift. Thus, from (30) and (50),

$$\vec{V}_{GC} = \vec{V}_G + \vec{V}_C$$

$$= \frac{\mathrm{r}_L \mathrm{v}_\perp}{2}\frac{\vec{\mathrm{B}}\times\nabla\mathrm{B}}{\mathrm{B}^2} - \frac{m\mathrm{v}_\parallel^2}{q\,\mathrm{B}^4}\Big[\big(\vec{\mathrm{B}}.\vec{\nabla}\big)\vec{\mathrm{B}}\Big]\times\vec{\mathrm{B}}$$

$$= \frac{v_\perp}{\Omega_C}\frac{v_\perp}{2}\frac{\vec{B}\times\nabla B}{B^2} - \frac{m v_\parallel^2}{q\,B^4}\Big[\big(\vec{B}.\vec{\nabla}\big)\vec{B}\Big]\times\vec{B}$$

$$= -\frac{1}{2}\frac{m\mathrm{v}_\perp^2}{q\,\mathrm{B}^3}(\nabla\mathrm{B})\times\vec{\mathrm{B}} - \frac{m\mathrm{v}_\parallel^2}{q\,\mathrm{B}^4}\Big[\big(\vec{\mathrm{B}}.\vec{\nabla}\big)\vec{\mathrm{B}}\Big]\times\vec{\mathrm{B}} \tag{51}$$

When volume currents are not present (in a vacuum field, for example) so that $\vec{\nabla}\times\vec{\mathrm{B}} = 0$, then using the vector identity

$$(\vec{\nabla}\times\vec{\mathrm{B}})\times\vec{\mathrm{B}} = \big(\vec{\mathrm{B}}.\vec{\nabla}\big)\vec{\mathrm{B}} - \vec{\nabla}\Big(\frac{1}{2}\mathrm{B}^2\Big), \tag{52}$$

We can write (51), as

$$\vec{\mathrm{v}}_{GC} = -\frac{m}{q\,\mathrm{B}^4}\Big(\mathrm{v}_\parallel^2 + \frac{1}{2}\mathrm{v}_\perp^2\Big)\Big(\nabla\frac{1}{2}\mathrm{B}^2\Big)\times\vec{\mathrm{B}} \tag{53}$$

4.11 Particle Trapping

We know that in a magnetic field that varies slowly inn space and time, the magnetic moment $\mu = \dfrac{(1/2)\, m\, v_\perp^2}{B}$ is a conserved quantity.

Consider a static inhomogeneous magnetic field, with no applied electric field. Then the particle kinetic energy K_1 is $\dfrac{1}{2}m(v_\perp^2 + v_{11}^2) = \text{constant}$.

• We denote the subscript zero the values of the particle and field parameters at the initial position which for convenience we shall locate at the point where the magnetic field strength is weakest. We have in general, by conservation of μ and K,

$$\frac{v_\perp^2}{B} = \frac{v_{\perp 0}^2}{B_0} \qquad\qquad (\mu \text{ is constant})$$

or $\qquad v_\perp^2 = \dfrac{B}{B_0} v_{\perp 0}^2$ $\hfill (54)$

Now from the conservation of kinetic energy K, we get

$$\frac{1}{2}mv_{z0}^2 + \frac{1}{2}mv_{\perp 0}^2 = \frac{1}{2}mv_z^2 + \frac{1}{2}mv_\perp^2$$

or $\qquad v_{z0}^2 + v_{\perp 0}^2 = v_z^2 + v_\perp^2$

Substituting for $v_\perp^2$ from equation (54), We have

$$v_{z0}^2 + v_{\perp 0}^2 = v_z^2 + \frac{B}{B_0}v_{\perp 0}^2$$

or $\qquad v_z^2 = v_{z0}^2 + v_{\perp 0}^2 - \dfrac{B}{B_0}v_{\perp 0}^2$

$$= v_0^2 - v_{\perp 0}^2 \cdot \frac{B}{B_0}$$

$$v_z^2 = v_0^2\left(1 - \frac{B}{B_0}\,\frac{v_{\perp 0}^2}{v_0^2}\right) \hfill (55)$$

In an inhomogeneous field, with increasing value of $B = B_T$ such that

$$1 - \frac{B_T}{B_0}\,\frac{v_{\perp 0}^2}{v_0^2} = 0 \qquad\qquad \text{i.e. } v_z^2 \text{ becomes zero}$$

Hence, when B reaches the value $B_T = \left(\dfrac{v_0^2}{v_\perp^2}\right)B_0,$ $\hfill (56)$

v_z is reduced to zero. Thereafter the particle reverses its path along the magnetic field. This is the phenomenon of magnetic trapping. Depending on their initial velocities, a certain portion of the charge particles in a inhomogeneous magnetic field will be restricted by this mechanism to the region of space where the magnetic field is weakest . Magnetic trapping is the basis of all mirror plasma confinement systems.

Let the maximum and minimum values of the magnetic field in a mirror confinement system be B_M and B_0 respectively.

It follows from (56) that all particles whose initial velocities are such that $B_T \leq B_M$ will be trapped. All particle with a large initial ratio of parallel to perpendicular velocity, $\dfrac{v_{z0}^2}{v_{\perp 0}^2} \geq \dfrac{B_M - B_0}{B_0}$, will escape. Mirror confinement systems are accordingly characterized by a loss-Cone distribution: that is a region in (v_x, v_y, v_z) space from which particles have escaped is formed of two cones, whose axes lie along the v_z axis and whose tips touch at the origin, as in Fig. (8)

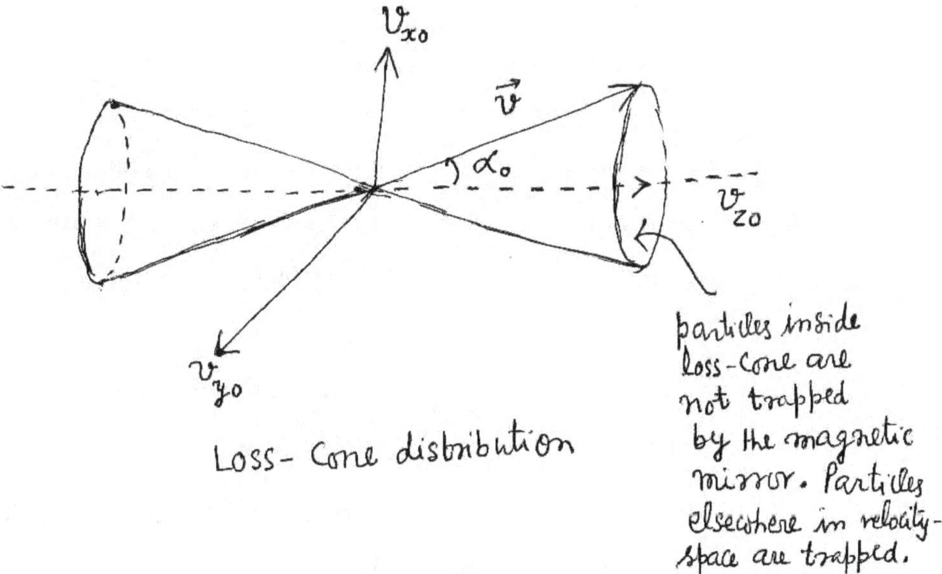

Figure 8: Loss-Cone distribution

4.12 Illustrative Examples

Example 1: A particle with charge q is emitted from the origin with momentum p, directed at an angle θ to a uniform magnetic field B which lies in the z-direction. At what point does the particle next intersect the z-axis.

Sol. At x=0, y=0, $z = v_{\parallel}T$, Substituting $T = \dfrac{2\pi\,m}{q\,B}$, and $v_{\parallel} = p\,\cos\theta$, we get

$$Z = \left(\dfrac{2\pi\,p}{q\,B}\right)\cos\theta.$$ The parallel velocity $\dfrac{p\,\cos\theta}{m}$ is unaffected by the magnetic field.

We know that the perpendicular motion is periodic, with angular velocity $\dfrac{q\,B}{m}$. It follows that after a time $\dfrac{2\pi m}{qB}$, the particle returns to x=0, and y=0, having travelled a distance $\left(\dfrac{2\pi m}{qB}\right)\left(\dfrac{p}{m}\right)\cos\theta$ along the z-axis.

4.13 Self Learning Exercise

Q.1 Calculate the cyclotron frequency and cyclotron radius for: (a) An electron in the Earth's ionosphere at 300 km altitude, where the magnetic flux density $B \simeq 0.5 \times 10^{-4}$ Tesla, considering that the electron moves at the thermal velocity $\left(\dfrac{kT}{m}\right)$, with T=1000K, Where K is Boltzmann's constant.

(b) A 50 MeV proton in the Earth's inner Van Allen radiation belt at about $1.5R_E$ (where R_E=6370 km is the Earth's radius) from the center of the Earth in the equatorial plane, considering $B \simeq 10^{-5}$ Tesla

(c) A 1MeV electron in the Earth's outer Van Allen radiation belt at about $4R_E$ from the centre of the Earth in the equatorial plane, where $B \simeq 10^{-7}$ Tesla.

Q.2 For an electron and an oxygen ion O^+ in the Earth's ionosphere at 300km altitude in the equatorial plane, where $B \simeq 0.5 \times 10^{-4}$ Tesla, calculate:

(a) The gravitational drift velocity $\vec{v}_g$

(b) The gravitational current density $\vec{J}_g$, considering $n_e = n_i = 10^{12}\,m^{-3}$. Assume that $\vec{g}$ is perpendicular to $\vec{B}$.

Q.3 Consider a system of two coaxial magnetic mirrors whose axis coincides with z axis, being symmetrical about the plane z=0, as show in Figure (9). Describe qualitatively the motion of a charged particle in this magnetic mirror system considering that at z=0 the particle has $v_{\parallel} = v_{\parallel}^{0}$ and $v_{\perp} = v_{\perp}^{0}$. What relation must exist between $\vec{B}_0 = B(z=0)\hat{z}$, $\vec{B}_m = B(z=\pm z_m)\hat{z}$ and α_0 (particle pitch angle at z=0) for particle to be reflected at Z_m?

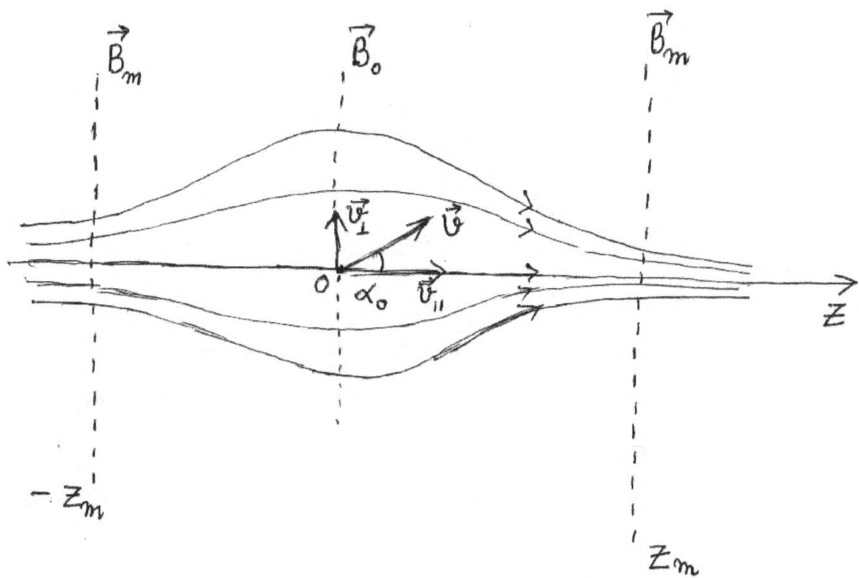

Figure 9: *Magnetic field line geometry for a system of two coaxial magnetic mirrors whose axis coincides with the z axis, being symmetrical about the plane z=0.*

4.14 Summary

In this chapter we have studied the motion of charged particle in uniform and non uniform magnetic field. We have also calculated drift velocity of charged particle in the combined fields of $\vec{E}$ and $\vec{B}$. Drift motion due to curvature of magnetic field lines has also been analyzed.

Some Important Formulae:

1. Electron cyclotron frequency $\Omega_{ce} = \dfrac{eB}{m_e} = 1.76 \times 10^{11} B \ \left(in \dfrac{rad}{sec}\right)$

    with B expressed in Tesla

2. Ion cyclotron frequency $\Omega_{ci} = \dfrac{ZeB}{m_i}$

3. Particle magnetic moment $\vec{\mu} = \dfrac{W_\perp \vec{B}}{B^2} = \dfrac{\frac{1}{2}mv_\perp^2}{B^2}\vec{B}$

4. Electron cyclotron radius $r_{ce} = \dfrac{v_{e\perp}}{\Omega_{ce}}$

5. Electron Drift velocity $\vec{\upsilon}_d = \dfrac{\vec{\Omega}_{ce} \times \vec{F}}{\Omega_{ce}^2}$

6. Curvature drift $\vec{\upsilon}_c = \dfrac{\vec{F}_c \times \vec{B}}{qB^2}$

4.15 Glossary

Magnetic Mirror Effect: As an electron moves into a region of higher magnetic field strength, the electron experiences a force that retards its longitudinal experiment motion. At some point of the increasing field vz comes to zero and the electron is reflected therefrom. This is called "Magnetic mirror effect".

"Magnetic moment" associated with orbiting electron: $\mu = \dfrac{\frac{1}{2}mv_\perp^2}{B}$

4.16 Answers to Self Learning Exercise

Ans.1: (a) $\Omega_C = \dfrac{eB}{m}$, $\dfrac{e}{m} = 1.76 \times 10^{11}\,\dfrac{C}{kg}$, $B = 5 \times 10^{-5}$

$$\Omega_C = 8.79 \times 10^6\,\dfrac{rad}{sec.}$$

$$r_L = \dfrac{mv_\perp}{eB},\ v_\perp = \sqrt{\dfrac{kT}{m}}$$

$$k = 1.38 \times 10^{-23}\,\dfrac{J}{K},\ T = 1000\,K$$

$$\therefore v_\perp = 1.2 \times 10^{-5}\,\dfrac{m}{s}\quad \therefore r_L = 1.3\ cm.$$

(b) $\Omega_C = \dfrac{qB}{M}$, mass of proton $= 1.67 \times 10^{-27\,kg}$

$B = 10^{-5}\,T$

$q = 1.6 \times 10^{-19}\,C$

and $\qquad r_L = \dfrac{MV_\perp}{qB} = \dfrac{\sqrt{2ME_K}}{qB}$

given $\qquad E_K = 50\,MeV = 50 \times 1.6 \times 10^{-19}\,J.$

Substitute these values in the formulae for Ω_C and r_L and get answer.

(c) 1 MeV electron, $\qquad\qquad B = 10^{-7}$ Tesla

1 MeV electron is relativistic as its kinetic energy is comparable to rest energy of the electron $(m_0 c^2 \sim 0.5\ MeV)$. As kinetic energy = Total energy – rest energy

or $E_K = \gamma\ m_0 c^2 - m_0 c^2$

$$E_K = (\gamma - 1)\ m_0 c^2$$

$$\therefore \gamma - 1 = \frac{E_K}{m_0 c^2}, \qquad E_K = MeV$$

or $\gamma = 1 + \dfrac{E_K}{m_0 c^2} = 1 + \dfrac{1}{0.5} = 3$

Thus $\qquad \Omega_0 = \dfrac{qB}{\gamma\ M_0} = \dfrac{1.76 \times 10^{11} \times 10^{-7}}{3}\ \dfrac{rad}{sec}$

$$= 5.8 \times 10^3\ \frac{rad}{sec}. \qquad \text{Also } r_L = \frac{\upsilon_\perp}{\Omega_0}.$$

(d) and (e) problems are to be attempted in similar way.

Ans.2: $\vec{v}_g = \dfrac{\vec{F} \times \vec{B}}{qB^2} = \dfrac{m}{q}\dfrac{\vec{g} \times \vec{B}}{B^2}$

This drift velocity depends on the ratio $\dfrac{m}{q}$ and therefore it is in opposite directions for electron and O^+ ion. Using $g_h \simeq g_s(1 - \dfrac{2h}{R}) = 9.8 \times \dfrac{58}{64}\dfrac{m}{s^2}$.

Thus v_g for electron is $= \dfrac{9.1 \times 10^{-31}}{1.6 \times 10^{-19}} \times \dfrac{9.8 \times 58}{5 \times 10^{-5} \times 64}\dfrac{m}{s}$

$$= \frac{9.1 \times 9.8 \times 58}{1.6 \times 5 \times 64} \times 10^{-31+24}\frac{m}{s}$$

$$v_{ge} \simeq 10.1 \times 10^{-7}\frac{m}{s}$$

$v^+_{gO^+}$ for O^+ ion is $\dfrac{M}{m} \times v_{ge} = 16 \times 1860 \times 10.1 \times 10^{-7}\dfrac{m}{s}$

$$= 3 \times 10^{-2} \, \frac{m}{s}$$

$$\vec{J} = n_e \, e \, \vec{v}_e + n_i \, e \, \vec{v}_i \quad ; \quad n_e = n_i$$

$$|\vec{J}| = 10^{12} \times 1.6 \times 10^{-19} \left[10.1 \times 10^{-7} + 3 \times 10^{-2} \right] \frac{A}{m^2}$$

$$\simeq 10^{12} \times 1.6 \times 10^{-19} \times 3 \times 10^{-2} \, \frac{A}{m^2}$$

$$= 4.8 \times 10^{-9} \, \frac{A}{m^2}$$

Ans.3: From energy conservation at $z = z_0$ and $z = z_m$:

$$\frac{1}{2} m (v_{\parallel}^{0^2} + v_{\perp 0}^2) = \frac{m}{2} (v_{\perp, z_m}^2 + 0), \tag{i}$$

using $v_{\parallel, z_m}$ is zero, as the particle is reflected there.

Also from the invariance of magnetic moment μ

$$\frac{v_{\perp 0}^2}{B_0} = \frac{v_{\perp, Zm}^2}{B_m}$$

$$\therefore \frac{B_m}{B_0} = \frac{v_{\perp, Zm}^2}{v_{\perp 0}^2},$$

From equation (i) , $v_{\perp, Zm}^2 = v_0^2$ (where $v_0^2 = v_{\parallel}^{0^2} + v_{\perp 0}^2$)

Thus, $\dfrac{B_m}{B_0} = \dfrac{v_0^2}{v_{\perp 0}^2} = \dfrac{1}{\sin^2 \alpha_0}$

$$\therefore \alpha_0 = \sin^{-1} \left[\frac{B_0}{B_m} \right]^{\frac{1}{2}}$$

4.17 Exercise

Q.1 Calculate the electron cyclotron frequency for the Ω_{ce} for the following magnetic fields: (a) the earth's magnetic field near a pole, $6 \times 10^{-5} T$,(b) the galactic field, $3 \times 10^{-10} T$, (c) a sunspot, $0.25 T$,

Q.2. Write down, in vector form, the relativistic equation of motion for a charged particle in the presence of a uniform magnetostatic field $\vec{B} = B_0 \hat{z}$ and show that its Cartesian components are given by

$$\frac{d}{dt}(\gamma v_x) = \frac{qB_0}{m} v_y$$

$$\frac{d}{dt}(\gamma v_y) = -\left(\frac{qB_0}{m}\right) v_x$$

$$\frac{d}{dt}(\gamma v_z) = 0 \quad \text{where } \gamma = \frac{1}{\sqrt{1-\beta^2}} \text{ and where } \beta = \frac{v}{c}.$$

Show that the velocity and trajectory of the charged particle are given by the same formulas as in the non relativistic case, but with Ω_c replaced by $\dfrac{|q|B_0}{m\,\gamma}$.

Q.3 Consider a system of N non-interacting electrons in a uniform magnetic field. Suppose that the initially, have an isotropic distribution of velocities, whose magnitude is v_0 for all electrons. If the magnetic field strength increases adiabatically with time from B_1 to $\alpha > 1$, with $B_2 = \alpha B_1$,calculate the change in energy of the system.

4.18 Answers to Exercise

Ans.1:(a) $\Omega_{ce} = \dfrac{e\,B}{m_e} = \dfrac{1.6\times10^{-19}\times6\times10^{-15}}{9.1\times10^{-31}} = 1.1\times10^{7}\,rad\ s^{-1}$

(b) 54 rad s^{-1}

(c) 4.5×10^{10} rad s^{-1}

Ans.3: Denoting velocities before and after the increase in magnetic field strength by subscripts 1 and 2 respectively: we have,

$$\frac{v_{\perp 1}^2}{B_1} = \frac{v_{\perp 2}^2}{B_2} \quad \text{(Adiabatic invariance of } \mu)$$

or $\quad v_{\perp 2}^2 = \dfrac{B_2}{B_1} v_{\perp 1}^2 = \propto v_{\perp 1}^2$ $\hfill$ (i)

whereas $\quad v_{\parallel 2}^2 = v_{\parallel 1}^2$ $\hfill$ (ii)

For a given electron, the initial and final energies are related by

$$K_2 = \frac{m}{2}\left(v_{\parallel 2}^2 = v_{\perp 2}^2\right) = \frac{m}{2}\left(v_{\parallel 1}^2 + \alpha v_{\perp 1}^2\right)$$

$$K_2 = K_1 + \frac{m}{2}(\alpha-1)\,v_{\perp 1}^2 \hfill \text{(iii)}$$

- We write $v_{\perp 1} = v_0 \sin\theta$, and since $K_1 = \dfrac{mv_0^2}{2}$,

We have a change in energy

$$\Delta K = K_2 - K_1$$

$$= \frac{m}{2}(\alpha - 1)v_0^2 \sin^2\theta$$

$$\Delta K = K_1(\alpha - 1)\sin^2\theta \quad \text{for the electron considered} \tag{iv}$$

- Initially, the electrons are uniformly distributed over an infinitesimally thin spherical shell in velocity space, with radius v_0. The number of electrons per unit area of this shell is $\dfrac{N}{4\pi v_0^2}$, and the number lying between θ and $\theta + d\theta$ is

$$\frac{N}{4\pi v_0^2} 2\pi v_0^2 \sin\theta \, d\theta = \frac{N}{2}\sin\theta \, d\theta$$

Each electron initially at θ undergoes a change in energy $\Delta K(\theta)$.

Therefore the total change in energy of the system is

$$\int_\theta^\pi \Delta K(\theta)\left(\frac{N}{2}\right)\sin\theta \, d\theta = \frac{NK_1(\alpha - 1)}{2}\int_0^\pi \sin^3\theta \, d\theta$$

$$= \frac{NK_1(\alpha - 1)}{3}$$

where NK_1 is the initial energy of the system.

References and Suggested Readings

1. Fundamentals of Plasma Physics: J.A. Bettencourt

2. Plasma Physics : F. F.Chen.

3. Plasma Dynamics : R.O.Dendy

4. Electrodynamics of Particles and Plasma : Clemmow and Dougherty

5. Plasma Physics :S. Chandra Chandrasekhar (Notes compiled by S.K. Trehan)

UNIT-5
Plasma Dielectric Constant, Magnetic Mirror Effect, Cyclotron Resonance

Structure of the Unit

5.0 Objectives

In this chapter our objectives are

- Introduction to the dielectric description of the plasma

- Derivation of the dielectric constant of the plasma

- Normal modes of an unmagnetized plasma

- Dispersion relation of electromagnetic wave in an unmagnetized plasma.

- Electrostatic mode

- Dielectric tensor of a cold magnetized plasma

- Cyclotron resonance

- Adiabatic invariants

- Magnetic mirror effect

- Magnetic heating of a plasma

5.1 Introduction

In this chapter we analyze the macroscopic response of any medium to an applied electric field. This response is determined by the sum of the macroscopic responses of the individual particles that make up the medium. Further, we analyze the motion of charged particles in the presence of time-varying fields. In the case of slowly varying magnetic filed we derive expressions for the quantities which remain invariant in such a case. These are called "Adiabatic invariants". Finally we discuss magnetic mirror effect and its application in the confinement of plasma.

5.2 Dielectric Description of Plasma

The macroscopic response of any medium to an applied electric field is determined by the sum total of microscopic responses of the individual particles of the medium.

In a conducting medium, the macroscopic response id determined at a microscopic level by the separation between positive and negative charges that is produced by the applied electric field.

- If the applied field varies with time, so too will the microscopic state of the medium : The separation between positive and negative charges will change with time, as will the electric field that is produced by their separation.

- Thus, particle currents and displacement currents are produced.

- At a macroscopic level, the particle current is described by the and conductivity tensor σ, where

$$J = \sigma . E \qquad (1)$$

σ is a macroscopic variable whose nature is determined by microscopic dynamics.

- According to Maxwell's equation states that $\vec{J}$ and the displacement current combine to act as the source of the magnetic field in the medium :

$$\vec{\nabla} \times \vec{H} = \vec{J} + \varepsilon_0 \frac{\partial \vec{E}}{\partial t}$$

Using (1), we have

$$\vec{\nabla} \times \vec{H} = \left(\varepsilon_0 \frac{\partial}{\partial t} + \vec{\sigma} . \right) \vec{E} \qquad (2)$$

Now using $\vec{B} = \mu_0 \vec{H}$ and $\mu_0 \varepsilon_0 = \dfrac{1}{c^2}$, we can write (2) as

$$\vec{\nabla} \times \vec{B} = (\mu_0 \varepsilon_0 \frac{\partial}{\partial t} + \mu_0 \sigma .) \vec{E} \qquad (3)$$

Then, if $\vec{E}$ varies as exp $(-i\omega t)$, then (3) gives

$$\vec{\nabla} \times \vec{B} = \frac{-i\omega}{c^2} \left(I + \frac{i}{\varepsilon_0 \omega} \sigma \right) . \vec{E} \qquad (4)$$

were I is the identity matrix.

It follows that all information about the macroscopic response of the medium to the the applied electric fields is contained in dielectric tensor, defined by

$$\varepsilon = I + \frac{i}{\varepsilon_0 \omega} \sigma \qquad (5)$$

and equation (4) can be written as

$$\vec{\nabla} \times \vec{B} - \frac{-i\omega}{c^2} \varepsilon . \vec{E} \qquad (6)$$

Operating on (6) with $\dfrac{\partial}{\partial t}$ and using Maxwell equation $\vec{\nabla} \times \vec{E} = -\dfrac{\partial \vec{B}}{\partial t}$

$$\vec{\nabla} \times \left(\frac{\partial \vec{B}}{\partial t} \right) = -\frac{i\omega}{c^2} . \frac{\partial}{\partial t} (\varepsilon . \vec{E})$$

$$\text{or } -\vec{\nabla} \times (\vec{\nabla} \times \vec{E}) = -\frac{i\omega}{c^2} (-i\omega) \varepsilon . \vec{E}$$

$$\nabla^2 \vec{E} - \nabla(\vec{\nabla}.\vec{E}) + \frac{\omega^2}{c^2}\varepsilon.\vec{E} = 0 \qquad (7)$$

- So far, our discussion in this section has been completely general, and applies to any conducting medium.

- The nature of the macroscopic quantity σ, and hence ε, for a plasma is determined at the microscopic level by the plasma particle dynamics.

As an example, we consider the motion of a plasma electron in the absence of external magnetic field :

$$m\frac{d\vec{v}}{dt} = -e\vec{E}, \text{ where } \vec{E} \text{ is the applied electric field and varies as}$$

exp (-iωt). Each plasma electron will respond with a velocity

$$\vec{v} = \left(\frac{e}{i\omega m}\right)\vec{E}$$

- The current density associated with n_0 such electrons per unit volume is

$$\vec{J} = -n_0 e\vec{v} = -\left(\frac{n_0 e^2}{i\omega m}\right)\vec{E} \qquad (8)$$

So that $\sigma = -\left(\dfrac{n_0 e^2}{i\omega m}\right)I$.

Then the definition of ε gives , (Eq. 5)

$$\varepsilon = \left(1 - \frac{\omega_{pe}^2}{\omega^2}\right)I , \qquad (9)$$

where we have $\qquad \omega_{pe}^2 = \dfrac{n_0 e^2}{m\varepsilon_0}$.

Eq.(9) is the dielectric tensor of the plasma.

5.3 Normal Modes of an Unmagnetized Plasma

We now calculate the normal modes – a macroscopic concept – of an unmagetized plasma. If $\vec{E}$ varies as exp $(i\vec{k}.\vec{r} - i\omega t)$, then equation (7) gives, after replacing $\vec{\nabla}$ operator by $i\vec{k}$ and $\dfrac{\partial}{\partial t}$ by $(-i\omega)$,

$$(i\vec{k})^2 \vec{E} - i\vec{k}\,(i\vec{k}.\vec{E}) + \frac{\omega^2}{c^2}\,\in.\vec{E} = 0$$

Now using from equation (9) $\in = \left(1 - \frac{\omega_{p_e}^{\;2}}{\omega^2}\right)I$, We finally get

$$(\omega^2 - \omega_{p_e}^{\;2} - c^2 k^2)\vec{E} + c^2\vec{k}(\vec{k}.\vec{E}) = 0 \tag{10}$$

There are two classes of normal modes. First, consider the case of transverse modes ,that have $\vec{k}.\vec{E} = 0$ and are accordingly electromagnetic. Then equation (10) gives

$$(\omega^2 - \omega_{p_e}^{\;2} - c^2 k^2)\vec{E} = 0 \tag{11}$$

This is compatible with non-zero $\vec{E}$ only if ω and k are related by

$$\omega^2 = \omega_p^{\;2} + c^2 k^2 \tag{12}$$

which is our first derivation of dispersion relation. It tells us that the frequency of any electromagnetic wave must exceed the electron plasma frequency.

In addition, if we try to launch an electromagnetic wave into the plasma with frequency $\omega < \omega_{p_e}$ eqn. (12) indicates that the wave will have an imaginary wave number k inside the plasma. The wave will therefore be evanescent, and unable to propagate through the plasma. Thus ω_{pe} plays the role of a cut-off frequency for electromagnetic waves in an unmagentized plasma.

This fact has a number of practical implications. For example, it determines the range of frequencies that can be used for different types of radio communication. In order to communicate with a satellite, one must choose a frequency that exceeds the plasma frequency of the ionospheric plasma. Otherwise, the signal will be reflected from the ionosphere, and will not reach the satellite. Conversely, may send a radio single to a distant point on the Earth's surface by choosing a frequency below the ionosphere to reflect the single in the required direction.

The second class of normal mode that satisfies eq.(10) is electrostatic, with $\vec{k}$ and $\vec{E}$ parallel. In this case, the dispersion relation is clearly

$$\omega^2 = \omega_{pe}^{\;2} \tag{13}$$

Thus electrostatic normal modes of plasma oscillate at the electron plasma frequency.

5.4 Adiabatic Invariants and Magnetic Mirror Effect

We consider the motion of a charged particle in the magnetic field that remains uniform but varies slowly in magnitude and direction. We shall show that when the conditions of motion are changed slowly, certain quantities called adiabatic invariants remain constant.

A charged particle in a uniform magnetic field is generally known to travel along a helical line. The projection of the trajectory on the plane perpendicular to the magnetic induction $\vec{B}$ is a circle with the radius,

$$\rho = \frac{m v_\perp}{qB} \tag{1}$$

where $v_\perp$ is the transverse component of the particle's velocity.

The motion along this circle is the rotation with the Larmor frequency

$$\omega_B = \frac{qB}{m} \tag{2}$$

The particle also travels along the lines of force with the constant velocity $v_\parallel$.

Generally, the magnetic field is not uniform .However ,in nonuniform fields typically considered in plasma physics, induction $\vec{B}$ has almost a constant value and sense at the distances of the order of Larmor radius of particles, that is, microscopic variations of the magnetic field are very small. We now investigate how such weak non-uniformity of the field affects the motion of particles.

At first, assume that the field varies along the field line. The trajectory of the particle traveling along this line noticeably changes its shape within a segment in which the magnetic induction B increases or decreases significantly.

When the particle travels towards the increasing field, the trajectory becomes more steep and it can be compared to a spring being compressed. When a particle travels towards a decreasing field, its trajectory becomes less steep.

This effect can be easily explained. A charged particle which rotates along the Larmor circle gives rise to a circular current so that is is equivalent to an elementary diamagnetic with the magnetic moment

$$\mu = \frac{W_\perp}{B}, \text{ where } W_\perp \text{ is the kinetic energy of the transverse motion.}$$

According to the Ampere's theory,

$$\mu = i \times \pi\rho^2 = \frac{e}{T} \times \pi\rho^2 = \frac{\frac{e}{2\pi\rho}}{v_\perp} \times \pi\rho^2$$

$$= \frac{ev_\perp}{2}\cdot\rho = \frac{e}{2}v_\perp \frac{mv_\perp}{eB}$$

$$\mu = \frac{1}{2}\frac{mv_\perp^2}{B} = \frac{W_\perp}{B} \tag{3}$$

- The magnetic field whose intensity varies along the lines of force acts upon a diamagnetic with the force

$$F = -\mu\frac{dB}{d\ell} \tag{4}$$

(here differentiation is done in the direction of the field).

The action of this force results in variation of the longitudinal velocity $v_\parallel$ according to the relationship

$$m\frac{dv_\parallel}{dt} = -\mu\frac{dB}{d\ell}$$

$$= -\frac{W_\perp}{B}\frac{dB}{d\ell} \tag{5}$$

On multiplying both sides of this equation by $v_\parallel$ yields

$$\frac{d}{dt}W_\parallel = -\frac{W_\perp}{B}\frac{dB}{dt} \tag{6}$$

When a particle travels in a magnetic field, $W_\perp + W_\parallel =$ constant.

Therefore $\qquad \dfrac{dW_\parallel}{dt} = -\dfrac{dW_\perp}{dt}$ $\hfill$ (7)

Substituting (7) into (6), we find

$$\frac{dW_\perp}{dt} = \frac{W_\perp^2}{B}\frac{dB}{dt}.$$

Form this we find

$$\frac{dW_\perp}{dt} = \frac{dB}{B} \quad ; \frac{W_\perp}{B} = \text{constant} \tag{8}$$

- This, the ratio $\dfrac{W_\perp}{B}$ is constant when a charged particle travels in the magnetic field whose intensity variation along the field lines is not too sharp. The constant $\dfrac{W_\perp}{B}$ is usually referred to as "adiabatic invariant" of motion.

This emphasizes the fact that the particle travels in a slowly changing magnetic field. The kinetic energy of the transverse motion is $W_\perp = W_0 \sin^2 \alpha$, where W_0 is the full energy of the particle and α is the angle between the velocity direction and the field line.

- Since W_0 is constant and $\dfrac{W_\perp}{B}$ is the adiabatic invariant, the ratio $\dfrac{\sin^2 \alpha}{B}$ is also an adiabatic invariant. This shows that the slope increases with increasing B and in the region with a stronger magnetic field the helical trajectory becomes more steep as discussed above.

Assume that at a certain trajectory point $\alpha = \alpha_0$ and $B = B_0$. Using these initial conditions, we can find the value of α at any trajectory point from the relationship

$$\frac{\sin^2 \alpha}{B} = \frac{\sin^2 \alpha_0}{B_0}, \text{ that is}$$

$$\sin \alpha = \sin \alpha_0 \sqrt{\frac{B}{B_0}} \tag{9}$$

When a particle traveling towards the increasing field comes to the point where $B = \dfrac{B_0}{\sin^2 \alpha_0}$, then the angle α becomes 90° and hence the longitudinal velocity $v_\parallel$ vanishes. This means that at this point the direction of the longitudinal motion is reversed. The particle is reflected from the region of high field towards the region of lower field.

Thus, high-field regions under certain conditions can act as some magnetic mirrors for charged particles. For instance, if the field increases in opposite directions from a certain intermediate region, a charged particle can be blocked between two magnetic mirrors ; it will oscillate along the field lines within a restricted space region. The particles with a large angle α ($\sin \alpha > \sqrt{(B_{min} B_{max})}$) will be confined in this region.

5.5 Dielectric Tensor of a Cold Magnetized Plasma

For calculating the dielectric tensor of the plasma in presence of electromagnetic field, we begin with the equations of motion for a single particle of species j (j = e for electron, i for ion) in an electromagnetic field :

$$m_j \frac{d}{dt}\vec{v}_j = q_j(\vec{E}+\vec{v}_j \times \vec{B})$$ (1)

along with the Maxwell Equations

$$\vec{\nabla}\times\vec{E} = -\frac{\partial \vec{B}}{\partial t}$$

$$\vec{\nabla}\times\vec{B} = \mu_0\left(\vec{J}+\varepsilon_0\frac{\partial \vec{E}}{\partial t}\right)$$ (2)

and the expression for the total current

$$\vec{J} = \sum_j n_j q_j \vec{v}_j$$

where the sum is over the species.

- Since the plasma has been presumed to be uniform and homogeneous in both space and time, we may Fourier transform these equations, or what is equivalent, assume that

$$\vec{E} = \vec{E}_1 e^{i(\vec{k}.\vec{r}-\omega t)}$$

$$\vec{B} = \vec{B}_0 + \vec{B}_1 e^{i(\vec{k}.\vec{r}-\omega t)}$$

$$\vec{v} = \vec{v}_1 e^{i(\vec{k}.\vec{r}-\omega t)}$$ (3)

and $\vec{B}_0$ is the static magnetic field and is taken to be in the Z-direction, and $|B_1|<<|\vec{B}_0|$ with these inserted into Eq. (1), we may rewrite that equation in linear from as

$$-i\omega m_j \vec{v}_{1j} = q_j(\vec{E}_1+\vec{v}_{1j}\times\vec{B}_0)$$ (4)

where the second order terms have been neglected because we have assumed the waves are of sufficiently low amplitude that the linear approximation is valid. The solution of Eq. (4) for the velocity is

$$v_{xj} = \frac{iq_j}{m_j(\omega^2-\omega_{cj}^2)}(\omega E_x + iE_j\omega_{cj}E_y)$$

$$v_{yj} = \frac{iq_j}{m_j(\omega^2 - \omega_{cj}^2)}(-iE_j\omega_{cj}E_x + \omega E_y)$$

$$v_{zj} = \frac{iq_j}{m_j\omega}E_z \tag{5}$$

Here we have introduced the definitions $\varepsilon_j = \dfrac{q_j}{|q_j|}$ to denote the sign of the charge

for species j and $\omega_{cj} \equiv \dfrac{|q_j|B_0}{m_j}$ is the cyclotron frequency for species j.

- The first two of these may be simplified by introducing rotating coordinates such that $v_{\pm} = v_x \pm iv_y$. Then we may write both of these components as

$$v_{\pm} = \frac{iq_j}{m_j(\omega \mp \varepsilon_j\omega_{cj})}E_{\pm} \tag{6}$$

Similarly, the current density may now be written as

$$J_{\pm} = i\varepsilon_0 \sum_j \frac{\omega_{pj}^2}{(\omega \mp \varepsilon_j\omega_{cj})}E_{\pm}$$

$$J_z = i\varepsilon_0 \sum_j \frac{\omega_{pj}^2}{\omega}E_z \tag{7}$$

where ω_{pj} is the plasma frequency for species j, given by

$$\omega_{pj}^2 = \frac{n_j q_j^2}{m_j \varepsilon_0} \tag{8}$$

If we now combine the plasma current and the displacement current such that

$$\vec{J} - i\omega\varepsilon_0\vec{E} \equiv -i\omega\varepsilon_0\vec{\vec{K}}.\vec{E} \tag{9}$$

Then the resulting equivalent dielectric tensor is given by

$$K = \begin{pmatrix} S & -iD & 0 \\ iD & S & 0 \\ 0 & 0 & P \end{pmatrix} = \begin{pmatrix} K_1 & K_2 & 0 \\ -K_2 & K_1 & 0 \\ 0 & 0 & K_3 \end{pmatrix}$$

where the Dielectric tensor elements are defined by

$$K_1 \equiv S = \frac{1}{2}(R+L) = 1 - \sum_j \frac{\omega_{pj}^2}{\omega^2 - \omega_{cj}^2}$$

$$iK_2 \equiv D \equiv \frac{1}{2}(R+L) = \sum_j \frac{\varepsilon_j \omega_{cj} \omega_{pj}^2}{\omega(\omega^2 - \omega_{cj}^2)}$$

$$K_3 \equiv P = 1 - \sum_j \frac{\omega_{pj}^2}{\omega^2}$$

$$K_1 + iK_2 \equiv R \equiv S + D = 1 - \sum_j \frac{\omega_{pj}^2}{\omega(\omega + \varepsilon_j \omega_{cj})}$$

$$K_1 - iK_2 \equiv L \equiv S - iD = 1 - \sum_j \frac{\omega_{pj}^2}{\omega(\omega - \varepsilon_j \omega_{cj})} \tag{11}$$

5.6 Dispersion Relation

The Maxwell equations are now written as

$$i\vec{k} \times \vec{E} = i\omega \vec{B} \tag{1}$$

$$i\vec{k} \times \vec{B} = -i\omega \varepsilon_0 \mu_0 \vec{K}.\vec{E}$$

with the resulting wave equation

$$\vec{n} \times (\vec{n} \times \vec{E}) + K.E = 0 \tag{2}$$

where $\vec{n} = \dfrac{\vec{k}c}{\omega}$ (3)

is the index of refraction vector whose direction is the direction of the wave vector $\vec{k}$ and whose magnitude is the index of refraction.

- If we now choose $\vec{n}$ to lie in the x-z plane, and since we have already chosen $\vec{B}_0$ to be in the Z-direction, then equation (2) becomes :

$$\begin{pmatrix} S - n^2 \cos^2\theta & -iD & n^2 \cos\theta \sin\theta \\ iD & S - n^2 & 0 \\ n^2 \cos\theta \sin\theta & 0 & P - n^2 \sin^2\theta \end{pmatrix} \begin{pmatrix} E_x \\ E_y \\ E_z \end{pmatrix} = 0 \tag{4}$$

Where θ is the angle between $\vec{k}$ and the Z-axis.

- In order to have a nontrivial solution, one requires that the determinant of coefficients vanish. This condition gives the dispersion relation.

$$An^4 - Bn^2 + C = 0 \tag{5}$$

where $A = S\sin^2\theta + P\cos^2\theta$

$$B = RL\sin^2\theta + PS(1+\cos^2\theta)$$

$$C = P\,R\,L \tag{6}$$

The solutions of (5) may be written in either of two forms, as a quadratic in n^2

$$n^2 = \frac{B \pm F}{2A}, \quad F^2 = B^2 - 4AC \tag{7}$$

where F^2 may be written in the form

$$F^2 = (RL - PS)^2 \sin^4\theta + 4P^2D^2\cos^2\theta \tag{8}$$

or in terms of the angle,

$$\tan^2\theta = \frac{P(n^2 - R)(n^2 - L)}{(Sn^2 - RL)(n^2 - P)} \tag{9}$$

The general condition for a resonance, where $n^2 \to \infty$ is given by (9) as

$$\tan^2\theta = -\frac{P}{S} \quad \text{General resonance condition} \tag{10}$$

and the general cutoff condition, where n =0, is given by

$$C = P\,R\,L = 0 \qquad \text{general cutoff condition}$$

We note some special cases : $\tag{11}$

1. Propagation parallel to $\vec{B}_0, \theta = 0$ (the numerator of (a) must vanish)

(a) $P = K_3 = 0$ (Plasma oscillations)

(b) $n^2 = R = K_1 + iK_2$ (wave with left-handed polarization)

2. Propagations perpendicular to $\vec{B}_0$, $\theta = \dfrac{\pi}{2}$ (the denominator of (9) must vanish)

(a) $n^2 = P = K_3$ (ordinary wave)

(b) $n^2 = \dfrac{RL}{S} = \dfrac{(K_1^2 + K_2^2)}{K_1}$ (Extra-ordinary wave)

5.7 Principal Solutions – Parallel Propagation

We first define the principal resonances to be those which occur at $\theta = 0$.

The general condition for a resonance $(n^2 \to \infty)$ is

$$\tan^2 \theta = -\frac{P}{S} \qquad (12)$$

Hence, for $\quad (\theta \to 0)$, we require $(S \to \infty)$,

Since, P = 0 is a cutoff. Since

$$S = \frac{1}{2}(R + L), \text{ this can be satisfied for}$$

either $R \to \infty$ (Electron-cyclotron resonance)

or $\quad L \to \infty \qquad$ (Ion cyclotron resonance)

- For a simple plasma of electrons and one ion species, the dispersion relation for the right-handed wave, $n_R^2 = R$, which propagation parallel to $\vec{B}_0$ is given by as

$$n_R^2 = R = 1 - \frac{\omega_{pi}^2}{\omega(\omega + \omega_{ci})} - \frac{\omega_{pe}^2}{\omega(\omega - \omega_{ce})} \qquad (13)$$

So the resonance is clearly at $\omega = \omega_{ce}$.

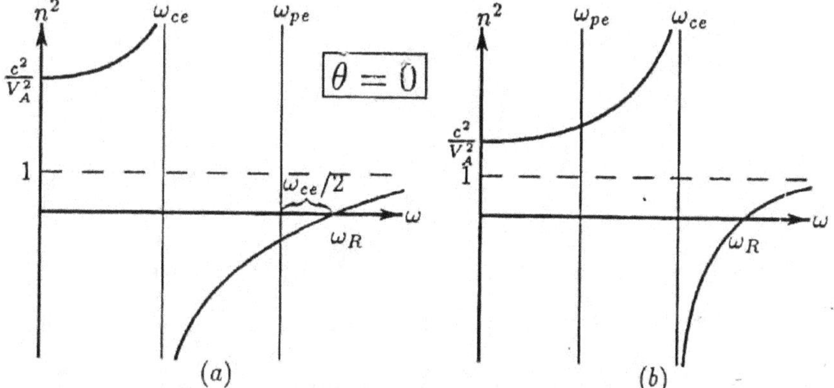

Dispersion relation for R-wave. (a) High density case, $\omega_{pe} \gg \omega_{ce}$. (b) Low density case, $\omega_{pe} \ll \omega_{ce}$.

Figure: Dispersion relation for R-wave.

5.8 Cyclotron Resonance

When the frequency ω of the applied electric field is equal to the electron cyclotron frequency then the electric field of right-handed circularly polarized wave speeds up the electron continuously and indefinitely in time. The electrons is able to absorb energy from the right-handed circular polarized wave. This phenomenon is called " Cyclotron resonance".

Example1

(a) Consider a right circularly polarized wave with electric field amplitude $\sqrt{2}E_R$, and a left circularly polarized wave with amplitude $\sqrt{2}\,E_L$, which have the same frequency ω. It both waves travel through a plasma along a uniform magnetic field which is oriented in the Z-direction, show that the total electric filed is

$$\vec{E} = [\hat{e}_x\{E_R\exp(ik_Rz) + E_L\exp(ik_Lz)\} \ + i\hat{e}_y[E_R\exp(ik_Rz) - E_L\exp(ik_Lz)]\exp(-i\omega t),$$

where k_R and k_L are wave numbers for right circularly polarized wave and left circularly polarized respectively. $\hat{e}_x$ and $\hat{e}_y$ are unit vectors along x and y axis respectively.

(b) Show that the ratio of the x and y components of the field, as a function of position Z, μ

$$\frac{E_x}{E_y} = -i\frac{\left[1+\left(\dfrac{E_L}{E_R}\right)\exp\{i(K_L - K_R)Z\}\right]}{\left[1-\left(\dfrac{E_L}{E_R}\right)\exp\{i(K_L - K_R)Z\}\right]}$$

(c) A linearly polarized wave can be written as the superposition of two oppositely circularly polarized waves that have equal amplitudes.

Show that, for the case considered, the plane of linear polarization rotated as the wave travels thought the plasma (Faraday rotation). Show that the plane of linear polarization has rotated through a right-angle once the wave has traveled a distance

$$L = \frac{\pi}{k_L - k_R}.$$

(d) For a high-frequency wave that has $\omega \gg \omega_{pe}$ and ω_{ce}, show that the angle of Faraday rotation is proportional to the product of electron density and magnetic field strength.

Sol:

(a) The vector amplitude of the right circularly polarized wave can be written $E_R(\hat{e}_x + i\hat{e}_y)$. Note that $\left|(\hat{e}_x + i\hat{e}_y)\right| = \sqrt{2}$

Similarly the vector amplitude of the left circularly polarized wave can be written $E_L(\hat{e}_x - i\hat{e}_y)$. These waves oscillate as $\exp(ik_R z - i\omega t)$ and $\exp(ik_L z - i\omega t)$ respectively.

Total electric field $\vec{E}$ is obtained by summing the two

$$\vec{E} = E_R(\hat{e}_x + i\hat{e}_y)\exp(ik_R z - i\omega t) + E_L(\hat{e}_x - i\hat{e}_y)\exp(ik_L z - i\omega t)$$

$$\vec{E} = \left[\hat{e}_x\{E_R\exp(ik_R z) + E_L\exp(ik_L z)\} + i\hat{e}_y\{E_R\exp(ik_R z) - E_L\exp(ik_L z)\}\right]\exp(-i\omega t)$$

(b) From the expression derived in (a), it follows that

$$\frac{E_x}{E_y} = -i\frac{\{E_R\exp(ik_R z) + E_L\exp(ik_L z)\}}{\{E_R\exp(ik_R z) - E_L\exp(ik_L z)\}}$$

Now dividing numerator and denominator by

$E_R\exp(ik_R z)$ we finally obtain

$$\frac{E_x}{E_y} = -i\frac{\left[1 + \left(\dfrac{E_L}{E_R}\right)\exp\{i(k_L - k_R)z\}\right]}{\left[1 - \left(\dfrac{E_L}{E_R}\right)\exp\{i(k_L - k_R)z\}\right]}$$

(c) We have $E_L = E_R$; in this case, multiplying top and bottom of the expression derived in (b) by $\exp\{i(k_R - k_L)z/2\}$, we obtain

$$\frac{E_x}{E_y} = -i\frac{\left[\exp\left\{i(k_R - k_L)\dfrac{z}{2}\right\} + \exp\left\{-i(k_R - k_L)\dfrac{z}{2}\right\}\right]}{\left[\exp\left\{i(k_R - k_L)\dfrac{z}{2}\right\} - \exp\left\{-i(k_R - k_L)\dfrac{z}{2}\right\}\right]}$$

$$\frac{E_x}{E_y} = \cot\left\{(k_L - k_R)\frac{z}{2}\right\}$$

This equation describes Faraday rotation .

If $\dfrac{E_x}{E_y}$ is infinite at $z = 0$, it has fallen to zero once the wave has traveled a distance

L such that

$$(k_L - k_R)\frac{L}{2} = \frac{\pi}{2},$$

or $\qquad L = \dfrac{\pi}{(k_L - k_R)}.$

(d) Since $\dfrac{\omega_{pe}^2}{\omega^2} \ll 1$ and using the dispersion relation for right-handed circularly polarized wave (R-wave) :

$$n^2 = R \,,\ R = 1 - \frac{\omega_{pe}^2}{\omega(\omega - \omega_{ce})}$$

or $\qquad \dfrac{c^2 k_R^2}{\omega^2} = 1 - \dfrac{\omega_{pe}^2}{\omega(\omega - \omega_{ce})}$

or $\qquad k_R^2 = \dfrac{\omega^2}{c^2}\left[1 - \dfrac{\omega_{pe}^2}{\omega^2\left(1 - \dfrac{\omega_{ce}}{\omega}\right)} \right]$

or $\qquad k_R \approx \dfrac{\omega}{c}\left[1 - \dfrac{1}{2}\dfrac{\omega_{pe}^2}{\omega^2}\dfrac{\omega_{pe}^2}{\omega^2\left(1 - \dfrac{\omega_{ce}}{\omega}\right)} \right]$

Similarly for L-wave,

$$k_L \approx \dfrac{\omega}{c}\left\{ 1 - \dfrac{\omega_{pe}^2}{2\omega^2\left(1 + \dfrac{\omega_{ce}}{\omega}\right)} \right\}.$$

Following (c), the angle of Faraday rotation is proportional to

$$k_L - k_R \approx \frac{\omega_{ce}}{c}\cdot\frac{\omega_{pe}^2}{\omega^2 - \omega_{ce}^2}$$

$$\approx \frac{\omega_{ce}}{c}\cdot\frac{\omega_{pe}^2}{\omega^2},$$

Since $\omega^2 \gg \omega_{ce}^2$,and ω_{ce} is proportional to magnetic field strength and ω_{pe}^2 to electron density, therefore Faraday rotation is proportional to the produced of magnetic field strength and plasma density.

5.10 Self Learning Exercise

Q.1 Calculate the electron cyclotron frequency ω_{ce} for the following magnetic fields :

(A)The earth's magnetic field near a pole, $6\times10^{-5}T$;

(B)The galactic filed, $3\times10^{-10}T$;

(C)A sunspot, 0.25T

5.11 Summary

In this chapter we have derived dispersion relation of electromagnetic waves in the absence of magnetic field and also in the presence of magnetic field. We have also explained the phenomenon of cyclotron resonance. Using the theory of adiabatic invariants we have also studied the magnetic mirror effect.

5.12 Glossary

Larmour frequency: The frequency of revolution of charged particle in a plane perpendicular to the magnetic field $\vec{B}$

Adiabatic invariant: The ratio $\dfrac{W_\perp}{B}$ is constant when a charged particle travels in the magnetic field whose intensity variation along the field lines is not too sharp.

5.13 Answers to Self Learning Exercise

Ans.1: $\omega_{ce}=\dfrac{eB}{m}=1.8\times10^{11}\left(\dfrac{B}{1Tesla}\right)rad\,s^{-1}$

    (A) $1.1\times10^{7}\,rad\,s^{-1}$

    (B) $54\,rad\,s^{-1}$;

    (C) $4.5\times10^{10}\,rad\,s^{-1}$

5.14 Exercise

Q.1 Suppose the magnetic field along the axis of a magnetic mirror is given by

$$B_z = B_0(1+\alpha^2 z^2),$$

(a) If an electron at z = 0 has a velocity given by $v^2 = 3v_\parallel^2 = 1.5v_\perp^2$, at what value of z is the electron reflected ?

(b) Write the equation of motion of the guiding centre for the direction parallel to the magnetic field.

(c) Show that the motion is sinusoidal, and calculated its frequency.

Q.2 A 20 keV deuteron in a large mirror fusion device has pitch angle $\theta = 45°$ at the mid-plane, where B=0.7T. Compute its Larmor radius.

Q.3 A 1-keV proton with $v_{\parallel} = 0$ is a uniform magnetic field B = 0.1 T is accelerated as B is slowly increased to 1.0T. It then makes an elastic collision with a heavy particles and charges direction so that $v_{\perp} = v_{\parallel}$. The B=field is then slowly decreased back to 0.1T. What is proton's energy now ?

5.15 Answers to Exercise

Ans.1:
$$\frac{Sin^2\theta}{B_{max}} = \frac{Sin^2\theta}{B_0}.$$

Using $Sin\theta_0 = \dfrac{v_\perp}{v} = \dfrac{v_\perp}{\sqrt{v_\perp^2 + v_\parallel^2}}$

$$= \frac{v_\perp}{\sqrt{v_\perp^2 + \frac{1}{2}v_\perp^2}} = \sqrt{\frac{2}{3}}$$

Thus $\dfrac{Sin^2}{B_{max}} = \dfrac{2}{3B_0}$;

Now is θ is $\dfrac{\pi}{2}$ at the location of B_{max} as the particle is reflected from there, we find

$$\frac{1}{B_{max}} = \frac{2}{3B_0}$$

or $\qquad B_{max} = \dfrac{3}{2}B_0$

or $B_0(1+\alpha^2 z^2) = \dfrac{3}{2}B_0$

$$\alpha^2 z^2 = \frac{1}{2} \qquad \Rightarrow z = \frac{1}{\sqrt{2}}\alpha$$

Equation of motion is

$$m\frac{dv_\parallel}{dt} = -\mu\frac{dB}{dl}$$

$$= -\frac{W_\perp}{B}\frac{dB}{dl}$$

$$\text{or } m\frac{d^2z}{dt^2} = -\frac{W_\perp}{B}\frac{dB}{dz}$$

$$= \frac{W_\perp}{B}\frac{d}{dz}(B_0(1+\alpha^2 z^2))$$

$$= -\frac{W_\perp}{B}B_0 2\alpha^2 z$$

$$\text{or } \frac{d^2z}{dt^2} + \frac{2W_\perp B_0}{mB}\alpha^2 z = 0.$$

Motion is SHM $\omega^2 = \dfrac{2W_\perp B_0}{mB}$.

Ans.2: $\alpha = 45^\circ$, $\therefore v_\parallel = v_\perp$.

Ans.3: $\dfrac{1}{2}\dfrac{mv_\perp^2}{B_{initial}} = \dfrac{\left(\dfrac{1}{2}mv_\perp^2\right)_{final}}{B_{final}}$

$$\text{or } \frac{1keV}{0.1} = \frac{\left(\dfrac{1}{2}mv_\perp^2\right)_{final}}{1}$$

$$\therefore \left(\frac{1}{2}mv_\perp^2\right)_{final} = 10keV$$

As $v_\parallel = v_\perp$ after collision ,therefore

$$\therefore \left(\frac{1}{2}mv_\perp^2\right)_{after\,collision} \approx 5keV .$$

Hence $\dfrac{5keV}{1T} = \dfrac{x}{0.1}$,

Thus x = 0.5 keV

References and Suggested Readings

1. Plasma Physics: F.F.Chen

2. Electrodynamics of particles and Plasmas: Clemmow and Dougherly

UNIT-6
Magneto Hydrodynamic (MHD) Equations

Structure of the Unit

6.0 Objectives

- Macroscopic equation for a conducting fluid

- MHD equations

- Diffusion phenomenon

- Ambipolar diffusion

6.1 Introduction

So far, we have based our study of plasmas on the single particle dynamics of the constituent electrons and ions. We have used these dynamics as the basis for a dielectric description of the plasma. This approach has been very effective. The understanding of any physical system is helped by knowledge of its normal models, and the dielectric approach has enabled us to formulate the normal modes of plasma over a wide range of frequencies, However, when we study normal modes, whose frequencies ω lie well below the lowest characteristic single particle frequency, the ions cyclotron frequency ω_{ci}, on these time scale, only the averaged, guiding centre positions of thee plasma particles are significant. In this regime of frequencies and length scales a fluid-like description of the plasma would be more appropriate. In the following we derive the basic equations for conducting fluid in a magnetic field.

6.2 Macroscopic Equations for a Conducting Fluid

- The plasma can be considered as a conducting fluid, without specifying its various individual species.

- If the plasma density is sufficiently high, we must take into account the fields produced by plasma motion. These fields, which are due to distribution and motion of plasma, affect the motion of plasma, that is, the plasma parameter and the filed produced by plasma are interrelated.

- The motion of plasma and variation of its parameters can be described by

(i) The continuity equation for the density of electrons and ions,

(ii) The equation for the mean momentum of electrons and ions,

(iii) Poisson's equation, and

(iv) The Maxwell equations.

Here we are considering plasma as composed of two distinct but intermingled fluids, the electron and ion fluids, coupled together by their opposing electrical charges.

At macroscopic level the plasma will be treated as electrically neutral. The plasma will be treated as a single, electrically neutral and perfectly conducting fluid. This reflects the fact that, on the timescales of interest, the high mobility of the electrons enables them both to keep the plasma electrically neutral and to cancel out any applied electric field. We shall need to distinguish between the macroscopic consequences of electron motion alone, which will support a current with volume density $\vec{J}$, and the macroscopic motion of the plasma as a whole, characterized by the bulk velocity $\vec{v}$, It should be noted that $\vec{J}$ and $\vec{v}$ are independent.

It is possible to visualize small elements of fluid moving in a given direction with velocity $\vec{v}$, while containing a current density $\vec{J}$ that is oriented in some other direction.

The mass density of the plasma fluid be denoted by ρ. At the macroscopic level, ρ is given by the product of the ion number density n_i and ion mass M- since M greatly exceeds the electron mass m.

- In the fluid equations we shall allow the single fluid representing the plasma to possess a pressure p, which implies non-zero temperature.

- We now proceed to setup the equations of ideal magnetohydrodynamics.

A. Continuity Equation:

The continuity equation is

$$\frac{\partial \rho}{\partial t} + \nabla \cdot (\rho \vec{v}) = 0. \tag{1}$$

This expresses the conservation of mass. To drive this equation consider some closed volume V of space, and denote its bounding surface by S ,the total mass of fluid contained in this volume is

$$M_f = \int_V \rho\, d^3x. \tag{2}$$

The change with time of the mass contained in a certain volume is determined by the derivative $\frac{\partial}{\partial t}\int \rho\, dV.$ On the other hand, the change in unit time, say, is determined by the quantity of mass which in unit time leaves the volume and goes to the outside or, conversely, passes to its interior. The quantity of mass which passes in unit time through the element $\vec{df}$ of the surface bounding our volume is equal to $\rho \vec{v}.\vec{df}$, where $\vec{v}$ is the velocity of the mass at the point in space where the element $\vec{df}$ is located. The vector $\vec{df}$ is directed, as always, along the external normal to the surface, that is, along the normal toward the outside of the volume under consideration. Therefore $\rho \vec{v}.\vec{df}$ is positive if mass leaves the volume, and negative if mass enters the volume.

The total amount of mass leaving the given volume per unit time is consequently $\oint \rho\vec{v}.d\vec{f}$,where the integral extends over the whole of the closed surface bounding the volume, from the equality of these two expressions,

we get
$$\frac{\partial}{\partial t}\int \rho dV = -\oint \rho\vec{v}.d\vec{f} \qquad\qquad (2)$$

The minus sign appears on the right, since the left side is positive if the total mass in the given volume increases. The equation (2) is the so-called "equation of continuity", expressing the conservation of mass in integral form. Noting that $\rho\vec{v}$ is the mass current density $\vec{j}$, we can rewrite (2) in the form

$$\frac{\partial}{\partial t}\int \rho dV = -\oint \vec{j}.d\vec{f} \qquad (\vec{j} = \rho\vec{v}) \qquad (3)$$

We also write this equation in differential form. To do this we apply Gauss's theorem to (2) :

$$\oint \rho\,\vec{v}.\,d\vec{f} = \int \text{div}\,(\rho\vec{v})\,dV$$

and we found $\int\left(\text{div}\,\rho\vec{v} + \frac{\partial \rho}{\partial t}\right)dV = 0.$

Since this must hold for integration over an arbitrary volume, integrand must be zero :

$$\frac{\partial \rho}{\partial t} + \text{div}\left(\rho\vec{v}\right) = 0 \qquad\qquad (4)$$

This is the equation of continuity in differential form.

Next, we require an equation for the time evolution of the fluid velocity $\vec{v}$. We wish to calculate the rate of change of the velocity of a particular fluid element in response to the forces acting on it.

• It will therefore be necessary to take a time derivative following the motion of the fluid element.

 For this consider any function g which depends both on spatial coordinates $\vec{x}$ and on time t. In the short interval of time Δt, a fluid element starting at $\vec{x}$ will move to a new position at $\vec{x} + \Delta t\,\vec{v}\,(\vec{x})$. The value g appropriate to the new coordinates of the fluid element is, by Taylor's theorem :

$$g(\vec{x}+\Delta t\ \vec{v}(\vec{x}),t+\Delta t)\simeq g\,(\vec{x},t)+\ +\Delta t\ \vec{v}(\vec{x}).\vec{\nabla}g(\vec{x},t)+\ +\Delta t\ .\frac{\partial}{\partial t}(g\ (\vec{x},t)) \tag{5}$$

In equation (5), the operator that is multiplied by Δt is the total or convective time derivative, following the motion of the fluid element, that we are seeking:

$$\frac{d}{dt}\equiv\frac{\partial}{\partial t}+\vec{v}.\vec{\nabla} \tag{6}$$

Thus in our equation for the time evolution of the fluid velocity, the left-hand side will be given by

$$\frac{d\vec{v}}{dt}\equiv\frac{\partial\vec{v}}{\partial t}+(\vec{v}.\vec{\nabla})\vec{v}. \tag{7}$$

- The right hand side is determined by the forces acting on the fluid element. First, the fluid element contains a current density $\vec{J}$, so that in a magnetic field it will be subject to the force $\dfrac{\vec{J}\times\vec{B}}{\rho}$ per unit mass. Here we have assumed that fluid element is large enough to be electrically neutral, so that there is no Lorentz force associated with its velocity $\vec{v}$. Second, there is a pressure force $-\left(\dfrac{1}{\rho}\right)\nabla p$ per unit mass. Hence

$$\rho\frac{d\vec{v}}{dt}=\vec{J}\times\vec{B}-\nabla p \tag{8}$$

Also, p and ρ are not independent. They are related by the standard equation of state, which for the behavior that we shall be considering is

$$pV^{\gamma}=\text{Constant, or}$$

$$\frac{d}{dt}\left(\frac{p}{\rho^{\gamma}}\right)=0, \tag{9}$$

where γ is the adiabatic exponent.

The three equations that we have considered so far, equation (1), (8), (9), govern the time evolution of the fluid quantities ρ ,$\vec{v}$ and p given a current $\vec{J}$ and magnetic field $\vec{B}$.

- In addition, we require complementary information: How $\vec{J}$ and $\vec{B}$ evolve in a given fluid. This is obtained as follows from Maxwell's equations and Ohm's law.

• Consider first Ohm's law for an element of a fluid that conductivity σ. we have

$$\vec{J} = \sigma\,\vec{E}'$$

Where $\vec{E}'$ is the electric field experienced by the fluid element in its frame. Now the fluid element is moving at velocity $\vec{v}$ with respect to the laboratory magnetic field $\vec{B}$. This motion gives rise to an electric field in the rest frame of the fluid element, which is described by the Lorentz Transformation:

$$\vec{E}' = \vec{E} + \vec{v} \times \vec{B}. \tag{11}$$

Hence $\vec{E}$ denotes the electric field in the laboratory frame, and we have neglected all higher order relativistic corrections because $v^2 \ll c^2$. combining (10), and (11), we have

$$\vec{J} = \sigma\,(\vec{E} + \vec{v} \times \vec{B}),$$

$$\text{or} \quad \frac{\vec{J}}{\sigma} = \vec{E} + \vec{v} \times \vec{B} \qquad . \tag{12}$$

• In a perfectly conducting fluid, σ tends to infinity. This limiting case defines the subject area known as ideal magneto hydrodynamics.

• The left hand side of equation (12) tends to zero, and we have

$$\vec{E} = -\vec{v} \times \vec{B}, \tag{13}$$

This equation tells us that the electric field $\vec{E}'$ in the rest frame of a conducting fluid is zero: This is evident from equation (11) and (13). The instantaneous response of the perfectly conducting fluid immediately cancels out any attempt to create a non-zero $\vec{E}'$.

• Now the laboratory is moving at velocity $-\vec{v}$ with respect to the rest frame of the fluid element, where the magnetic field $\vec{B}' = \vec{B}$, when terms of order $\dfrac{v^2}{c^2}$ are neglected. Then the Lorentz transformation back to the laboratory frame gives

$$\vec{E} = \vec{E}' - \vec{v} \times \vec{B}' = -\vec{v} \times \vec{B}, \tag{14}$$

Since $\vec{E}' = 0$.

• Thus the existence of non-zero $\vec{E}$ has two physical origins. First, there is the assumed infinite conductivity of the fluid. Second, there is the fact that the division of an electromagnetic field into electric and magnetic components is frame independent. $\vec{E}'$ is zero, and $\vec{v} \times \vec{B}$ is non-zero, and this is sufficient to make $\vec{E}$ non-zero.

• We now include Faraday's law of electromagnetic induction equation (Maxwell equation) :

$$\vec{\nabla} \times \vec{E} = -\frac{\partial \vec{B}}{\partial t}$$

Taking the curl of equation (13), gives

$$\frac{\partial \vec{B}}{\partial t} = \vec{\nabla} \times (\vec{v} \times \vec{B}). \tag{15}$$

We have now eliminated $\vec{E}$, and obtained in equation (15) an expression for the time evolution of $\vec{B}$ in terms of $\vec{v}$ and $\vec{B}$ itself.

• Equation (15) is a combined statement of Ohm's law and Faraday's law for a perfectly conducting fluid in a magnetic field. It leads to the concept of magnetic flux freezing. To understand this important physical phenomenon, we return to the equation (15):

$$\frac{\partial \vec{B}}{\partial t} = \text{Curl}(\vec{v} \times \vec{B})$$

We expand the right hand side, using the fact that $\text{div}\,\vec{B} = 0$: (use formula $\vec{\nabla} \times (\vec{a} \times \vec{b}) = (\vec{b}.\vec{\nabla})\vec{a} - (\vec{a}.\vec{\nabla})\vec{b} + \vec{a}\,\text{div}\,\vec{b} - \vec{b}\,\text{div}\,\vec{a}$.)

$$\frac{\partial \vec{B}}{\partial t} = (\vec{B}.\,\text{grad})\,\vec{v} - (\vec{v}.\,\text{gard})\,\vec{B} - \vec{B}\,\text{div}\,\vec{v},$$

Substituting from the equation of continuity (equation 4):

$$\frac{\partial \rho}{\partial t} + \text{div}(\rho\,\vec{v}) = 0$$

$$\text{or} \quad \frac{\partial \rho}{\partial t} + \rho\,\text{div}\,\vec{v} + \vec{v}.\,\text{grad}\,\rho = 0$$

$$\text{or} \quad \text{div}\,\vec{v} = -\frac{1}{\rho}\frac{\partial \rho}{\partial t} - \frac{\vec{v}.\,\text{grad}\,\rho}{\rho},$$

We obtain after a simple rearrangement of terms

$$\left(\frac{\partial}{\partial t} + \vec{v}.\,\text{grad}\right)\left(\frac{\vec{B}}{\rho}\right) \equiv \frac{d}{dt}\left(\frac{\vec{B}}{\rho}\right) = \left(\frac{\vec{B}}{\rho}.\,\text{grad}\right)\,\vec{v}$$

or $\qquad \dfrac{d}{dt}\left(\dfrac{\vec{B}}{\rho}\right)=\left(\dfrac{\vec{B}}{\rho}.\,\text{grad}\right)\vec{v}$ $\qquad\qquad\qquad$ (16)

We now consider some "fluid line", i.e. a line which moves with the fluid particles composing it. Let $\delta\vec{\ell}$ be an element of length of this line; we shall determine how $\delta\vec{\ell}$ varies with time. if $\vec{v}$ is the fluid velocity at one end of the element $\delta\vec{\ell}$, then the fluid velocity at the other end is $\vec{v}+(\delta\vec{\ell}.\text{grad})\,\vec{v}$. During a time interval dt, the length of $\delta\vec{\ell}$ therefore changes by $dt(\delta\vec{\ell}.\text{grad})\,\vec{v}$, i.e. $d\,(\delta\vec{\ell})/dt=(\delta\vec{\ell}.\text{grad})\,\vec{v}$.

• We see that rates of change of the vectors $\delta\vec{l}$ and $\vec{B}/\rho$ are given by identical formulae. Hence it follows that, if these vectors are initially in the same direction, they will remain parallel and their length will remain in the same ratio. In other words, if two infinitely close fluid particles are on the same line of force at any time, then they will always be on the same line of force, and the value of $\dfrac{\vec{B}}{\rho}$ will be proportional to the distance between particles.

• Passing now from particles as an infinitesimal distance apart to those at any distance apart .We conclude that every line of force moves with the fluid particles which lie on it. We can picture this by saying (in the limit $\sigma\to\infty$) the lines of magnetic force are "frozen" in the fluid and more with it.

The quantity $\dfrac{B}{\rho}$ varies at every point proportionally to the extension of the corresponding "fluid line". if the fluid may be supposed incompressible $\rho=$ constant, and the field B varies as the extension of the lines of force.

6.3 Complete set of Magnetohydrodynamic(MHD)Equations

We collect here the following set of simplified MHD equations:

1. Equation of continuity:

$$\dfrac{\partial\rho}{\partial t}+\nabla.(\rho\vec{v})=0$$

2. Equation of Motion of conducting fluid:

$$\rho\left[\dfrac{\partial\vec{v}}{\partial t}+(\vec{v}.\vec{\nabla})\,\vec{v}\right]=\vec{J}\times\vec{B}-\nabla p$$

3. Maxwell's equations :

$$\vec{\nabla} \times \vec{E} = -\frac{\partial \vec{B}}{\partial t}$$

4. $\qquad \vec{\nabla} \times \vec{B} = \mu_0 \vec{J}$

5. Generalized Ohm's law

$$\vec{J} = \sigma(\vec{E} + \vec{v} \times \vec{B}) - \frac{\sigma}{\rho} \vec{J} \times \vec{B}$$

In this set of equations, viscosity and thermal conductivity are neglected. Also from equation (4) we have $\vec{\nabla} . \vec{J} = 0$

6.4 Diffusion of Charged Particles in a Weakly Ionized Gas

• Diffusion of particles in a gas and plasma are due to the gradients of the macroscopic parameters. These gradients give rise to fluxes that finally equalize the microscopic parameters over the plasma volume.

• We first consider the simplest diffusion phenomenon due to the transport of particles. If the density of a given particle species is not spatially uniform and this gas state is hydrodynamically stable, there is a directional flow of particle tending to equalize in space the density of the given particle species.

• If the density of the given species varies slightly over the mean free path, that is, the density gradient is small, the diffusion flux density $\vec{j}$ is proportional to that gradient:

$$\vec{j} = -D \ \text{grad} \ N \qquad\qquad (1)$$

• The factor D in equation (1) is called the diffusion coefficient.

• We analyze the restoration of the equilibrium gas density after a density gradient has been established. Inserting (1) into the continuity equation viz $\frac{\partial N}{\partial t} + \text{div} \ \vec{j} = 0$, where $\vec{j} = N \vec{v}$, we obtain:

$$\frac{\partial N}{\partial t} = -\text{div} \ \vec{j}$$

$$= -\text{div}(-D \ \text{grad} \ N)$$

$$= D \ (\nabla . \nabla N)$$

$$\frac{\partial N}{\partial t} = D\,(\nabla^2 N) \qquad\qquad (2)$$

Let us denote by L the characteristic distance at which there occurs a noticeable variation of the gas density. Then equation (2) yields the characteristic time of the density variation,

$$\tau_L \sim \frac{L^2}{D} \sim \frac{L^2}{v\lambda},$$ where λ is the mean free path and v is the characteristic velocity.

6.5 Ambipolar Diffusion

We now take up the case of diffusion of charged particles in a weakly ionized gas. The degree of ionization is assumed to be so small that collisions between charged particles may be neglected in comparison with those between charged particles and neutral atoms.

Even under these conditions, the diffusion of the two types of charged particles (electrons and ions) is not independent, because an electric field arises in the diffusion process.

• It may be noted that electrons have a considerably larger diffusion coefficient than the ions ;therefore ,the electrons are spreading over the gas volume considerably faster than the ions. This results in a disturbance of quasineutrality of the plasma and in the emergence of electric fields in the plasma created by the charged particles.

• The electric field $\vec{E}$ created due to the spatial distribution of charged particles satisfies Poisson's equation: $\operatorname{div} \vec{E} = 4\pi e(N_i - N_e)$ $\qquad\qquad (1)$

Where N_e and N_i are the densities of the electrons and ions (ions are assumed to be singly charged).

• The flux densities of the electrons and ions in the system are the sums of the diffusion flux density and the flux density due to the electric field:

$$\vec{j}_e = -D_e \ \operatorname{grad} N_e - K_e N_e \vec{E} \qquad\qquad (2)$$

$$\vec{j}_i = -D_i \ \operatorname{grad} N_i - K_i N_i \vec{E} \qquad\qquad (3)$$

In the above equation D_e and D_i are the diffusion coefficients of the electrons and ions in the gas, K_e and K_i are the mobilities ($K_e = \frac{v_e}{E}$, $K_i = \frac{v_i}{E}$) of the electrons and ions respectively, v_e and v_i are electron and ion drift velocities respectively,

Let us now consider such a mode of development that the plasma remains quasineutral is the process of motion. This is the case for relatively high densities of charged particles; the separation of charges gives rise to large fields which prevent further separation and preserve the quasineutrality of the plasma. This phenomenon is termed the "ambipolar diffusion" In this case $N_e = N_i = N$ So that $\Delta N \equiv |N_e - N_i| \ll N$ and the fluxes of the electrons and ions are the same.

• We now analyze equation (2) and (3) for the electron and ion flux densities in the case of ambipolar diffusion.

• For the electron flux to be equal to the ion flux, the first term in equation (2) must almost cancel the second term: $-D_e \, \text{grad} \, N_e = K_e \, N_e \, \vec{E}$.

This means that the electric field strength must be

$$\vec{E} = -\left(\frac{D_e}{K_e}\right) \times \frac{(\text{grad} \, N_e)}{N_e} \tag{4}$$

$$= -\left(\frac{D_e}{K_e}\right) \times \frac{(\text{grad} \, N)}{N}$$

The ratio of the mobility K_e to the diffusion coefficient is known as the Einstein relation and is given by

$$\frac{K_e}{D_e} = \frac{e}{T_e} \tag{5}$$

• Thus the electric field $\vec{E}$ must be

$$\vec{E} = -\left(\frac{T_e}{e}\right) \times \frac{(\text{grad} \, N)}{N} \tag{6}$$

• Inserting this into equation (3), we find the flux density of the charged particles:

$$\vec{j}_i = -D_i \, \text{grad} \, N - K_i \, N_i \, \vec{E}$$

$$= -D_i \, \text{grad} \, N - \left(\frac{e \, D_i}{T_i}\right) N \left(\frac{-D_i}{K_e}\right) \times \left(\frac{\text{grad} \, N}{N}\right) \tag{7}$$

Now substituting $K_e = \frac{(-e)D_e}{T_e}$ (Einstein's relation) in equation (7), we find

$$\vec{j}_i = -D_i \, grad \, N - \left(\frac{e \, D_i}{T_i}\right) N \left(\frac{-D_e \, T_e}{(-e) \, D_e}\right) \times \frac{gradN}{N}$$

$$= -\left(D_i + D_i \frac{T_e}{T_i}\right) gradN$$

$$= -D_a \, grad \, N \tag{8}$$

where $D_a = D_i \left(1 + \frac{T_e}{T_i}\right)$ \hfill (9)

Above eq. is known as the ambipolar diffusion coefficient. In the above analysis, we made use of the Einstein relation viz. $K = e\dfrac{D}{T}$, and assumed that the electron T_i thus, the ambipolar diffusion of the charged particles is a diffusion-like motion with the time parameter corresponding to the ions.

Assuming $T_e \approx T_i$, then the ambipolar diffusion coefficient is twice that of the ions, Thus, during a time $\sim \tau_i$ (τ_i is the characteristic ion diffusion time), the electrons and ions diffuse together ($\delta N_e \approx \delta N_i$) with a diffusion coefficient twice that of the ions; this process is called "ambipolar diffusion." Half of the coefficient is due to the intrinsic diffusion of the ions, and half to the electric field resulting from the accelerating electrons.

6.6 Illustrative Examples

Example 1 (a) solve the "diffusion equation"

$$\frac{\partial n(\vec{r},t)}{\partial t} = D\nabla^2 n(\vec{r},t)$$

by the method of separation of variables, let

$$n(\vec{r},t) = S(\vec{r})T(t)$$

and show that

$$T_k(t) = T_0 \exp(-D k^2 t)$$

$$(\nabla^2 + k^2)S(\vec{r}) = 0$$

Where k^2 is the separation constant and T_0 is constant.

(b) Assuming that S depends only on the x coordinate, show that
$$S(x) = C(k) \exp(i k x)$$
where k can be either positive or negative, and that

$$n(x,t) = \int_{-\infty}^{\infty} C(k) \exp(i k x - D k^2 t) \, dk$$

$$n_0(x) = \int_{-\infty}^{\infty} C(k) \exp(i k x) \, dk$$

where $n_0(x) = n(x,0)$ is the know "initial" density distribution.

(c) Using Fourier transform theory, show that

$$C(k) = \frac{1}{2\pi} \int_{-\infty}^{\infty} n_0(x) \exp(-i k x) \, dx \,, \text{ and consequently, that}$$

$$n(x,t) = \frac{1}{2(\pi D t)^{1/2}} \int_{-\infty}^{\infty} n_0(x') \exp\left[\frac{-(x-x')^2}{4Dt} \right] dx'$$

(d) Taking as initial condition

$$n_0(x) = \exp\left(\frac{-x^2}{x_0^2} \right)$$

show that $\qquad n(x,t) = \left(\frac{\tau_D}{\tau_D + 4t} \right)^{1/2} \exp\left[\frac{-x^2}{x_0^2} \left(\frac{\tau_D}{\tau_D + 4t} \right) \right]$

Where $\tau_D = \dfrac{x_0^2}{D}$ is a characteristic time for diffusion to smooth out the density n.

Sol.(a) Diffusion equation is

$$\frac{\partial n(\vec{r},t)}{\partial t} = D \nabla^2 n(\vec{r},t)$$

We solve this equation by method of separation of variables:

$$\text{Let } n(\vec{r},t) = S(\vec{r}) T(t);$$

substituting in diffusion equation, we find

$$S \frac{\partial T}{\partial t} - D\, T \nabla^2 S = 0$$

$$\text{or} \qquad \frac{1}{T} \frac{\partial T}{\partial t} - \frac{D}{S} \nabla^2 S = 0$$

Let $\dfrac{1}{D\,T}\dfrac{\partial T}{\partial t} = -k^2$, then $-\dfrac{1}{S}\nabla^2 S = +k^2$

Integrating we obtain $T = T_0\, e^{-D k^2 t}$

and $\nabla^2 S + k^2 S = 0$

or $(\nabla^2 + k^2)S = 0$

(b) $\quad n(x,t) = \displaystyle\int_{-\infty}^{\infty} S(x)T_k(t)\,dk$

$$= \int_{-\infty}^{\infty} C(k)\exp(i\,k\,x)\ \text{to}\ \exp(-D k^2\, t)\,dk$$

$$n(x,t) = \int_{-\infty}^{\infty} C(k)\exp(i\,k\,x - D k^2\, t)\,dk$$

$T_0 = \text{Constant} = 1$ (by Choice)

and $n_0(x) = n(x, 0)$ is obtain by putting t=0 in the above expression

$$n_0(x) = \int_{-\infty}^{\infty} C(k)\exp(i\,k\,x)\,dk$$

(c) Substituting from Fourier transform theory

$$C(k) = \frac{1}{2\pi}\int_{-\infty}^{\infty} n_0(x')\exp(-i\,k\,x')\,dx'$$

In the expression for n(x, t)

$$n(x,t) = \frac{1}{2\pi}\int_{-\infty}^{\infty} dx'\, n_0(x')\int_{-\infty}^{\infty} e^{-i k x' + i k x - D k^2 t}\,dk$$

carrying out this simple integration and using the well know integral $\displaystyle\int_{-\infty}^{\infty} e^{-\alpha x^2}\,dx = \sqrt{\dfrac{\pi}{\alpha}}$, we finally obtain :

$$n(x,t) = \frac{1}{2\sqrt{\pi D t}}\int_{-\infty}^{\infty} n_0(x')\exp\left[\frac{-(x-x')^2}{4\,Dt}\right]$$

(d) Substituting $n_0(x') = \exp\left(\dfrac{-x'^2}{x_0^2}\right)$ and carrying out this simple integration we

obtain the required expression.

Example 2 Consider the solution of the diffusion equation by separation of variables in the linear geometry of the plasma slab indicated in figure given below. Show that solution of the equation

$$\frac{d^2 S(x)}{d x^2} + k^2 S(x) = 0$$

That satisfy the boundary conditions S=0 at $x = \pm L$,are

$$S(x) = \sum_m a_m \cos\left[(m + \frac{1}{2}) \frac{\pi\, x}{L} \right]$$

$$S(x) = \sum_m b_m \sin(\frac{m\,\pi\, x}{L})$$

Explain why the solution as a sine series is not a physically acceptable solution for this case. Consequently, from n(x,t) = S(x) T(t), show that the number density can be expressed as:

$$n(x,t) = \sum_m a_m \exp\left[- D\pi^2 (m + \frac{1}{2})^2 \frac{t}{L^2} \right] \cos\left[\pi (m + \frac{1}{2}) \frac{x}{L} \right]$$

Therefore, the decay time constant for the mth mode is

$$\tau_m = \left[\frac{L}{\pi (m + \frac{1}{2})} \right]^2 \frac{1}{D}$$

This result shows that the higher modes decay faster than the lower ones.

How are the coefficients a_m determined in terms of n_0 (x) ?

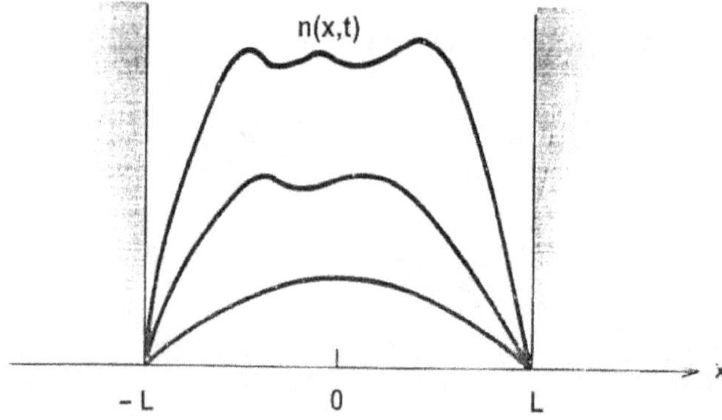

Figure : Geometry of the plasma slab for the solution of the diffusion equation considered in illustrative example 2.

Sol. Diffusion equation is

$$\frac{\partial n(x,t)}{\partial t} = D\frac{\partial^2 n(x,t)}{\partial x^2} \tag{1}$$

Let $\quad n(x,t) = S(x)\, T(t)$

Substituting in (1) one fields

$$S\frac{dT}{dt} = DT\frac{d^2S}{dx^2}$$

Dividing by DST gives

$$\frac{1}{DT}\frac{dT}{dt} - \frac{1}{S}\frac{d^2S}{dx^2} = 0$$

The variables are separated.

$$\text{Let}\quad \frac{1}{DT}\frac{dT}{dt} = -k^2 \tag{2}$$

$$-\frac{1}{S}\frac{d^2S}{dx} = +k^2 \tag{3}$$

Integration of equation (2) gives

$$T = T_0 \,\exp\left(-k^2 D\, t\right),\ \text{where } T_0 \text{ is a constant.}$$

Equation (3) gives $\dfrac{d^2S}{dx^2} + k^2 S = 0$

The solution for $S(x)$ is

$$S(x) = C_1 \cos k\, x + C_2 \sin k\, x$$

Using the boundary condition

$$S(x) = 0, \qquad x = \pm L$$

$$0 = C_1 \cos k\, L + C_2 \sin k\, L$$

If $\quad C_2 = 0,$ then

$$kL = \left(m + \frac{1}{2}\right)\pi, \qquad m \text{ is integer}$$

$$\therefore k = \left(m + \frac{1}{2}\right)\frac{\pi}{L}$$

$$\therefore S(x) = \sum_m a_m \cos\left(m + \frac{1}{2}\right)\frac{\pi}{L}x \tag{4}$$

If $C_1 = 0$, then the solution with the boundary condition $S(x)=0,\ x = \pm L$ gives

$$\therefore S(x) = \sum_m b_m \sin\left(\frac{m \pi x}{L}\right) \tag{5}$$

Taking account the symmetry of the problem $[N_e(x) = N_e(-x)]$ the solution for $S(x)$ is given by (4).

Therefore

$$n(x,t) = \sum_m a_m \exp\left[-D k^2 t\right] \cos\left(m + \frac{1}{2}\right)\frac{\pi x}{L}$$

Substituting the value of $k = \left(m + \frac{1}{2}\right)$,we find

$$n(x,t) = \sum_m a_m \ exp\left[-D\left(m + \frac{1}{2}\right)^2 \frac{\pi^2}{L^2} t\right]\cos\left(m + \frac{1}{2}\right)\frac{\pi x}{L}.$$

Therefore, the decay constant for m^{th} mode is given by

$$D\left(m + \frac{1}{2}\right)^2 \frac{\pi^2}{L^2}, t \sim 1$$

or $\qquad t = \dfrac{L}{\pi\left(m + \frac{1}{2}\right)^2 D} \equiv \tau_m \ (\text{say})$

or $\qquad \tau_m = \left[\dfrac{L}{\pi\left(m + \frac{1}{2}\right)}\right]^2 \cdot \dfrac{1}{D}$

Therefore higher modes decay faster than the lower modes.

Example 3 Consider a fluid whose conductivity σ is not infinite.

(a) Show the MHD equation in this case satisfies

$$\frac{\partial \vec{B}}{\partial t} = \vec{\nabla} \times (\vec{v} \times \vec{B}) + \frac{1}{\mu_0 \sigma}\nabla^2 \vec{B}$$

(b) Show that, when the evolution of the magnetic field is dominated by the new, non-ideal term, the field decays in a way described by a diffusion equation. What is the relation between the characteristic timescale and length scale of this process ?

(c) Show that the decay of the magnetic field is due to the dissipation of energy through Joule heating.

Sol.(a)
$$\vec{\nabla} \times \vec{B} = \mu_0 \vec{J} \qquad (1)$$

$$\vec{J} = \sigma(\vec{E} + \vec{v} \times \vec{B}) \qquad (2)$$

Substituting (2) into (1) gives

$$\vec{\nabla} \times \vec{B} = \mu_0 \sigma\left[\vec{E} + \vec{v} \times \vec{B}\right] \qquad \text{(Ohm's law)}$$

Taking the curl of this equation, and dividing by $\mu_0\sigma$, we obtain

$$\frac{1}{\mu_0\sigma} \vec{\nabla} \times (\vec{\nabla} \times \vec{B}) = \vec{\nabla} \times \vec{E} + \nabla \times (\vec{v} \times \vec{B})$$

or
$$\frac{1}{\mu_0\sigma}\left[\vec{\nabla}(\vec{\nabla}.\vec{B}) - \nabla^2\vec{B}\right] = -\frac{\partial\vec{B}}{\partial t} + \vec{\nabla} \times (\vec{v} \times \vec{B})$$

$$\Rightarrow \frac{\partial\vec{B}}{\partial t} = \nabla \times (\vec{v} \times \vec{B}). \qquad (3)$$

represents MHD equation of non-ideal MHD fluid $(\sigma \neq \infty)$.

(b) If the ideal term on the right hand side of (3) is negligible, we are left with a diffusion equation:

$$\frac{\partial\vec{B}}{\partial t} = \frac{1}{\mu_0\sigma}\nabla^2\vec{B}$$

Each component of $\vec{B}$ satisfies the diffusion equation.

Let $\qquad B(x,t) = X(x)\,\theta(t).$ Then we have

$$X(x)\,\theta'(t) = \frac{1}{\mu_0\sigma}X''(x)\theta(t)$$

Where the prime denotes differentiation with respect to x or t as appropriate.

Dividing both sides by $X(x)\,\theta(t)$

$$\frac{\theta'(t)}{\theta(t)} = \frac{1}{\mu_0\sigma}\frac{X''(x)}{X(x)}$$

The left and right hand sides of the equation are functions of different independent variables. Equality for all values of x and t is possible only if both sides of the equation are, in fact, constant.

We write $\quad \dfrac{\theta'(t)}{\theta(t)} = \dfrac{1}{T}$

where T is a fixed quantity which necessarily has time as its dimension, similarly, we write

$$\frac{X''(x)}{X(x)} = \frac{1}{L^2}$$

Where L is a fixed quantity which necessarily has length as its dimension. Substituting back, we obtain the relation.

$$T = \mu_0 \sigma L^2$$

6.7 Self Learning Exercise

Q.1 Show that the solution of the diffusion equation in the case of cylindrical geometry (see figure)

$$\frac{d^2 S(r)}{d r^2} + \frac{1}{r}\frac{dS(r)}{dr} + k^2 S(r) = 0$$

can be written in terms of Bessel functions $J_m(kr)$, Explain how k must be determined so that $n(\vec{r}, t)$ satisfies the boundary condition $n = 0$ at $r = R_0$.

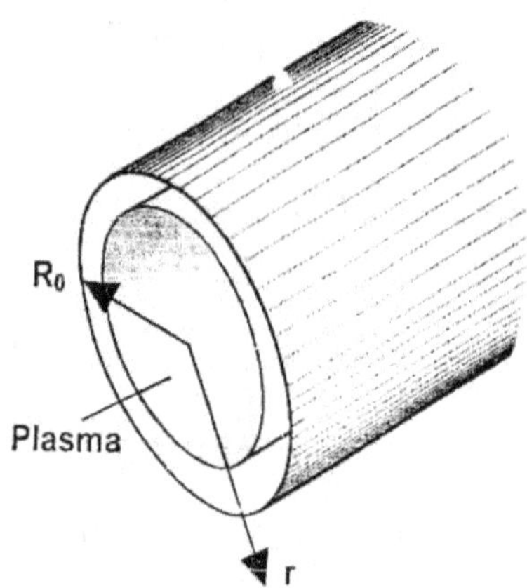

Figure: Cylindrical geometry of plasma column for the solution of the diffusion equation considered

Q.2 Prove that total time derivative following the motion of the fluid element is given by

$$\frac{d}{dt} \equiv \frac{\partial}{\partial t} + \vec{v}.\vec{\nabla}$$

6.8 Summary

In this chapter we have treated plasma as a conducting fluid. The behavior of the plasma is completely described by a complete set of magnetohydrodynamic (MHD) equations. We have also discussed the phenomenon of diffusion in a weakly ionized plasma. This leads to an interesting phenomenon of " Ambipolar diffusion."

6.9 Glossary

Ambipolar diffusion:

diffusion of electrons and ions together in steady state.

"Frozen" in magnetic field:

In the limit of infinite conductivity of plasma fluid ,the lines of magnetic force are frozen in fluid and move with it. This is called "Frozen" in magnetic field.

6.10 Answers to Self Learning Exercise

Ans.1 : S(r) satisfies the Bessel equation.

$$S(r) = J_0(k\, r)$$

The boundary condition $\qquad$ n=0 at $r = R$

$\qquad$ implies $\qquad$ $S(r = R_0) = 0$, This gives

$$J_0(k\, R_0) = 0$$

This gives $\qquad$ $k\, R_0 = 2.405$ $\quad$ or $\qquad$ $k = \dfrac{2.405}{R_0}$

Hence the solution is $S(r) = J_0(2.405\,\dfrac{r}{R_0})$.

The complete solution for $n(\vec{r}, t)$ is

$$n(\vec{r}, t) = S(\vec{r})\, T(t)$$

there $\qquad$ $T(t) = T_0\, e^{-D k^2 t}$

$\qquad$ an $\qquad$ $S(r) = J_0(\dfrac{2.405\ r}{R_0})$.

Ans.2 : Consider any function g which depends on spatial position x and on time t. In time interval Δt, a fluid element starting at $\bar{x}$ will move to a new position at $\vec{x} + \Delta t \vec{v}(\vec{x})$. The value of g appropriate to the new coordinates of the fluid element is, by Taylor's theorem

$$g\,(\bar{x} + \Delta t\,\vec{v}(\bar{x}), t + \Delta t) \simeq g(\bar{x}, t) + \Delta t\,\vec{v}(\bar{x}).\nabla g\,(\bar{x}, t) + \Delta t\,\frac{\partial}{\partial t} g(\bar{x}, t).$$

or
$$\frac{g\,(\bar{x} + \Delta t\,\vec{v}(\bar{x}), t + \Delta t) - g(\bar{x}, t)}{\Delta t} = \left(\vec{v}.\nabla + \frac{\partial}{\partial t}\right) g(\bar{x}, t)$$

or
$$\frac{d}{dt}\,\vec{g}\,(\bar{x}, t) = \left(\vec{v}.\vec{\nabla} + \frac{\partial}{\partial t}\right) g(\bar{x}, t)$$

or equivalently

$$\frac{d}{dt} = \frac{\partial}{\partial t} + \vec{v}.\vec{\nabla}$$

6.11 Exercise

Q.1   Write full set of ideal MHD equations.

Q.2   What is ambipolar diffusion? derive and expression the coefficient of ambipolar diffusion. solve ambipolar diffusion equation when the fluid is confined in rectangular box.

Q.3   Explain the concept of magnetic flux freezing.

Q.4   Deduce the equation of continuity of an ideal fluid.

Q.5   Derive generalized Ohm's law.

References and Suggested Readings

1. Electrodynamics of continuous media : Landau and Lifshitz

2. Plasma Physics : J.A. Bittencourt

UNIT-7
Some Basic Plasma Phenomena
and Pinch Effect

Structure of the Unit

To learn

• Electron plasma oscillations and Plasma Sheath

• Floating negative potential & Electric Potential on the wall

• Plasma Probe (Langmuir Probe)

• Characteristics of I-V curve of an electrostatic plane probe immersed in a plasma

• Pinch effect and Equilibrium of a perfectly conducting fluid (Plasma)

• Bennet Pinch

• Kink instability& Sausage instability

7.1 Introduction

 One of the fundamental properties of a plasma is its tendency to maintain electric charge neutrality on a macroscopic scale under equilibrium conditions .When this macroscopic charge neutrality is disturbed, so as to produce a significant imbalance of charge, large Coulomb forces come into play, which tend to restore the macroscopic charge neutrality. Since these Coulomb forces cannot be naturally sustained in the plasma, it breaks into high-frequency electron plasma oscillations, which enable the plasma to maintain on average its electrical neutrality, The frequency of these oscillations is usually very high, and since the ions (in view of their higher mass) are unable to follow the rapidity of the electron oscillations, their motion is often neglected. In the following we will examine the mechanism by which the plasma strives to shield its interior from a disturbing electric field. When a material body is immersed in a plasma, the body acquires a net negative charge and therefore a negative potential with respect to the plasma potential. In the region near the wall of the body there is a boundary layer, known as "Plasma sheath" in which the electron and the ion number densities are different. We shall study the mechanism of formation of plasma-sheath."

In this chapter we consider the equilibrium of a plasma column at rest in a constant magnetic field. This study is important from the point of view of plasma confinement by a magnetic field in controlled thermonuclear research. The confinement is produced by an azimuthal (θ) self-magnetic field, due to an axial current in the plasma generated by an applied electric field. $\vec{J} \times \vec{B}$ force, acting on the plasma, forces the column to contract radially. This radial constriction of   the plasma column is known as the "Pinch effect".

7.2 The Debye Shielding Problem

To examine the mechanism by which the plasma strives to shield its interior form a disturbing electric field, consider a plasma whose equilibrium state is pretended by an electric field due to external charged particle. Let this test particle' has a positive test charge $+\,Q$ and it is located at the origin of spherical coordinate system. We are interested in determining the electrostatic potential $\phi(\vec{r})$ that is established near the test charge Q, due to the combined effects of the test charge and the distribution of charged particles surrounding it. Since the positive test charge Q attracts the negatively charged particles and repels the positively charged ones, the number densities of electrons $n_e(\vec{r})$ and of the ions $(n_i(\vec{r}))$ will be slightly different near the origin (test particle), whereas at large distances from the origin the electrostatic potential vanishes, so that

$$n_e(\infty) = n_i(\infty) = n_0,$$

Since this is a steady- state problem under the action of a conservative electric field, we have

$$\vec{E}(\vec{r}) = -\nabla\phi(\vec{r}) \tag{1}$$

$$n_e(\vec{r}) = n_0 \exp\left[\frac{e\,\phi(\vec{r})}{kT}\right] \tag{2}$$

$$n_i(\vec{r}) = n_0 \exp\left[\frac{-e\,\phi(\vec{r})}{kT}\right] \tag{3}$$

where we have assumed that the electrons and ions have the same temperature T.

The total electric charge density $\rho(\vec{r})$, including the test charge Q can be expressed as

$$\rho(\vec{r}) = -\,e\left[n_e(\vec{r}) - n_i(\vec{r})\right] + Q\delta(\vec{r}) \tag{4}$$

Using (2) and (3)

$$\rho(\vec{r}) = -\,e\,n_0\left\{\exp\left[\frac{e\,\phi(\vec{r})}{kT}\right] - \exp\left[\frac{-e\,\phi(\vec{r})}{kT}\right]\right\} + Q\delta(\vec{r}) \tag{5}$$

Making use of the following Maxwell equation,

$$\vec{\nabla}.\vec{E} = \frac{\rho(\vec{r})}{\epsilon_0} \tag{6}$$

and $\vec{E} = -\nabla\phi$, we obtain the following differential equation:

$$\nabla^2\phi(\vec{r}) = \frac{-e\,n_0}{\epsilon_0}\left\{\exp\left[\frac{e\,\phi(\vec{r})}{kT}\right] - \exp\left[\frac{-e\,\phi(\vec{r})}{kT}\right]\right\} + \frac{Q\delta(\vec{r})}{\epsilon_0} \tag{7}$$

which allows the evolution of the electrostatic potential $\phi(\vec{r})$.

We now assume that the perturbing electrostatic potential is weak so that the electrostatic potential energy is much less than the mean thermal energy, that is

$$e\phi(\vec{r}) \ll kT \tag{8}$$

Under this assumption we can use the approximation

$$\exp\left[\pm\frac{e\phi(\vec{r})}{kT}\right] \simeq 1 \pm \frac{e\phi(\vec{r})}{kT} \tag{9}$$

Therefore the equation (7) simplifies to

$$\nabla^2\phi(\vec{r}) - \frac{2}{\lambda_D^2}\phi(\vec{r}) = -\frac{Q\delta(\vec{r})}{\epsilon_0} \tag{10}$$

where λ_D denotes the Debye length

$$\lambda_D = \left(\frac{\epsilon_0\,kT}{n_0 e^2}\right)^{\frac{1}{2}} = \frac{v_{th}}{\omega_{pe}} \tag{11}$$

where $v_{th} = \left(\dfrac{kT_e}{m_e}\right)^{\frac{1}{2}}$ and ω_{pe} is the plasma -frequency $\left(\dfrac{n_0 e^2}{m_e\,\epsilon_0}\right)^{\frac{1}{2}}$

Since the problem has spherical symmetry, therefore ϕ depends only on the radial distance r measured from the position of the test particle, it being independent of the spatial orientation of $\vec{r}$ i.e. θ and Φ .

$$\frac{1}{r^2}\frac{d}{dr}\left(r^2\frac{d}{dr}\phi(\vec{r})\right) - \frac{2}{\lambda_D^2}\phi(\vec{r}) = 0 \tag{12}$$

In order to solve this equation we note initially that for an isolated particle of charge $+Q$, in free space, the electric field is

$$\vec{E}(\vec{r}) = \frac{1}{4\pi\,\epsilon_0}\frac{Q}{r^2}\hat{r}, \tag{13}$$

So that the electrostatic Coulomb potential ϕ_c due to this isolated charged $\phi_c(\vec{r})$ particle in free space is

$$\phi_c(r) = \frac{1}{4\pi\,\epsilon_0}\frac{Q}{r} \tag{14}$$

Hence we note that in the very close proximity of the test particle the electrostatic potential should be the same as that for an isolated particle in free space. Hence it is reasonable to seek the solution of (12) in the form

$$\phi(r) = \phi_c(r)\, F(r) = \frac{Q}{4\pi\,\epsilon_0}\,\frac{F(r)}{r} \tag{15}$$

where the function $F(r)$ must be such that $F(r) \to 1$ when $r \to 0$

Further, $\phi \to 0$ when $r \to \infty$.

Substituting (15) into (12) yields the following differential equation for $F(r)$:

$$\frac{d^2 F(r)}{dr^2} = \frac{2}{\lambda_D^2}\, F(r) \tag{16}$$

Diff. Equation (16) has the solution for $F(r)$ as

$$F(r) = A\exp\left(\frac{\sqrt{2}\,r}{\lambda_D}\right) + B\exp\left(\frac{-\sqrt{2}\,r}{\lambda_D}\right) \tag{17}$$

The condition that $\phi(r) \to 0$ as $r \to \infty$, requires A=0, Also the condition that $F(r) \to 1$ as $r \to 0$ requires B=1, Therefore, the solution of (12) is

$$\phi(r) = \phi_c(r)\exp\left(\frac{-\sqrt{2}\,r}{\lambda_D}\right) = \frac{Q}{4\pi\,\epsilon_0\,r}\exp\left(\frac{-\sqrt{2}\,r}{\lambda_D}\right) \tag{18}$$

Eqn. (18) represents Debye potential, The charge Q of the test particle is neutralized by the charge distribution surrounding the test particle. The charge density is (from eq. (5) and (9)

$$\rho(r) = -2\frac{n_0 e^2 \phi(\vec{r})}{kT} + Q\delta(\vec{r}) \tag{19}$$

Substituting for $\phi(\vec{r})$ from (18), then

$$\rho(r) = -\frac{Q}{2\pi r\,\lambda_D^2}\exp\left(\frac{-\sqrt{2}\,r}{kT}\right) + Q\,\delta(\vec{r}) \tag{20}$$

The neutralization of the test particle takes place effectively on account of the charged particles inside the Debye sphere. In the neighborhood of the test particle the electron number density is larger than the ion number density, on account of the fact that the positive test particles attracts the electrons and repels the ions. Therefore, in the close neighborhood of the test particle there is an imbalance of charge and consequently, an electric field, the shielding of this electric field is effectively completed over a distance of the order of $\sim \lambda_D$. Thus, for macroscopic neutrality, it is necessary that the plasma characteristic dimension L be much greater than λ_D.

7.3 Plasma Sheath

When a material body is immersed in a plasma, the body acquires a net negative charge and therefore a negative potential with respect to the plasma potential. In the region near the wall of the body there is a boundary layer, known as the "Plasma sheath" in which the electron and the ion number densities are different. Inside the plasma sheath the potential increases monotonically from a negative value on the wall to the value corresponding to the unperturbed plasma. The thickness of the plasma sheath is found to be of the order of a Debye length.

Physical Mechanism For The Formation of The Plasma Sheath:

We can understand the physical mechanism responsible for the formation of the plasma sheath as follows:

• The charged particles in the plasma that strike the wall because of their random thermal motions are for the most part lost to the plasma.

• The ions generally recombine at the wall an return to the plasma as neutral particles, whereas the electrons may either recombine there or enter the conduction band if the surface is a metal.

• As it is already known that the number of particles that hit the surface per unit area and per unit time, from one side only, for the case of an isotropic velocity distribution, is given by

$$\Gamma_\alpha = \frac{n_\alpha}{4} <v>_\alpha \tag{1}$$

where $<v>_\alpha$ is the average particle speed for a species. But

$$<v>_\alpha = \left(\frac{8}{\pi}\right)^{\frac{1}{2}} \left(\frac{kT_\alpha}{m}\right)^{\frac{1}{2}} \tag{2}$$

So that $\quad \Gamma_\alpha = n_\alpha \left(\frac{kT_\alpha}{2\pi m_\alpha}\right)^{\frac{1}{2}} \tag{3}$

Γ_α is known as random particle flux. If is evident from this result that if initially the electron and the ion number densities are equal, then the random particle flux for the electrons $\left(\Gamma_e\right)$ greatly exceeds that for the ions $\left(\Gamma_i\right)$ .This is because the electron thermal velocity $\left(\dfrac{T_e}{m_e}\right)^{\frac{1}{2}}$ is much greater than $\left(\dfrac{T_i}{m_i}\right)^{\frac{1}{2}}$. Therefore the wall in contact with plasma rapidly accumulates a negative charge, since initially more electrons reach the wall than positive ions. The negative potential repels the electrons and attracts the ions so that electron flux diminishes and the ion flux increases. Eventually, the negative potential at the wall becomes large enough in magnitude to equalize the rate at which electrons and ions hit the surface. At this "floating negative potential" the wall and the plasma reach a dynamical equilibrium such that the net current at the wall is zero.

7.4 Electric Potential on the Wall

Let the electric potential $\phi(x)$ at the wall (x = 0) be given by

$$\phi(0) = \phi_w \tag{1}$$

Let us choose the reference potential inside the plasma, at a very large distance from the wall, equal to zero,

$$\phi(\infty) = 0 \tag{2}$$

The electrons and ions are assumed to be in thermodynamic equilibrium at the same temperature T, with the negative potential on the wall, at $x \to \infty$, the plasma is unperturbed and $n_e \approx n_i \approx n_0$ .The electron and ion number densities can be expressed as

$$n_e(\vec{r}) = n_0 \exp\left[\frac{e\phi(\vec{r})}{kT}\right] \tag{3}$$

$$n_i(\vec{r}) = n_0 \exp\left[-\frac{e\phi(\vec{r})}{kT}\right] \tag{4}$$

Under equilibrium conditions there must be no charge build up at the wall (x=0), so that

$$J_e(0) = J_i(0) \tag{5}$$

where J_e and J_i represent electron and ion current densities respectively, using the formula $\vec{J} = n\,e\,\vec{v},$ equation (5) gives

$$\left(\frac{1}{m_e}\right)^{\frac{1}{2}} \exp\left(\frac{e\,\phi_w}{kT}\right) = \left(\frac{1}{m_i}\right)^{\frac{1}{2}} \exp\left(\frac{-e\,\phi_w}{kT}\right) \tag{6}$$

or $\quad \exp\left(\dfrac{-2e\,\phi_w}{kT}\right) = \left(\dfrac{m_i}{m_e}\right)^{\frac{1}{2}}$ $\hfill (7)$

From (7), We obtain

$$\phi_w = -\left(\dfrac{kT}{4e}\right)\ln\left(\dfrac{m_i}{m_e}\right) \hfill (8)$$

From(8) we note that the magnitude of the potential energy near the wall $|e\,\phi_w|$ is of the same order as the average thermal energy kT of the particles in the plasma, Since $\dfrac{|e\,\phi_w|}{kT} = \dfrac{1}{4}\ln\left(\dfrac{m_i}{m_e}\right)$,for a hydrogen ion, for examples, $\dfrac{|e\,\phi_w|}{kT}$ is approximately equal to 2, whereas for heavier ions it may be close to 3.

7.5 Plasma Probe (Langmuir Probe)

The plasma probe is a device used to measure the temperature and density of a plasma.The plasma probe (electrostatic probe) was originally developed by Langmuir Its physical mechanism of its operation can be explained using the theory of "Plasma sheaths". A conducting probe, or electrode, is immersed in a plasma and the current that flows through it is measured for various potentials applied to the probe.The temperature and number density of the electrons can be obtained from the characteristics of the resulting current-potential curve. When the surface of the probe is a plane, the current-potential curve has a shape like that in figure 1.

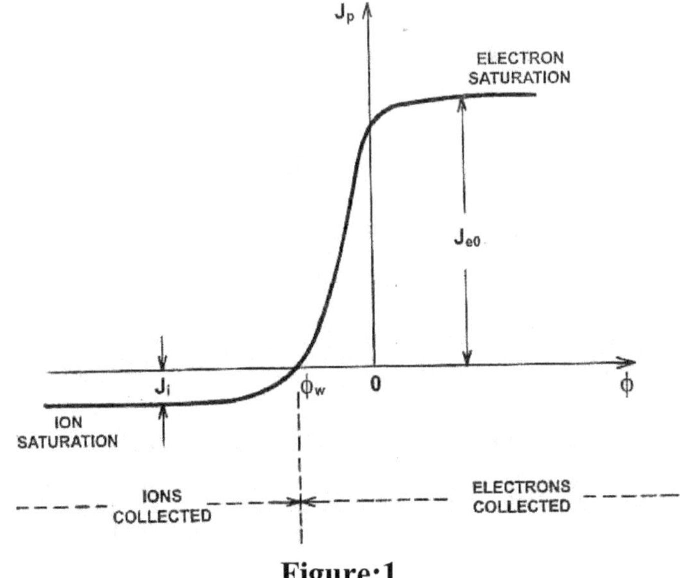

Figure:1

The probe, when inserted in the plasma, is surrounded by a plasma sheath that shields the major portion of the plasma from the disturbing probe field.The thickness of the sheath is of the order of a Debye length.

• When no current flows through the electrode, it stays at the negative floating potential ϕ_w , which is the wall potential.

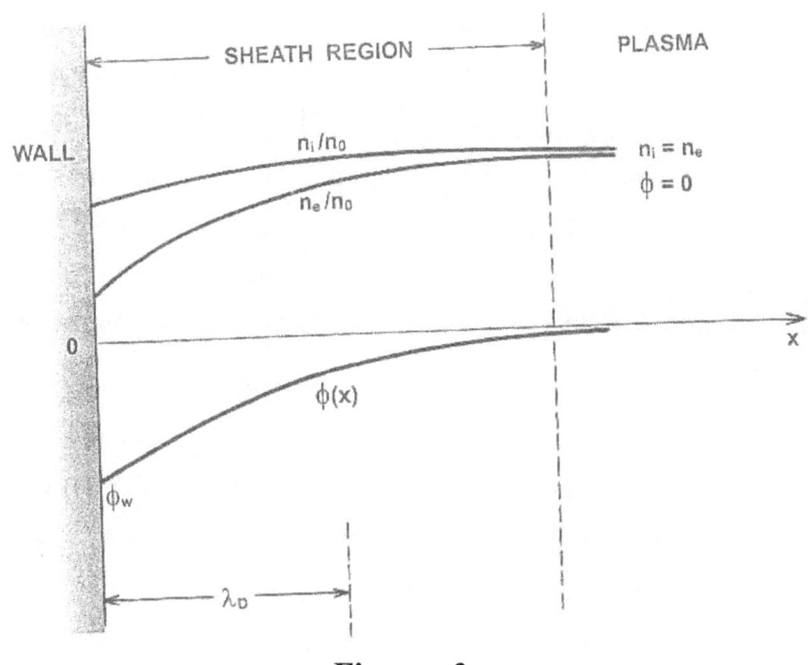

Figure : 2

• Under these equilibrium conditions, the number of electrons reaching the probe per unit time is equal to the number of positive ions reaching the probe per unit time.

• We assume the current to be positive when it flows in the direction away from the probe, The current associated with the electron flow is directed away from the probe and therefore it is considered as positive. Consequently, the electric current associated with the flow of ions is negative.

• Under equilibrium conditions there is no net current flowing through the probe and its potential is the floating potential ϕ_w .

• When the probe potential is made more negative than ϕ_w , the electron current is reduced due to increased repulsive force imposed by the probe electric field on the electrons.

• As the potential is made more negative, the contribution to the electric current arising from the electrons will eventually become negligible and the total electric current asymptotically approaches a constant negative value, corresponding to the electric current density J_i associated with the flow of ions only.

• The ions that reach the edge of the plasma sheath fall into the potential well and their current is practically unaffected if the potential is made even more negative.

• On the other hand, when the probe potential is increased from the negative value ϕ_w, more electrons reach the probe than ions per unit time due to the decrease in the repulsion force on the electrons, and the net electric current becomes positive.

• When the electric potential is zero, that is when the probe is at the same potential as the plasma, there is no electric field near the electrode and, since the average thermal velocity of the electrons is much greater than that of ions, the electron current densities $J_{e0}(for\ \phi = 0)$ is much greater than the ion current density.

• If the potential is made sufficiently positive, a situations arises in which the current associated with the ions become negligible, but all the electrons that reach the edge of the sheath are collected by the probe. The electron current density reaches a fairly constant value for sufficiently high positive value of ϕ.

 The plateau region is the probe current-potential curve is called the region of saturation of the electron current.

• For higher positive values of ϕ there are complications in the i- v curve.

• An approximate expression for the magnitude of the electron density, away from the region of saturation is given by

$$J_e = J_{e0}\ \exp\left(\frac{e\phi}{kT_e}\right) \tag{1}$$

where J_0 is the electron current density when electric potential is zero.

• Since for $\phi = 0$ electron flux $\lceil_e = \frac{n_e}{4} <v>_e$,

we find $\qquad J_e = e\,n_e \left(\frac{kT_e}{2\pi m_e}\right)^{\frac{1}{2}}, \tag{2}$

where n_e is the electron number density in the unperturbed plasma region, Note that J_i is constant in the negative potential region,

Thus, we can express the probe current density in the region where $\phi < 0$, as

$$J_p = J_{e0}\ \exp\left(\frac{e\phi}{kT_e}\right) - J_i \quad \text{(for } \phi < 0 \text{)} \tag{3}$$

From this equation we can deduce the result

$$T_e = \frac{e}{k}\left\{\frac{d}{dt}\Big[\ln(J_p + J_i)\Big]\right\}^{-1} \tag{4}$$

This expression can be used to determine the electron temperature as follows. First, the electrode is made sufficiently negative with reference to the plasma potential, so that the current that flows through the probe is due to the ions only. The measurement of this current gives directly the value of J_i. Then, the current-potential characteristic curve of the probe is measured and a plot of $\ln(J_p + J_i)$ as a function of ϕ is made. This curve has a straight-line section corresponding to the probe potential less than the plasma potential, and the slope of this straight line gives the value of $\dfrac{d}{d\phi}\Big[\ln(J_p + J_i)\Big]$ which, when substituted in (4) gives the electron temperature in the plasma. After the electron temperature T_e has been determined, we can evaluate the electron number density from (2), which can be

written as $n_e = \dfrac{J_{e0}}{e}\left(\dfrac{2\pi m_e}{KT_e}\right)^{\frac{1}{2}}$. The value of J_{e0} is determined by measuring the probe current corresponding to the plateau (electron saturation) region of the current potential characteristic of the probe.

7.6 The Equilibrium of a Perfectly Conducting Fluid (Plasma) at Rest in a Constant Magnetic Field

The equilibrium of a perfectly conducting fluid (plasma) at rest in a constant magnetic field is described by the equations (obtained from fluid equations)

$$grad\ p\ =\ \vec{J} \times \vec{B}, \tag{1}$$

where $\vec{J}$ is the current density, $\vec{B}$ is the magnetic field, and p is the plasma kinetic pressure.

$$\vec{\nabla} \times \vec{B} = \mu_0 \vec{J} \tag{2}$$

$$\vec{\nabla}.\vec{B} = 0 \tag{3}$$

For simplicity, the current density, the magnetic field, and the plasma kinetic pressure are assumed to depend only on the distance from the cylinder axis. Here we shall consider some general properties of equilibrium configurations derivable from these equations.

Scalar multiplication of (1) by $\vec{B}$ and by $\vec{J}$ gives

$$(\vec{B}.grad)p = 0, \quad (\vec{J}.grad)p = 0; \tag{4}$$

That is, the pressure gradient is zero along the lines of magnetic induction $\vec{B}$ and along the current lines. Thus both sets of lines lie on surfaces

$$p(x, y, z) = \text{constant,} \tag{5}$$

called "Magnetic surface".

These isobaric surface, for which p= constant, are concentric cylinders.

We now consider an infinite cylindrical column of conducting fluid with an axial current density $\vec{J} = J_z(r)\,\hat{z}$ and a resulting azimuthal magnetic induction $\vec{B} = B_\theta(r)\,\hat{\theta}$, as depicted in Figure 3

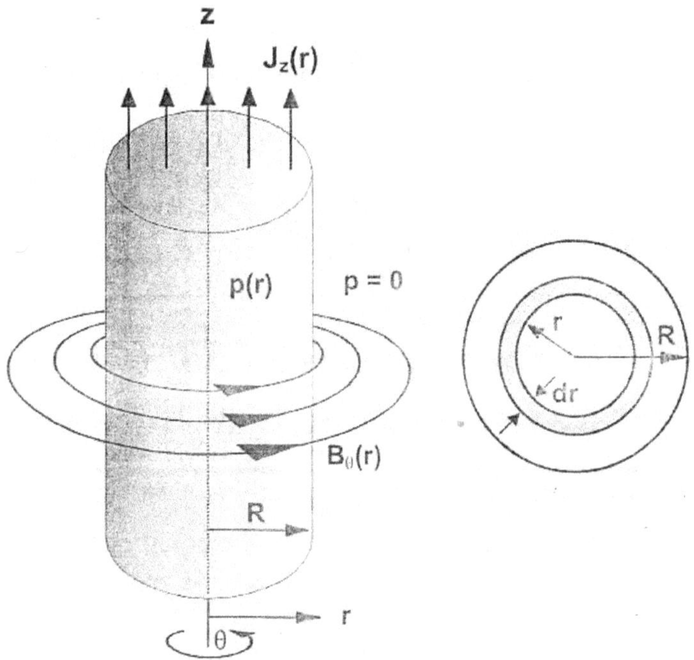

Figure 3: *Schematic diagram illustrating the various parameters relevant to the study of the equilibrium longitudinal pinch configuration.*

The radial component of (1) gives $\dfrac{d\,p(r)}{dr} = -J_z(r)B_\theta(r)$ $\tag{6}$

Inside a cylinder of general radius r, the total enclosed current $I_z(r)$ is

$$I_z(r) = \int_0^r J_z(r) 2\pi\, r\ \mathrm{dr} \tag{7}$$

From (7) we get

$$\frac{d\, I_z(r)}{dr} = 2\pi\, r\, J_z(r) \tag{8}$$

Ampere's law in integral form relates $B_\theta(r)$ to the total enclosed current:

$$B_\theta(r) = \frac{\mu_0}{2\pi r} I_z(r) = \frac{\mu_0}{r} \int_0^r J_z(r)\, r\ \mathrm{dr} \tag{9}$$

A number of results can be obtained even without specifying the precise form of $J_z(r)$. If the conducting fluid lies almost entirely inside r=R, then the magnetic induction $B_\theta(r)$ outside the plasma is

$$B_\theta(r) = \frac{\mu_0\, I_0}{2\pi r} \quad (r \geq R) \tag{10}$$

where $I_0 = \int_0^r J_z(r) 2\pi r dr = I_z(R)$ (11)

which is the total current flowing inside the cylindrical plasma column.

Thus the substitution of $B_\theta(r)$ and $J_z(r)$ from (9) and (8) respectively in (6) results in

$$\frac{dp(r)}{dr} = -\frac{\mu_0}{4\pi^2 r^2} I_z(r) \frac{d\, I_z(r)}{dr}, \tag{12}$$

which can be written as

$$4\pi^2 r^2 \frac{dp(r)}{dr} = -\frac{d}{dr}\left[\frac{1}{2}\mu_0 I_z^2(r) \right] \tag{13}$$

Integrating this equation from r=0 to r=R

$$4\pi^2 r^2\ p(r)\big|_0^R - 4\pi \int_0^R 2\pi r\ p(r)\ \mathrm{dr} = -\frac{1}{2}\mu_0 I_0^2 \tag{14}$$

where $I_0 = I_z(R)$ is the total current flowing through the entire cross-section of plasma column and obviously, $I_z(0) = 0$. Considering the plasma column to be confined to the range $0 \leq r < R$, it follows that p(r) is zero for $r \geq R$, and finite for $0 \leq r < R$, so that the first term in the left-hand side of (14) vanishes. Therefore, we find that

$$I_0^2 = \frac{8\pi}{\mu_0} \int_0^R 2\pi r\, p(r)\, \mathrm{dr} \tag{15}$$

If the partial pressures of the electrons and ions are governed by the ideal gas law,

$$p_e(r) = n(r)kT_e \tag{16}$$

$$p_i(r) = n(r)kT_i \tag{17}$$

assuming that T_e and T_i are constants throughout the plasma column, we have

$$p(r) = p_e(r) + p_i(r) = n(r)k(T_e + T_i) \tag{18}$$

therefore, (15) becomes

$$I_0^2 = \frac{8\pi}{\mu_0} k(T_e + T_i) \int_0^R 2\pi r \ n(r) \ dr \tag{19}$$

which can be rewritten as

$$I_0^2 = \frac{8\pi}{\mu_0} k(T_e + T_i) N_l \tag{20}$$

where $N_l = \int_0^R 2\pi r \ n(r) \ dr \tag{21}$

is the number of particles per unit length of the plasma column. Eq. (19) is known as the "Bennett relation".

It gives the total current that must be discharged through the plasma column in order to confine a plasma at a specified temperature and a given number of particles (N_l) per unit length.

For example we consider $N_l = 10^{19}$ m^{-1}, $T_e + T_i \simeq 10^8 \, K$

$$\mu_0 = 4\pi \times 10^{-7} \frac{H}{m}, \quad k = 1.38 \times 10^{-23} \frac{J}{K}, \text{then } I_0 \sim \text{One million amperes.}$$

 To obtain the radial distribution of p(r) in terms of $B_\theta(r)$,it is convenient to start from (6) and proceed as follows.

First, we note that from Maxwell equation $\vec{\nabla} \times \vec{B} = \mu_0 \vec{J}$ we have, in cylindrical coordinates, with only radial dependence,

$$\frac{1}{r}\frac{d}{dr}\left[r\, B_\theta(r)\right] = \mu_0 \, J_z(r) \tag{22}$$

from which we get

$$J_z(r) = \frac{1}{\mu_0}\frac{d\, B_\theta(r)}{dr} + \frac{1}{\mu_0}\frac{B_\theta(r)}{r} \tag{23}$$

Substituting the expression for $J_z(r)$ from (23) into (6), we get

$$\frac{dp(r)}{dr} = -\frac{1}{2\mu_0 r^2}\frac{d}{dr}\left[r^2 B_\theta^2(r)\right] \tag{24}$$

Now, integrating this equation from r=0 to a general radius r,

$$p(r) = p(0) - \frac{1}{\mu_0}\int_0^r \frac{1}{r^2}\frac{d}{dr}\left[r^2 B_\theta^2(r)\right]dr \tag{25}$$

In particular, since for r=R we have $p(R) = 0$,

$$p(0) = \frac{1}{\mu_0}\int_0^R \frac{1}{r^2}\frac{d}{dr}\left[r^2 B_\theta^2(r)\right]dr \tag{26}$$

and substituting this result into (25),

$$p(r) = \frac{1}{2\mu_0}\int_r^R \frac{1}{r^2}\frac{d}{dr}\left[r^2 B_\theta^2(r)\right]dr \tag{27}$$

The average pressure $\bar{p}$ inside the cylinder can be related to the total current I_0 and the column radius R without knowing the detailed radial dependence. The average of the kinetic pressure inside the column is defined by

$$\bar{p} = \frac{1}{\pi R^2}\int_0^R 2\pi r\, p(r)\, dr \tag{28}$$

Simplifying this expression by an integration by parts, gives

$$\bar{p} = -\frac{1}{R^2}\int_0^R r^2 \frac{dp(r)}{dr}\, dr \tag{29}$$

Since the integrated term is zero, because $p(R) = 0$.

Replacing $\dfrac{dp(r)}{dr}$ using (24), we get

$$\bar{p} = -\frac{B_\theta^2(R)}{2\mu_0} = \frac{\mu_0^2 I_0^2}{8\pi^2 R^2} \tag{30}$$

This result shows that the average kinetic pressure in the equilibrium plasma column is balanced by the magnetic pressure at the boundary. From (7), (9), and (27) we can deduce the radial distribution for $I_z(r)$, $B_\theta(r)$, and $p(r)$ if we know the radial dependence of $J_z(r)$.

As a simple example consider the case in which the current density $J_z(r)$ is constant for $r < R$.

Taking $J_z = \dfrac{I_0}{\pi R^2}$ in (9), we obtain for $r < R$

$$B_\theta(r) = \frac{\mu_0 I_0}{\pi R^2 r} \int_0^r r \, dr = \frac{\mu_0 \, I_0 \, r}{2\pi \, R^2} \qquad (r < R) \tag{31}$$

Substituting this result into (27) we obtain a parabolic dependence for the pressure versus radius,

$$p(r) = \frac{1}{2\mu_0} \int_0^R \frac{1}{r^2} \frac{d}{dr}\left(\frac{\mu_0^2 \, I_0^2 \, r^4}{4\pi^2 \, R^4} \right) dr$$

$$= \frac{\mu_0 \, I_0^2}{4\pi^2 \, R^2}\left(1 - \frac{r^2}{R^2} \right) \tag{32}$$

Note that, in this case, the axial pressure $p(0)$ is twice the average pressure $\bar{p}$ given in (30). The radial dependence of the various quantities for this example is shown in Figure.

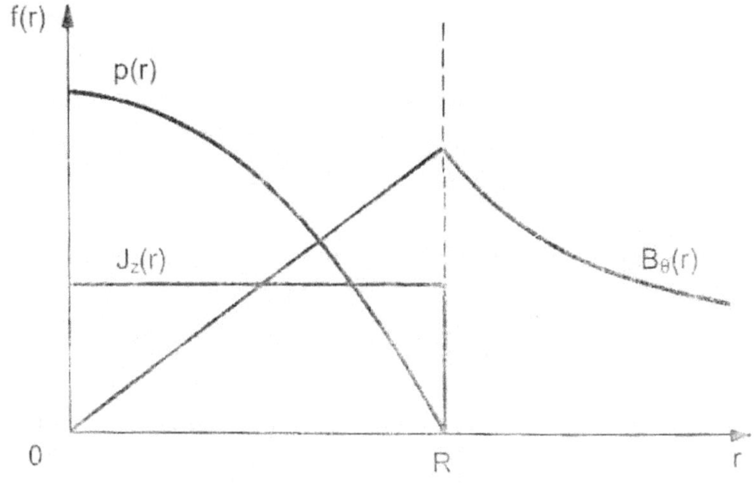

Figure 4

Another radial distribution $J_z(r)$ that is of interest in the investigation of the equilibrium pinch is the one in which the current density is confined to a very thin layer on the surface of the column. This model is appropriate for a highly conducting fluid. In a perfectly conducting plasma, the current cannot penetrate the plasma and exists only on the column surface. This surface current density can be conveniently represented by a Dirac delta function at r=R. In this case there is no magnetic field inside the plasma and $B_\theta(r)$ exists only for $r > R$.

From (10) the magnetic induction is given by

$$B_\theta(r) = \frac{\mu_0 I_0}{2\pi \, r} \qquad (r > R) \tag{33}$$

where I_0 is the total axial current. Therefore, from (25) we have

$$p(r) = p(0) \qquad (0 < r < R) \tag{34}$$

So that the plasma kinetic pressure is constant inside the cylindrical column and equal to the average value given in (30).Thus, for a perfectly conducting plasma column, the magnetic induction vanishes inside the column and falls of as $\dfrac{1}{r}$ outside the column. The plasma kinetic pressure outside it. The pinch effect, in this special case, can be thought of as due to an abrupt build up of the magnetic pressure $\dfrac{B_\theta^2}{2\mu_0}$ in the region external to the plasma column.

7.7 Illustrative Examples

Example1. Consider a cylindrical plasma, extending along the z-axis, which carries a steady electric current whose density can be written $J(r)\hat{e}_z$, where r denotes distance from the z-axis.

(a) obtain the magnetic field as a function of r.

(b) obtain an expression for $J(r)$, in terms of the magnetic field and its gradient.

(c) Show that the $\vec{J} \times \vec{B}$ force is directed radially inward, and has magnitude

$$\frac{1}{2\mu_0 r^2}\frac{d}{dr}(r^2 B^2(r))$$

(d) If the plasma is in magnetohydrodynamic equilibrium, obtain an expression for the pressure $p(r)$, assuming that the pressure vanishes at the plasma boundary r=a.

Sol. (a) Clearly, the magnetic field is azimuthal:

$B = B(r)\hat{e}_\theta$, where $\hat{e}_\theta$ is the azimuthal unit vector.

Using Ampere's law

$$\oint_{Loop} \vec{B}.d\vec{l} = \mu_0 \text{ times the total current through the loop.}$$

or $\qquad B(r)2\pi r = \mu_0 \int_0^r J(r')2\pi r'dr'$ Hence $\qquad B(r) = \dfrac{\mu_0}{r}\int_0^r J(r')r'dr'$

(b) Take the derivative of the preceding expression with respect to r:

$$\frac{dB}{dr} = -\frac{\mu_0}{r^2}\int_0^r J(r')r'dr' + \frac{\mu_0}{r}J(r)r$$

Then $\quad \mu_0 J(r) = \dfrac{B(r)}{r} + \dfrac{dB}{dr}$

(c) $\quad \vec{J} \times \vec{B} = J(r)\hat{e}_z \times B(r)\hat{e}_\theta = -J(r)\,B(r)\hat{e}_r$

The is an inward radial force ,whose magnitude is

$$J(r)B(r) = \left(\frac{1}{\mu_0}\frac{B(r)}{r} + \frac{1}{\mu_0}\frac{dB}{dr} \right) B(r)$$

$$= \frac{1}{\mu_0}\left(\frac{B^2(r)}{r} + B\frac{dB}{dr} \right)$$

Using the result of (b) it is easy to check that this expression is equivalent to the more compact one given in the question.

(d) At equilibrium, we know that $\vec{J} \times \vec{B} = \nabla p$. That is, using the previous result

$$\frac{dp}{dr} = (\vec{J} \times \vec{B})_r = -J_z(r)\,B_\theta(r)$$

$$= -\left[\frac{1}{\mu_0}\left(\frac{B}{r} + \frac{dB}{dr} \right)B \right]$$

$$= -\frac{1}{\mu_0}\left[\frac{B^2}{r} + \frac{1}{2}\frac{d}{dr}(B^2) \right]$$

or $\quad \dfrac{dp}{dr} = \dfrac{-1}{2\mu_0 r^2}\dfrac{d}{dr}\left\{ r^2 B^2(r) \right\}$

Integrating this expression outwards from $r' = r$ to $r' = a$ where the pressure is zero, we obtain $\quad p(r) = \dfrac{1}{2\mu_0}\displaystyle\int_r^a \dfrac{1}{r'^2}\dfrac{d}{dr'}\left(r'^2 B^2(r') \right)dr'$

where $B(r)$ follows from (a) above.

7.8 The Bennet Pinch

In this case a stream of electrons and a stream of ions are moving through each other along the z-axis. In view of the fact that the ion mass is much larger than the electron mass, the drift velocity of the ions is much smaller than that of the electrons and can therefore be neglected on a first approximation.

Thus, we consider the current density to be given by

$$\vec{J}(r) = -e\,\text{n}\,(\text{r})\bar{\text{u}}_e \tag{1}$$

Since the applied electric field is in the z-direction, we have

$$\vec{J}(r) = J_z(\mathrm{r})\hat{z} \quad \text{and} \quad \vec{u}_e = -u_{ez}\hat{z}.$$

Therefore, $J_z(\mathrm{r}) = e\,n\,(\mathrm{r})\,u_{ez}$ (2)

Using the equation (previously derived) $\dfrac{d}{dr}p(r) = -J_z(r)B_\theta(r)$ and using $p = n(r)k(T_e + T_i)$, we get

$$k(T_e + T_i)\frac{dn(r)}{dr} = -\,e\,n\,(r)\,u_{ez}B_\theta(r) \tag{3}$$

if we multiply this equation by $\dfrac{r}{\big[n\,(r)\,k(T_e + T_i)\big]}$ and differentiate it with respect to r, we obtain

$$\frac{d}{dr}\left[\frac{r}{n\,(r)}\frac{dn\,(r)}{dr}\right] = \frac{-e\,u_{ez}}{k\,(T_e + T_i)}\frac{d}{dr}\big[r\,B_\theta(r)\big] \tag{4}$$

Making use of Maxwell equation $\vec{\nabla}\times\vec{B} = \mu_0\vec{J}$ we have, in cylindrical coordinates, with only radial dependence $\dfrac{1}{r}\dfrac{d}{dr}\big[r\,B_\theta(r)\big] = \mu_0 J_z(r)$

or $\quad \dfrac{d}{dr}\big[r\,B_\theta\,(r)\big] = \mu_0\,e\,u_{ez}\,r\,n(r)$ (5)

and using this result in (4),

$$\frac{d}{dr}\left[\frac{r}{n\,(r)}\frac{dn\,(r)}{dr}\right] + \left[\frac{\mu_0 e^2 u_{ez}^2}{k(T_e + T_i)}\right]r\,n\,(r) = 0 \tag{6}$$

Equation (6) is a non linear differential equation. Bennet obtained the solution of this nonlinear equation subjected to the boundary condition that n(r) is symmetric about the z-axis, where r=0, and is a smoothly varying function of r, so that

$$\left[\frac{d}{dr}n(r)\right]_{r=0} = 0 \tag{7}$$

The solution of (6), subjected to the boundary condition (7), is known as the Bennet distribution and is given by

$$n(r) = \frac{n_0}{(1 + n_0\,b\,r^2)^2} \tag{8}$$

where $n_0 = n(0)$ which is the number density on the axis,

and $\quad b = \dfrac{\mu_0\,e^2\,u_{ez}^2}{8k(T_e + T_i)}$ (9)

which has the dimension of length. The radial dependence of the number density n(r) is sketched in Figure (5). It can be used to determine $B_\theta(r)$.

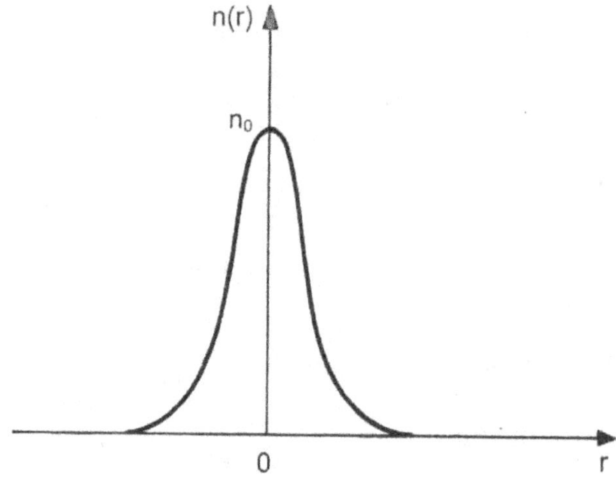

Figure 5: *The Bennet distribution for the particle number density n(r) in an equilibrium pinched plasma column.*

The Bennet distribution shows that particles are present up to infinity but, n(r) falls off every rapidly with increasing values of r, we can consider, for all practical purposes ,that the plasma is essentially confined symmetrically in a small cylindrical region about the z-axis. Using (8) we obtain the number of particles N_I (R) per unit length contained in a cylindrical column of radius R,

$$N_I(R) = \int_0^R n(r)2\pi\ r\,d\,r\ =\ 2\pi n_0 \int_0^R \frac{r}{(1+n_0\ b\ r^2)}\,dr \tag{10}$$

Evaluating this integral gives

$$N_I(R) = \frac{n_0\pi\ R^2}{(1+n_0\ b\ R^2)} \tag{11}$$

Since particles are present up o infinity, the total number of particles per unit length can be obtained from (11) by taking the limit as $R \to \infty$, which gives

$$N_I(\infty) = \frac{\pi}{b} \tag{12}$$

Let α denotes the fraction of the number of particles per unit length that is contained in a cylinder of radius R, that is

$$\alpha = \frac{N_I(R)}{N_I(\infty)} = \frac{b}{\pi} N_I(R) \tag{13}$$

Making use of (11), one finds

$$(n_0 b)^{\frac{1}{2}} R = \left(\frac{\alpha}{1-\alpha}\right)^{\frac{1}{2}} \tag{14}$$

Therefore, if 90% of the plasma particles are confined within the cylindrical plasma column of radius R, that is $\alpha = 0.9$, we must have

$$(n_0 b)^{\frac{1}{2}} R = 3 \tag{15}$$

Thus, even though the particles extend up to infinity, the major portion of them lies in a small neighborhood around the z-axis. $(n_0 b)^{\frac{1}{2}} R$ is the normalized radius of the cylindrical plasma column.

7.9 Instabilities in a Pinched Plasma Column

Although it is possible to achieve an equilibrium state for a plasma confinement with the pinch effect, this equilibrium state is not possible. A small departure from the cylindrical geometry of the equilibrium state results in the growth of the original perturbations with time and in the   disintegration of the plasma column. The growth of instabilities is the reason why it is difficult to sustain reasonably long-lived pinched plasma in the laboratory.

7.10 The Sausage Instability

In nature as well as in the laboratory, plasmas are generally confined by magnetic fields bent into bounded volumes. The magnetic tension associated with the bent magnetic fields constitutes a source of free energy for instabilities. We consider the simplest confinement Z-pinch geometry. In a Z-pinch, the current is made to run through a plasma between two electrodes. The magnetic field generated by the current provides a force that can accelerate the plasma inwards.

If a z-pinched plasma is squeezed somewhere between the two electrodes, however, the current density and therefore the confining magnetic field will increase at that point. This has the effect of squeezing the plasma further "Sausage instability", can be eliminated by applying a longitudinal magnetic field B_z (Figure 6) Because of conservation of flux, the perturbation of the longitudinal field for a constriction δa of the radius of column ,a , is given by considering that the magnetic flux $(\phi_m = B_z \pi r^2)$ through the cross sectional area of the column remains constant:

$$\phi_m = B_z \pi r^2 = \text{constant} \tag{16}$$

Therefore $\quad d\phi_m = \pi\, r^2 dB_z + B_z\, 2\pi\, r\, dr = 0$

$$\therefore dB_z = -\frac{B_z\, 2\pi\, r\, dr}{\pi\, r^2} \tag{17}$$

or if the plasma column radius is a,

Then $\quad \delta B_z = 2B_z \dfrac{\delta a}{a}$ (during constriction $\delta r = -\delta a$) $\tag{18}$

The azimuthal field perturbation can be calculated by considering that the azimuthal flux density B_θ external to the plasma column is given by Ampere's law as

$$r\, dB_\theta(r) = \text{Constant}. \tag{19}$$

Therefore, the azimuthal magnetic flux density, at the constricted surface increases by amount

$$d\, B_\theta = -B_\theta \frac{dr}{r}, \tag{20}$$

The azimuthal field perturbation at r=a is

$$\delta\, B_\theta = B_\theta \frac{\delta a}{a} \qquad (-dr = \delta a \text{ is constriction}) \tag{21}$$

The column is stable if the restoring pressure caused by the longitudinal field compression exceeds the increase in the azimuthal field pressure, or if

$$\delta\,(B_z^2) > \delta\,(B_\theta^2) \tag{22}$$

We see that a longitudinal field has a strong stabilizing influence. This stabilizing influence is ineffective to present shear-Alfvin instabilities.

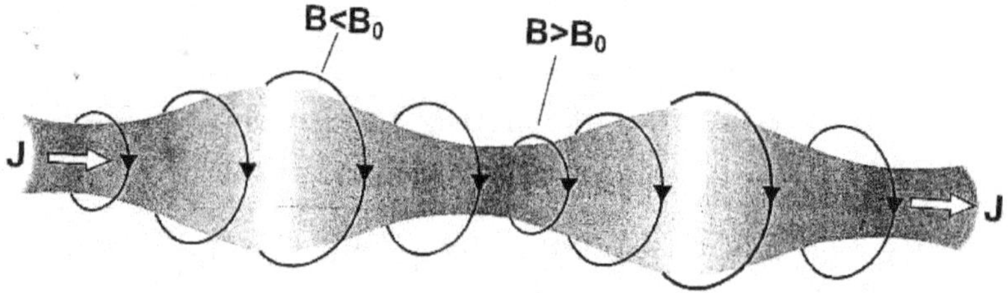

Figure 6: *The sausage instability in a z pinch plasma*

7.11 The Kink Instability

Another type of instability of the pinched plasma column is the so-called "Kink instability". The Kink distortion consists of a perturbation in the form of a bend or kink in the column, but with the disturbed column maintaining its uniform circular cross section, as shown in Fig. (8).Consider the case of long thin, current carrying plasma wire in a longitudinal magnetic field. Twisting this plasma into a corkscrew introduces an angle between the external magnetic field and the plasma current. This result in a force $\vec{J} \times \vec{B}$ (Figure 7)

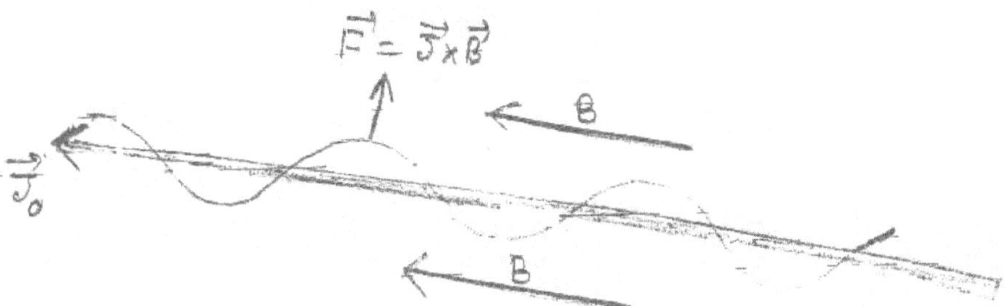

Figure 7: *Kink instability of a current carrying plasma filament*

If the pitch of the twisted plasma matches that of the unperturbed magnetic field, the perturbation will amplify as the plasma tries to align the magnetic field produced by the current it is carrying with the externally applied magnetic field. The thin current carrying plasma is thus unconditionally unstable. This is called the "Kink " instability. Usually there may be several kinks along the column length. In the neighborhood of the column, where the kink has developed, the magnetic field lines are brought closer together on the concave side, and separated on the convex side, so that the external magnetic pressure is increased on the concave side and decreased on the convex side.

Therefore, the changes in the external magnetic pressure are in such a way as to accentuate the distortion still further. This type of distortion is therefore unstable.

The kink instability can be obstructed by the application of a longitudinal magnetic field within the plasma column, as in the case of the sausage instability. In the kink distortion, the longitudinal magnetic field lines frozen inside the plasma column are stretched and the increase tension acting along the longitudinal magnetic field lines opposes the external forces. The net result is the stabilization of the column.

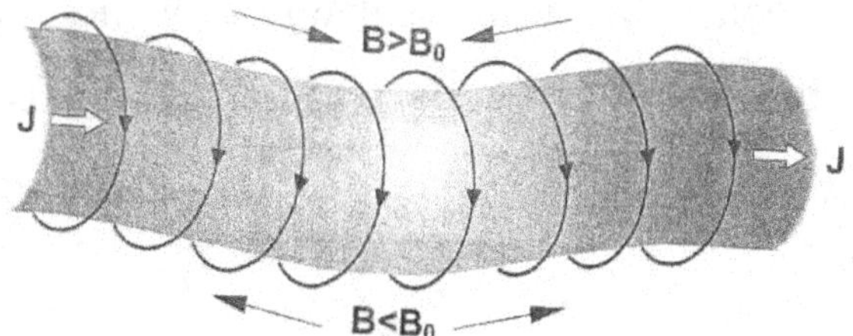

Figure 8: *The Kink instability*

7.12 Plasma Confinement in a Magnetic field

The subject of plasma confinement by magnetic fields is of considerable interest in the theory of controlled thermonuclear fusion. We now pay attention to the consequences of the fundamental equations characterizing the state of a plasma in an electromagnetic field. Here we will deal only with quasi-stationary states without touching upon the question of their stability. From the fluid equation of motion

$$\rho_m \left[\frac{\partial}{\partial t} + \vec{u}.\vec{\nabla} \right] \vec{u} = \vec{J} \times \vec{B} - \nabla p$$

or $\qquad \rho_m \dfrac{D \vec{u}}{D t} = \vec{J} \times \vec{B} - \nabla p \qquad\qquad\qquad\qquad (1)$

where $\dfrac{D}{D t} \equiv \dfrac{\partial}{\partial t} + \vec{u}.\vec{\nabla}$ represents the total time derivative operator. In the static case $\dfrac{D}{D t} \equiv 0$, thus (1) reduces to

$$\vec{J} \times \vec{B} = \nabla p \qquad\qquad\qquad\qquad (2)$$

In this case, the pressure gradient is perpendicular to $\vec{B}$ .Hence, the plasma pressure must be constant along the lines of magnetic force. For the case where the lines of force are straight and parallel to each other, it is possible to obtain an equation relating the pressure to the field strength. To these equations we must add Maxwell curl equation, in the following reduced form,

$$\vec{\nabla} \times \vec{B} = \mu_0 \vec{J} \qquad\qquad\qquad\qquad (3)$$

From equations (2) and (3) it follows that in a quasi-stationary state

$$grad\ p = \frac{1}{\mu_0} \left[curl\ \vec{B} \right] \times \vec{B} \qquad\qquad\qquad\qquad (4)$$

Let the z axis be parallel to the lines of force. Then $\vec{B}$ will have only the component B_z, whose value is a function of x and y and is independent of z (due to the condition div $\vec{H} = 0$).From (4), the components of the pressure gradient are given by

$$\frac{\partial p}{\partial x} = -\frac{1}{\mu_0} B_z \frac{\partial B_z}{\partial x} = -\frac{1}{2\mu_0} \frac{\partial}{\partial x}(B_z^2),$$

$$\frac{\partial p}{\partial y} = -\frac{1}{\mu_0} B_z \frac{\partial B_z}{\partial y} = -\frac{1}{2\mu_0} \frac{\partial}{\partial y}(B_z^2).$$

Therefore $\quad \dfrac{\partial}{\partial x}\left(p+\dfrac{B_z^2}{2\mu_0}\right)=\dfrac{\partial}{\partial y}\left(p+\dfrac{B_z^2}{2\mu_0}\right)=0$

Consequently, $p+\dfrac{B^2}{2\mu_0}=$ constant $\hspace{4cm}$ (5)

This result is interesting from a number of viewpoints. First, it follows that a plasma is diamagnetic, since the field strength within the plasma is less than that outside. Second, the equation (5) is the mathematical expression of the idea of confining a plasma by means of a magnetic field. The magnetic pressure outside the plasma is greater than the magnetic pressure inside by the value of a kinetic pressure inside by the value of the kinetic pressure of the plasma. Relatively speaking, the plasma pressure attains its greatest value p_{max} when the field inside the plasma becomes zero; that is:

$$p_{max}=\frac{B_0^2}{2\mu_0}$$ where B_0 is the field strength just outside the plasma boundary.

7.13 Self Learning Exercise

Q.1 What are "magnetic surfaces"?

Q.2 What is the electron plasma frequency ? Prove that electron plasma frequency $f_p=q\sqrt{n}$, where n is electron density is the unit of $\dfrac{1}{m^3}$.

Q.3 Consider a cylindrically symmetric plasma, using cylindrical coordinates (r, θ, z), with symmetry in the θ and z- directions. A purely axial current $\vec{J}=\hat{z}\,J(r)$, maintains the azimuthal magnetic field $B=\hat{\theta}\,B(r)$. Use MHD equations derive the equilibrium relation between $B(r), J\ (r)$ and the pressure p(r). Also compute the magnetic curvature for this configuration .

(In the plasma confinement literature, this configuration is called a z- pinch, since a plasma current along the z-axis creates the confining magnetic field).

7.14 Summary

In this chapter we have examined the mechanism by which plasma tries to shield its interior from a disturbing electric field. It is found that the shielding of this electric field is effectively completed over a distance of $\sim\lambda_D$ (Debye length). We have also discussed the physical mechanism for the formation of the plasma sheath. In order to measure the temperature and density of the plasma ,we have discussed the physics of plasma probe (Langmuir probe).In this chapter we have discussed the equilibrium of plasma column at rest in a constant magnetic field. We have also discussed the radial constriction of the plasma column commonly known as the "pinch effect" .We have found that for a perfectly conducting plasma column, the magnetic induction vanishes inside the column   and falls off as $\dfrac{1}{r}$ outside the column. The plasma kinetic pressure is constant inside the column.

7.15 Glossary

Plasma sheath: In the region near the wall of the body there is a boundary layer, known as the "Plasma sheath" in which the electron and the ion number densities are different.

7.16 Answers to Self Learning Exercise

Ans.1: These are isobaric surfaces which satisfy the conditions

$$\left(\vec{B}.grad\right)p = 0, \left(\vec{J}.grad\right)p = 0,$$

That is ,the pressure gradient is zero along the lines of magnetic induction $\vec{B}$ and along the current lines. Thus both sets of lines lie on surfaces

$$p(x,y,z) = \text{constant}$$

called "magnetic surfaces"

7.17 Exercise

Q.1   What is pinch effect?

Q.2   What are the instabilities in a pinched plasma column ?

Q.3   Estimate the maximum plasma pressure that can be confined by a magnetic field B_0

Q.4   What is Z pinch ?

Q.5   Why a plasma is diamagnetic ?

Q.6   Explain the reason for acquiring a negative charge by a body in contact with plasma.

7.18 Answers to Exercise

Ans.3: $p_{max} = \dfrac{B_0^2}{2\mu_0}$

Ans.4: A plasma column carrying a plasma current along the z axis is called z pinch .A purely axial current $\vec{J} = J(r)\hat{z}$ maintains the azimuthal magnetic field $\vec{B} = B(r)\hat{\theta}$ and this magnetic field confines the plasma .

Ans.5: From the equilibrium condition

$$p + \frac{B^2}{2\mu_0} = \frac{B_0^2}{2\mu_0}$$

It follows that the field strength within the plasma is less than the outside . In other words the magnetic field inside the plasma is reduced than its value outside. For this reason we can say that plasma is diamagnetic.

References and Suggested Readings

1. Fundamental of Plasma Physics by J. A. Bittencourt

2. Classical Electrodynamics by D. Jackson

UNIT-8
MHD Waves

Structure of the Unit

8.0 Objectives

To learn

• The fluid model of the plasma.

• Fluid approximation.

• Force and motion in ideal magneto hydrodynamics.

• Alfven velocity .

• MHD waves

• Propagation parallel and perpendicular to magnetic field.

• Dispersion relation of Alfven waves.

• Magnetosonic waves

8.1 Introduction

So far, we have based our study of plasmas on the single particle dynamics of the constituent electrons and ions. We have used this dynamics as the basis for a dielectric description of the plasma. This approach has been very effective. The understanding of any physical system is aided by knowledge of its normal modes, and the dielectric approach has enabled us to formulate the normal modes of plasmas over a very wide range of frequencies. However, to study phenomena whose frequencies lie well below the lowest characteristic single particle frequency, the ion cyclotron frequency ω_{ci}, an alternative approach might be fruitful. when $\omega \ll \omega_{ci}$, the time scales in question are much longer than those of any ion or electron oscillation associated with cyclotron or space charge motion. On these time scales, only averaged, guiding centre positions of the plasma particles are significant. It is, thus, clear that we have entered regime of frequencies and length scales where a macroscopic, fluid-like description of the plasma would be more appropriate.

The question now arises, what sort of fluid model should we develop ? What features can we distinguish, that will make the model apply specifically to plasmas?

One set of answers is provided by the two-fluid description of plasmas, where the plasma is taken as composed of two distinct but intermingled fluids, the electron and ion fluids, coupled together by their opposing electrical charges.

Here, we shall adopt a simpler approach to the fact that at a macroscopic level the plasma is composed of oppositely charged particles of unequal mass, but that at a macroscopic level it is electrically neutral. The plasma will be treated as a single, electrically neutral and perfectly conducting fluid. This reflects the fact that, on time scales of interest, the high mobility of electrons enables them both to keep the plasma electrically neutral and to cancel out any applied electric field.

• The most fundamental type of wave motion that propagates in a compressible, non conducting fluid is that of longitudinal "Sound Waves". For these waves the variations in pressure (p) and in density (ρ_m), associated with the fluid compressions and rarefactions, obey the adiabatic energy equation commonly used in thermodynamics

$$p \, \rho_m^{-\gamma} = \text{constant}$$

where γ denotes the ratio of the specific heats at constant pressure and at constant volume.

Differentiating (1) we obtain

$$\nabla p = \frac{\gamma p}{\rho_m} \nabla \rho_m = V_s^2 \nabla \rho_m$$

$$\text{where } V_s = \left(\frac{\gamma p}{\rho_m} \right)^{\frac{1}{2}} = \left(\frac{\gamma kT}{m} \right)^{\frac{1}{2}}$$

is the wave propagation speed, known as the "adiabatic sound speed".

Longitudinal sound waves propagate in a compressible, non conducting fluid with speed

$$V_s = \left(\frac{\gamma p}{\rho_m} \right)^{\frac{1}{2}}$$

8.2 Alfven Waves

In the case of a compressible, conducting fluid immersed in a magnetic field, other types of wave motion are possible.

We are now going to examine time-dependent phenomena described by the complete set of hydromagnetic equations.

As in electromagnetic theory we are looking for periodic wave solutions. However, hydromagnetic set of equations is nonlinear. Nevertheless we shall examine the possibility of obtaining wave equations.

Since these equations are nonlinear we shall linearize the hydromagnetic equations by considering small amplitudes.

•Looking at the hydromagnetic equations, the existence of some wave solutions is immediately apparent. With $\vec{B} = 0$ these equations reduce to the regular fluid equations, and yield after the linearization the well-known sound waves.

•Apart from these trivial solutions one suspects that the coupling of fluid and electromagnetic phenomena might result in new types of waves. To simplify our investigations we restrict ourselves first to an incompressible fluid with infinite conductivity.

The appropriate system of equations, which governs the behavior of this type of fluid, with the assumptions involved are

$$\frac{\partial \rho_m}{\partial t} + \nabla.(\rho_m \vec{u}) = 0 \tag{1}$$

$$\rho_m \frac{\partial \vec{u}}{\partial t} + \rho_m (\vec{u}.\vec{\nabla})\vec{u} = -\nabla p + \vec{J} \times \vec{B} \tag{2}$$

$$\nabla p = V_s^2 \nabla \rho_m \tag{3}$$

$$\vec{\nabla} \times \vec{B} = \mu_0 \vec{J} \tag{4}$$

$$\vec{\nabla} \times \vec{E} = -\frac{\partial \vec{B}}{\partial t} \tag{5}$$

$$\vec{E} + \vec{u} \times \vec{B} = 0 \tag{6}$$

To reduce this system of equations we combine equations (2) to (4) in the form

$$\rho_m \frac{\partial \vec{u}}{\partial t} + \rho_m (\vec{u}.\vec{\nabla})\vec{u} = -V_s^2 \ \nabla \rho_m + \frac{1}{\mu_0}(\vec{\nabla} \times \vec{B}) \times \vec{B} \tag{7}$$

as well as (5) and (6) in the form

$$\nabla \times (\vec{u} \times \vec{B}) = \frac{\partial \vec{B}}{\partial t} \tag{8}$$

• Under equilibrium conditions, the fluid is assumed to be spatially uniform with constant density ρ_{m0}, the equilibrium velocity is considered zero, and throughout the fluid the magnetic induction $\vec{B}_0$ is uniform and constant.

• In order to deduce the dispersion relation for small amplitude waves, consider small- amplitude departures from the equilibrium values, so that

$$\vec{B}(\vec{r},t) = \vec{B}_0 + \vec{B}_1(\vec{r},t) \tag{9}$$

$$\rho_m(\vec{r},t) = \rho_{m0} + \rho_{m1}(\vec{r},t) \tag{10}$$

$$u(\vec{r},t) = \vec{u}_1(\vec{r},t) \tag{11}$$

Substituting (9) to (11) into (1), (7), and (8), and neglecting second-order terms, one obtains the linearized equations in the small first-order quantities:

$$\frac{\partial \rho_{m1}}{\partial t} + \rho_{m0}(\vec{\nabla}.\vec{u}_1) = 0 \tag{12}$$

$$\rho_{m0}\frac{\partial \vec{u}_1}{\partial t} + V_s^2 \nabla \rho_{m1} + \frac{1}{\mu_0}\vec{B}_0 \times (\vec{\nabla} \times \vec{B}_1) = 0 \tag{13}$$

$$\frac{\partial \vec{B}_1}{\partial t} - \nabla \times (\vec{u}_1 \times \vec{B}_0) = 0 \tag{14}$$

Equation (12) to (14) can be combined to yield an equation for $\vec{u}_1$, alone. For this purpose, we first differentiate (13) with respect to time, obtaining

$$\rho_{m0}\frac{\partial^2 \vec{u}_1}{\partial t^2} + V_s^2 \nabla\left(\frac{\partial \rho_{m1}}{\partial t}\right) + \frac{1}{\mu_0}\vec{B}_0 \times \left[\vec{\nabla} \times \left(\frac{\partial \vec{B}_1}{\partial t}\right)\right] = 0$$

Next, using (12) and (14), we can write (15) as

$$\frac{\partial^2 \vec{u}_1}{\partial t^2} - V_s^2\, \vec{\nabla}(\vec{\nabla}.\vec{u}_1) + \vec{V}_A \times \left\{\vec{\nabla} \times \left[\vec{\nabla} \times \left(\vec{u}_1 \times \vec{V}_A\right)\right]\right\} = 0$$

where we have introduced the vector Alfven velocity.

$$\vec{V}_A = \frac{\vec{B}_0}{(\mu_0 \rho_{mo})^{\frac{1}{2}}} \tag{17}$$

• Without loss of generality, we can consider plane wave solutions for (16) in the form

$$\vec{u}_1(\vec{r},t) = \vec{u}_1 \exp(i\,\vec{k}.\,\vec{r} - i\,\omega\,t) \tag{18}$$

Thus, in (16) we can replace the operator ∇ by $i\,\vec{k}$ and $\dfrac{\partial}{\partial t}$ by $-i\omega$, so that

$$-\omega^2 \vec{u}_1 + V_s^2(\vec{k}.\vec{u}_1)\vec{k} - \vec{V}_A \times \left\{\vec{k} \times \left[\vec{k} \times \left(\vec{u}_1 \times \vec{V}_A\right)\right]\right\} = 0 \tag{19}$$

• Since for any three vectors $\vec{A}$, $\vec{B}$, and $\vec{C}$ we have the vector identity

$$\vec{A} \times (\vec{B} \times \vec{C}) = (\vec{A}.\vec{C})\vec{B} - (\vec{A}.\vec{B})\vec{C} \tag{20}$$

We can rearrange (19) to read

$$-\omega^2 \vec{u}_1 + (V_s^2 - \vec{V}_A^2)(\vec{k}.\vec{u}_1)\vec{k} +$$
$$(\vec{k}.\vec{V}_A)\left[(\vec{k}.\vec{V}_A)\vec{u}_1 - (\vec{V}_A.\vec{u}_1)\vec{k} - (\vec{k}.\vec{u}_1)\vec{V}_A\right] = 0 \tag{21}$$

Although this expression appears to be somewhat involved, it leads to remarkable simple solutions for waves propagating in the directions parallel or perpendicular to the magnetic field.

8.3 Propagation Perpendicular To The Magnetic Field

When the wave vector $\vec{k}$ is perpendicular to the magnetic induction $\vec{B}_0$, we have $\vec{k}.\vec{V}_A = 0$,and simplifies to

$$-\omega^2 \vec{u}_1 + (V_s^2 + \vec{V}_A^2)(\vec{k}.\vec{u}_1)\vec{k} = 0 \tag{22}$$

from which we obtain

$$\vec{u}_1 = \frac{1}{\omega^2}(V_s^2 + V_A^2)(\vec{k}.\vec{u}_1)\vec{k} \tag{23}$$

Therefore, $\vec{u}_1$, is parallel to $\vec{k}$, so that $\vec{k}.\vec{u}_1 = ku_1$, and the solution for $\vec{u}_1$, is a longitudinal wave with the phase velocity

$$\frac{\omega}{k} = \left(V_s^2 + V_A^2\right)^{\frac{1}{2}} \tag{24}$$

The magnetic field associated with this longitudinal wave can be obtained from (14):

$$\frac{\partial \vec{B}_1}{\partial t} - \vec{\nabla} \times (\vec{u}_1 \times \vec{B}_0) = 0$$

Taking $\qquad \vec{B}_1(\vec{r}, t) = \vec{B}_1 \exp (i\,\vec{k}.\,\vec{r} - i\omega t) \tag{25}$

One gets $\qquad -\omega\vec{B}_1 - \vec{k} \times (\vec{u}_1 \times \vec{B}_0) = 0 \tag{26}$

or $\qquad \omega\vec{B}_1 - \vec{u}_1(\vec{k}.\vec{B}_0) + \vec{B}_0(\vec{k}.\vec{u}_0) = 0$

or $\qquad \vec{B}_1 = \frac{\vec{B}_0(\vec{k}.\vec{u}_1)}{\omega} = \frac{\vec{B}_0 u_1}{\left(\dfrac{\omega}{k}\right)} \tag{27}$

The electric field associated with this wave is to be given by

$$\vec{E} = -\vec{u}_1 \times \vec{B}_0 \tag{28}$$

This wave is somewhat similar to an electromagnetic wave, since the time-varying magnetic field is perpendicular to the direction of propagation, but parallel to the magnetostatic field, whereas the time- varying electric field is perpendicular to both the direction of propagation and the magnetostatic field.

It is a "Longitudinal" wave, however, since the velocity of mass flow and the fluctuating mass density associated with the wave motion are both in the wave propagation direction. For these reasons, this wave is called the magnetosonic wave. The phase velocity $\dfrac{w}{k} = (V_s^2 + V_A^2)^{\frac{1}{2}}$ is independent of frequency, so that it is a nondispersive wave.

The magnetosonic wave produces compressions and rarefactions in the magnetic field lines without changing their direction. Since the fluid is perfectly conducting, the lines of the force and the fluid move together.

• The restoring forces operating in the magnetosonic wave are the fluid pressure gradient and the gradient of the compressional stresses between the magnetic field lines.

• When the fluid pressure is much greater than the magnetic pressure, the effect of the magnetic field is negligible, so that $\dfrac{w}{k} = V_s$ and the wave becomes essentially an acoustic wave (sound wave).

On the other hand, when the magnetic field is very strong so that the magnetic pressure is much larger than the fluid pressure, then the phase velocity of the magnetosonic wave becomes equal to the Alfven wave velocity V_A.

The magnetosonic wave mode is also known as the compressional Alfven wave or fast Alfven wave.

8.4 Propagation Parallel To The Magnetic Field

For waves propagating along the magnetic field $(\vec{k}\,||\,\vec{B}_0)$, we have $\vec{k}.\vec{V}_A = kV_A$ .We now take up the general dispersion relation viz. equation (21):

$$-\omega^2 \vec{u}_1 + (V_s^2 + V_A^2)(\vec{k}.\vec{u}_1)\vec{k} +$$
$$+(\vec{k}.\vec{V}_A)\left[(\vec{k}.\vec{V}_A)\vec{u}_1 - (\vec{V}_A.\vec{u}_1)\vec{k} - (\vec{k}.\vec{u}_1)\vec{V}_A\right] = 0$$

This equation now simplifies to

$$\left(k^2 V_A^2 - \omega^2\right)\vec{u}_1 + \left(\frac{V_s^2}{V_A^2} - 1\right)k^2(\vec{u}_1.\vec{V}_A)\vec{V}_A = 0 \qquad (29)$$

In this case, there are two types of wave motion possible.

For $\vec{u}_1$ parallel to $\vec{B}_0$ and $\vec{k}$, we find from (29) that a longitudinal mode is possible, with the phase velocity

$$\frac{\omega}{k} = V_S \qquad (30)$$

This is an ordinary longitudinal sound wave, in which velocity of mass flow is in the propagation direction. There is no electric field, electric current density, or magnetic field associated with this wave.

A transverse wave, with $\vec{u}_1$ perpendicular to $\vec{B}_0$ and $\vec{k}$, is other possibility. In this case $\vec{u}_1.\vec{V}_A = 0$ and (29) gives for the phase velocity of this transverse wave, known as the "Alfven wave",

$$\frac{w}{k} = V_A \qquad (30)$$

where

$$V_A = \left(\frac{B_0^2}{\mu_0 \rho_m}\right)^{\frac{1}{2}} \qquad (31)$$

Since the phase velocity is independent of frequency , there is no dispersion.

The magnetic field associated with the Alfven wave is found, from (14) and (26), to be given by

$$\vec{B}_1 = -\frac{B_0}{(\omega / k)}\vec{u}_1 \qquad (32)$$

Hence, the magnetic field disturbance is normal to the original magnetostatic induction $\vec{B}_0$. The small component $\vec{B}_1$, when added to $\vec{B}_0$, gives the lines of force a sinusoidal ripple (see figure...). The associated electric field is given by (28).

The Alfven wave involves no fluctuations in the fluid density or pressure, although both the fluid and the magnetic field lines oscillate back and forth laterally, in the plane normal to $\vec{B}_0$.

The magnetic energy density of this wave motion $\left(\dfrac{B_1^2}{2\mu_0}\right)$ is equal to the kinetic energy density of the fluid motion $\left(\dfrac{\rho_{m0}u_1^2}{2}\right)$. This equipartition of energy is easily verified:

$$\frac{B_1^2}{2\mu_0} = \frac{B_0^2\, u_1^2}{2\mu_0\left(\dfrac{\omega}{k}\right)^2}$$

$$= \frac{B_0^2\, u_1^2}{2\mu_0 V_A^2}$$

$$\Rightarrow \frac{B_1^2}{2\mu_0} = \frac{1}{2}\rho_{m0}\, u_1^2 \tag{33}$$

• The Alfven wave mode is also known as the "shear Alfven wave" or the "slow Alfven wave".

8.5 Illustrative Examples

Example 1: Find the velocity of an Alfven wave in mercury in a magnetic field $B_0 = 100$ gauss.

Sol.
$$V_A = \frac{B_0}{\left(\mu_0\rho_m\right)^{\frac{1}{2}}}$$

$$B_0 = 100\text{ gauss} = 100\times10^{-4}\,\text{Tesla}$$

$$\rho_m = 13.6\times10^3\,\frac{\text{kg}}{\text{m}^3},$$

$$\mu_0 = 4\pi\times10^{-7}\,\frac{\text{Wb}}{\text{A.m}}$$

Substituting in the above formula gives

$$V_A = 7.6\,\frac{\text{cm}}{\text{sec}}$$

Example 2. Calculate the speed of an Alfven wave for the following cases:

(a) In the Earth's ionosphere, considering that $n_e = 10^5 \, cm^{-3}$, $B = 0.5$ gauss, and that the positive charge carriers are atomic oxygen ions.

(b) In the solar corona, assuming $n_e = 10^6 \, cm^{-3}$, $B = 10$ gauss, and that positive charge carriers are protons.

(c) In the interstellar space, considering $n_e = 10^7 \, m^{-3}$ and $B = 10^{-7}$ Tesla, the positive charge carriers are protons.

Sol. (a) $\quad V_A = \dfrac{B_0}{(\mu_0 \rho_m)^{\frac{1}{2}}}$, given $B_0 = 0.5 \times 10^{-4}$ Tesla

$$\rho_m = n \times \text{mass of one oxygen ion}$$

$$= 10^5 \times 16 \times 1.67 \times 10^{-24} \, \frac{g}{cm^3}$$

$$= 16 \times 1.67 \times 10^{-19} \times 10^3 \, \frac{kg}{m^3}$$

$$= 26.72 \times 10^{-16} \, \frac{kg}{m^3}$$

Substituting in the formula for V_A, we find

$$V_A = \frac{0.5 \times 10^{-4}}{\left[4\pi \times 10^{-7} \times 26.72 \times 10^{-16}\right]^{\frac{1}{2}}} \, \frac{m}{s}$$

$$= 8.8 \times 10^5 \, \frac{m}{sec}$$

In a similar way find V_A for (b) and (c).

8.6 Self Learning Exercise

Q.1 What are sound waves and what is their speed ?

Q.2 Prove the equipartition of kinetic and magnetic (wave) energy in a small-amplitude Alfven wave.

Q.3 Find the equations and phase velocity of an Alfven wave in an incompressible fluid propagating along the magnetic field in

(i) longitudinal mode i.e. $\vec{u}_1$ is parallel to $\vec{B}_0$

(ii) a transverse mode with $\vec{u}_1$ perpendicular to $\vec{B}_0$.

Q.4 What are Alfven waves ? Derive a formula for the speed of Alfven waves.

8.7 Summary

The behavior of conductive fluids is described by the set of hydromagnetic equations.

$$\frac{\partial \rho_m}{\partial t} + \nabla \cdot (\rho_m \vec{u}) = 0$$

$$\rho_m \frac{\partial \vec{u}}{\partial t} + \rho_m (\vec{u} \cdot \vec{\nabla}) \vec{u} = -\nabla p + \vec{\tau} \times \vec{B}$$

$$\nabla p = V_s^2 \nabla \rho_m$$

$$\vec{\nabla} \times \vec{B} = \mu_0 \vec{J}$$

$$\vec{\nabla} \times \vec{E} = -\frac{\partial \vec{B}}{\partial t}$$

$$\vec{E} + \vec{u} \times \vec{B} = 0$$

The conductivity is assumed to be large enough so that the displacement current can be neglected compared to the conduction current. On the fluid-vacuum boundary, the boundary condition

$$p + \frac{B^2}{2\mu_0} = 0 \quad \text{apply.}$$

For the special case $\sigma \to \infty$, the fluid-vacuum system conserves the sum of kinetic and potential energy

$$K + W = \text{constant}$$

where
$$K = \int \frac{1}{2} \rho_{m0} v^2 dV$$

$$W = \int \left(p + \frac{B^2}{2\mu_0} \right) dV$$

where integration extends over the entire fluid-plus-vacuum volume.

Fluid elements lying along a magnetic field line can be thought of as being mechanically connected by the (elastic) field line. An oscillation of fluid element transverse to the magnetic field propagates along the field lines, with the Alfven velocity

$$V_A = \frac{B_0}{(\mu_0 \, \rho_m)^{\frac{1}{2}}}$$

Compressional acoustic waves (except when propagating along the field) couple to the magnetic field, giving rise to magnetoacoustic waves. Such waves, when propagating perpendicular to the field with the phase velocity

$$\frac{\omega}{k} \equiv V_{MA} = (V_s^2 + V_A^2)^{\frac{1}{2}}$$ can be easily interpreted ; compression (and expansion) of a fluid element is linked with compression (and expansion) of the flux frozen therein, adding to $p_0\gamma$ the quantity $\left(\dfrac{B_0^2}{2\mu_0}\right) \times 2$, in accordance with the notion of magnetic pressure and $\gamma = 2$ for its two-dimensional equation of state.

8.8 Glossary

Two fluid description of plasmas:

In this description the plasma is taken as composed of two distinct but intermingled fluids, the electron and ion fluids.

Linearization:

Equations retaining first order quantities and neglecting second and higher order terms.

Hydromagnetic waves:

These are the waves in a compressible ,conducting fluid immersed in a magnetic field.

Fast Alfven wave: The magnetosonic wave mode is also known as the compressional Alfven wave or fast Alfven wave.

8.9 Answers to Self Learning Exercise

Ans.1: Sound waves are longitudinal that propagate in a compressible , nonconducting fluid. The adiabatic sound speed is given by

$$V_S = \left(\frac{\gamma p}{\rho_m}\right)^{1/2}$$

Where V_s is the sound speed

γ is the ratio of the specific heats $\dfrac{C_P}{C_V}$

ρ_m = mass density of conducting fluid

p = pressure

Ans.2: The magnetic energy density of this wave motion is $\dfrac{B_1^2}{2\mu_0}$.

The equipartition of energy is verified by nothing that

$$\frac{B_1^2}{2\mu_0} = \frac{B_0^2 u_1^2}{2\mu_0\left(\dfrac{\omega}{k}\right)^2} = \frac{B_0^2 u_1^2}{2\mu_0 V_A^2}$$

$$= \frac{1}{2}\rho_{m0}\, u_1^2$$

That is magnetic energy density $\dfrac{B_1^2}{2\mu_0}$ is equal to the kinetic energy density of the fluid motion.

8.10 Exercise

Q.1 What is a magnetosonic wave and what is the phase velocity of the magnetosonic wave ?

Q.2 What is an Alfven wave and what is the phase velocity of Alfven wave?

Q.3 What are magnetohydrodynamic (MHD) waves ?

Q.4 Find the equations for magnetosonic waves propagating normal to $\vec{B}_0$. Show that the phase velocity is $V = (V_s^2 + V_A^2)^{\frac{1}{2}}$

Q.5 What are Alfven waves? Using the fluid equations appropriate to an incompressible fluid with infinite conductivity find the dispersion relation for Alfven waves propagating perpendicular to the magnetic field.

Q.6 Find the dispersion relation of Alfven waves propagating parallel to the magnetic field.

Q.7 Derive expression for speed of longitudinal sound wave,

Q.8 Consider an incompressible fluid with infinite conductivity . Write complete and appropriate system of equations that governs the behavior of this type of fluid.

8.11 Answers to Exercise

Ans.1: Magnetosonic waves are longitudinal waves in which the velocity of mass flow and the fluctuating mass density associated with the wave motion are both in the wave propagation direction. The phase is given by

$$\frac{\omega}{k} = \left(V_s^2 + V_A^2 \right)^{\frac{1}{2}}$$

Where V_S is sound velocity and V_A is Alfven velocity. $\dfrac{\omega}{k}$ is the phase velocity of magnetosonic wave, ω is the frequency and k is the wavenumber of magnetosonic waves.

Ans.2: These are the waves propagating in a homogeneous conducting medium in a uniform constant magnetic field.

The physical velocity of propagation of waves is called the group velocity (Alfven speed) and is given by

$$\frac{\partial \omega}{\partial \vec{k}} = \vec{V}_A = \frac{\vec{B}_0}{\left(\mu_0 \rho_m \right)^{1/2}}$$

In Alfven wave an oscillation of fluid elements transverse to the magnetic field propagates along the field lines with the Alfven speed V_A.

Ans.3: If a conducting fluid moves in a magnetic field, electric fields are induced in it and electric currents flow. Conversely ,the currents themselves modify the magnetic field.

Thus we have a complex interaction between the magnetic field and the fluid dynamic phenomena and this flow is examined by combining the field equations with those of fluid dynamics. MHD waves are small disturbances that propagate in a homogeneous conducting medium in a uniform constant magnetic field.

Ans.4: When the wave vector $\vec{k}$ is perpendicular to $\vec{B}_0$, then $\vec{k}.\vec{V}_A = 0$. equation (21) simplifies to

$$-\omega^2 \vec{u}_1 + \left(V_s^2 + V_A^2\right)\left(\vec{k}.\vec{u}_1\right)\vec{k} = 0$$

Therefore $\vec{u}_1 = \dfrac{1}{\omega^2}\left(V_s^2 + V_A^2\right)\left(\vec{k}.\vec{u}_1\right)\vec{k}$.

Thus $\vec{u}_1$ is parallel to $\vec{k}$, so that $\vec{k}.\vec{u}_1 = k u_1$ and the solution for $\vec{u}_1$ is a longitudinal wave with the phase velocity

$$\frac{\omega}{k} = \left(V_s^2 + V_A^2\right)^{\frac{1}{2}}$$

References and Suggested Readings

1. Physics of high temperature plasmas (Second Edition) by George Schmidt (Academic press)

2. Fundamentals of Plasma physics (Third Edition) By J.A. Bitten Court (Springer)

3. Plasma Dynamics by R. O. Dendy (Clarendom Press. Oxford).

UNIT-9
Dispersion Relation, Wave Propagation In Magnetized Cold Plasma

Structure of the Unit

9.0 Objectives

To learn

- Derivation of dielectric constant of unmagnetized plasma.

- Derivation of dispersion relation of electromagnetic wave propagating through an unmagnetized plasma.

- Derivation of dielectric tensor of a cold magnetized plasma.

- Highly frequency waves in a cold magnetized plasma.

- Dispersion relation of left circularly polarized wave.

- Dispersion relation of right circularly polarized wave.

- Dispersion relation when the wave vector $\vec{k}$ is perpendicular to the magnetic field direction.

9.1 Introduction

In this chapter we discuss the theory of wave propagation in a cold homogeneous plasma, immersed in a magnetic field.

There are two main different methods of approach that are normally used in analyzing the problem of wave propagation in plasmas. In one of them, the plasma is characterized as medium having either a conductivity or a dielectric constant and the wave equation for this medium is derived from Maxwell equations. In the presence of an externally applied magnetostatic field, the plasma is equivalent to an anisotropic dielectric characterized by dielectric tensor or dyad.

• In another approach, Maxwell equations are solved simultaneously with the fluid equations describing the particle motions. In this case we do not explicitly derive a wave equation, and expressions for the dielectric or conductivity dyad are not obtained directly. Instead, we obtain a "dispersion relation", which relates the wave number k to the wave frequency ω. All the information about the propagation of a given mode is contained in the appropriate dispersion relation. We shall adopt the first method.

9.2　Dielectric　Tensor　of　The　Plasma　(Without External Magnetic Field)

The macroscopic response of any medium is to an applied electric field is determined by the sum of the microscopic responses of the individual particles that make up the medium. In a conducting medium, macroscopic level by the separation between positive and negative charges that is produced by the applied electric field. If the applied field varies with time, so too will the microscopic state of the medium:

The separation between the positive and negative charges will change with time, as will the electric field that is produced by their separation. Thus particle currents and displacement currents are produced.At a macroscopic level, the particle current is described by the conductivity tensor σ, where

$$\vec{J} = \sigma.\vec{E} \tag{1}$$

That is, σ is a macroscopic variable whose nature is determined by microscopic dynamics,

Maxwell's equation $\vec{\nabla} \times \vec{H} = \vec{J} + \varepsilon_0 \dfrac{\partial \vec{E}}{\partial t}$ states that $\vec{J}$ and the displacement current combine to act as the source of the magnetic field in the medium .Using (1), we have

$$\vec{\nabla} \times \vec{H} = \left(\varepsilon_0 \frac{\partial}{\partial t} + \sigma \right) \vec{E} \tag{2}$$

• Now recalling equation $\vec{B} = \mu_0 \vec{H}$ and $\mu_0 \varepsilon_0 = \dfrac{1}{c^2}$ and if varies $\vec{E}$ as $\exp(-i\omega t)$,

then equation (2) gives $\qquad \vec{\nabla} \times \vec{B} = \dfrac{-i\omega}{c^2} \left(I + \dfrac{i}{\varepsilon_0 \ \omega} \sigma \right) \vec{E} \tag{3}$

Where I is the identity matrix. It follows that all information about the macroscopic response of the medium to applied electric fields is contained in the dielectrics tensor defined by

$$\varepsilon = I + \frac{i}{\varepsilon_0 \omega} \sigma \tag{4}$$

and equation (3) can now be written

$$\vec{\nabla} \times \vec{B} = \frac{-i\omega}{c^2} \varepsilon E \tag{5}$$

•Operating on equation (5) with $\dfrac{\partial}{\partial t}$, we find

$$\vec{\nabla} \times \frac{\partial \vec{B}}{\partial t} = (-i\omega) \left(\frac{-i\omega}{c^2} \right) \varepsilon.E$$

$$-\vec{\nabla} \times \left(\nabla \times \vec{E} \right) = -\frac{\omega^2}{c^2} \varepsilon.\vec{E}$$

Using the vector formula

$$\vec{\nabla} \times \left(\vec{\nabla} \times \vec{E} \right) = \nabla \left(\vec{\nabla}.\vec{E} \right) - \nabla^2 \vec{E} \text{ , One obtains}$$

$$\nabla^2 \vec{E} - \nabla(\nabla.\vec{E}) + \frac{\omega^2}{c^2} \varepsilon.E = 0 \qquad (6)$$

- The nature of the macroscopic quantity σ , and hence ε for a plasma is determined at the microscopic level by the plasma particles dynamics. As the simplest example, consider equation of motion of a plasma electron in the absence of an magnetic field.

$$m\ddot{x} = -e\,\vec{E}, \text{ where } \vec{E} \text{ is the applied electric field varying as } \exp(-i\omega t).$$

Each plasma electron will respond with a velocity

$$\vec{v} = \left(\frac{e}{i\,\omega\,m}\right)\vec{E} \qquad (7)$$

- The current density associated with n_0 such electrons per unit volume is

$$\vec{J} = -n_0\,e\,\vec{v} = -\frac{n_0\,e^2}{i\,\omega\,m}\,\vec{E} \qquad (8)$$

So that $\sigma = -\left(\dfrac{n_0 e^2}{i\,\omega\,m}\right) I$. Then the definition equation (4) of ε gives

$$\varepsilon = I + \frac{i}{\varepsilon_0 \omega}\left(\frac{-n_0\,e^2}{i\,\omega\,m}\right) I$$

$$\varepsilon = \left(1 - \frac{\omega_{pe}^2}{\omega^2}\right) I \qquad (9)$$

where we have used $\omega_{pe}^2 = \dfrac{n_0 e^2}{m\,\varepsilon_0}$

- Now we calculate the normal modes- a macroscopic concept of an unmagnetized plasma. If $\vec{E}$ varies as $\exp(i\vec{k}.\vec{r} - i\omega t)$ equations (6) and (9) combined to give (by replacing the operator $\vec{\nabla}$ by $i\vec{k}$)

$$(\omega^2 - \omega_{pe}^2 - c^2 k^2)\vec{E} + c^2 \vec{k}(\vec{k}.\vec{E}) = 0 \qquad (10)$$

- There are two classes of normal mode. First, consider the case of transverse mode, that have $\vec{k}.\vec{E} = 0$ and are accordingly electromagnetic. Then equation (10) gives

$$(\omega^2 - \omega_{pe}^2 - c^2 k^2)\vec{E} = 0 \qquad (11)$$

This is compatible with non-zero $\vec{E}$ only if ω and $\vec{k}$ are related by

$$\omega^2 = \omega_{pe}^2 + c^2 k^2 \qquad (12)$$

which is our first derivation of a dispersion relation. It tells us that the frequency of any electromagnetic wave in an unmagnetized plasma must exceed the electron plasma frequency. In addition, if we try to launch an electromagnetic wave into the plasma with frequency $\omega < \omega_{pe}$ equation (12) indicates that the wave will have an imaginary wave number k inside the plasma. The wave will therefore be evanescent, and unable to propagate through the plasma.

Thus ω_{pe} plays the role of a cutoff frequency for electromagnetic waves in an unmagnetized plasma.

This fact has a number of practical applications. For example, it determines the range of frequencies that can be used for different types of radio communication.

In order to communicate with a satellite, one must choose a frequency that exceeds the plasma frequency of the ionospheric plasma. Otherwise the signal will be reflected from the ionosphere, and will not reach satellite.

Conversely, we may send a radio signal to a distant point on the Earth's surface by choosing a frequency below the ionospheric plasma frequency, and use ionosphere to reflect the signal in the required direction.

The second class of normal mode that satisfies (10) is electrostatic, with $\vec{k}$ and $\vec{E}$ parallel. In this case, the dispersion relation is clearly

$$\omega^2 = \omega_{pe}^2 \tag{13}$$

Thus the electrostatic normal modes of a plasma oscillate at the electron plasma frequency .

9.3 Dielectric Tensor of A Cold Magnetized Plasma

The effect on particle dynamics of the magnetic component of an electromagnetic wave can be shown to be negligible so long as the particle velocities are non-relativistic and $\omega \neq \omega_{ce}$.

• In this case, we only consider the effect of the rapidly oscillating electric field. The motion of an electron parallel to the static uniform magnetic field is governed

by
$$\dot{v}_z = \frac{-e}{m} \vec{E}_z(t) \tag{14}$$

which is the same as we have used while considering the motion in electric field only. This is because the magnetic field does not affect motion parallel to it. The perpendicular equation of motion,

$$\dot{\vec{v}}_\perp = -\frac{e}{m}\vec{E}_\perp(t) + \vec{\omega}_{ce} \times \vec{v}_\perp(t) \tag{15}$$

follows from Lorentz force equation: Differentiating (15) with respect of time, and substituting from (15) for $\vec{\omega}_{ce} \times \dot{\vec{V}}_\perp$, we obtain

$$\ddot{\vec{v}}_\perp = -\frac{e}{m}\dot{\vec{E}}_\perp + \vec{\omega}_{ce} \times \dot{\vec{v}}_\perp$$

$$= -\frac{e}{m}\dot{\vec{E}}_\perp + \vec{\omega}_{ce} \times \left(\frac{-e}{m}\vec{E}_\perp + \vec{\omega}_{ce} \times \vec{v}_\perp \right)$$

$$= \frac{-e}{m}\left(\dot{\vec{E}}_\perp + \vec{\omega}_{ce} \times \vec{E}_\perp \right) + \vec{\omega}_{ce}\left(\vec{\omega}_{ce}.\vec{v}_\perp \right) - \vec{v}_\perp \omega_{ce}^2$$

Here $\vec{\omega}_{ce} \times \left(\vec{\omega}_{ce} \times \vec{v}_\perp \right) = \vec{\omega}_{ce}\left(\vec{\omega}_{ce}.\vec{v}_\perp \right) - \vec{v}_\perp \left(\vec{\omega}_{ce}.\vec{\omega}_{ce} \right)$

Now, using the fact that $\vec{\omega}_{ce}$ and $\vec{v}_\perp$ are perpendicular to each other and hence $\vec{\omega}_{ce}.\vec{v}_\perp$ is equal to zero, we obtain

$$\ddot{\vec{v}}_\perp + \vec{\omega}_{ce}^2 \vec{v}_\perp = \frac{-e}{m}\left(\dot{\vec{E}}_\perp + \vec{\omega}_{ce} \times \vec{E}_\perp \right) \tag{16}$$

In the absence of perpendicular electric field, equation (16) reduces to equation which describes cyclotron motion, whose solution we shall denote by $\vec{v}_{\perp 0}(t)$

Now we write $\vec{E}_\perp(t) = \text{Re}(\tilde{E}_\perp e^{-i\omega t})$ and
$$\vec{v}_\perp(t) = \vec{v}_{\perp 0}(t) + \tilde{v}_\perp e^{-i\omega t}$$

Then equation (16) becomes

$$(\omega_{ce}^2 - \omega^2)\tilde{v}_\perp = \frac{-e}{m}\left(-i\omega\tilde{E}_\perp + \vec{\omega}_{ce} \times \tilde{E}_\perp \right) \tag{17}$$

Combining equation (14) and (17)

$$\tilde{v}_x = -\frac{e}{m}\left(\frac{-i\omega}{\omega_{ce}^2 - \omega^2} \right)\tilde{E}_x + \frac{e}{m}.\frac{\omega_{ce}}{\omega_{ce}^2 - \omega^2}\tilde{E}_y + 0\tilde{E}_z$$

$$\tilde{v}_y = \frac{e}{m}\left(\frac{-\omega_{ce}}{\omega_{ce}^2 - \omega^2} \right)\tilde{E}_x + \frac{e}{m}.\frac{i\omega}{\omega_{ce}^2 - \omega^2}\tilde{E}_y + 0\tilde{E}_z$$

$$\tilde{v}_z = 0\,\tilde{E}_x + 0.\tilde{E}_y + \frac{e}{m}\left(-\frac{i}{\omega} \right)\tilde{E}_z$$

and hence

$$\vec{v}(t) = \vec{v}_{\perp 0}(t) + \frac{e}{m}\begin{bmatrix} \dfrac{i\omega}{\omega_{ce}^2 - \omega^2} & \dfrac{\omega_{ce}}{\omega_{ce}^2 - \omega^2} & 0 \\[2ex] -\dfrac{\omega_{ce}}{\omega_{ce}^2 - \omega^2} & \dfrac{i\omega}{\omega_{ce}^2 - \omega^2} & 0 \\[2ex] 0 & 0 & \dfrac{-i}{\omega} \end{bmatrix}\begin{bmatrix} \tilde{E}_x \\[2ex] \tilde{E}_y \\[2ex] \tilde{E}_z \end{bmatrix}$$

$$\vec{v}(t) = \vec{v}_{\perp 0}(t) + \mathrm{Re}\left[\hat{O}.\tilde{E}\, e^{-i\omega t} \right] \tag{18}$$

$$\text{where } \hat{O} \equiv \frac{e}{m}\begin{bmatrix} \dfrac{i\omega}{\omega_{ce}^2 - \omega^2} & \dfrac{\omega_{ce}}{\omega_{ce}^2 - \omega^2} & 0 \\[2ex] -\dfrac{\omega_{ce}}{\omega_{ce}^2 - \omega^2} & \dfrac{i\omega}{\omega_{ce}^2 - \omega^2} & 0 \\[2ex] 0 & 0 & \dfrac{-i}{\omega} \end{bmatrix} \tag{19}$$

All information about the particle drift arising form $\vec{E}_\perp$ is contained in $\hat{O}$ setting $E_y = 0$, we see that

$$\tilde{v}_{\perp x}^2 + \tilde{v}_{\perp y}^2\left(\frac{\omega}{\omega_{ce}}\right)^2 = \left(\frac{e}{m}\right)^2 \frac{\omega^2}{(\omega_{ce}^2 - \omega^2)^2} E_x^2 \tag{20}$$

From equation (20) we see that particle motion in the x y plane follows an ellipse, whose mean radius and eccentricity depend on the size of the electric field and the value of $\dfrac{\omega}{\omega_{ce}}$. This motion is superimposed on oscillation along the magnetic field in the harmonic field E_z .

In equation (18), the response of the electron to the wave field is given by the second term on the right. This is the term on which we wish to concentrate, and we shall accordingly adopt a model in which the plasma is cold so that the electrons have no thermal motion, so we have $v_\perp = 0$ and Larmor radius $r_L = \dfrac{v_\perp}{\omega_{ce}} = 0$ and consequently in equation (18) $v_{\perp 0}(t) = 0$.

Then the current $\vec{J}(t)$ associated with the response to an electromagnetic wave of the n_0 electrons contained in unit volume of magnetized plasma is

$$\vec{j}(t) = -n_0 e \vec{v}(t)$$

$$= -n_0 e\ \hat{O}.\vec{E}(t) = \vec{\sigma}.\vec{E} \tag{21}$$

This defines the conductivity tensor $\vec{\sigma}$ for a cold magnetized plasma.

• We now obtain the magnetic field which is set up in the plasma by the oscillating electric field and by the response to it of the electrons from (5). The dielectric tensor ε in equation (5) for a cold magnetized plasma follows from (4), (21), and (19):

$$\varepsilon = I + \frac{i}{\varepsilon_0 \omega}\sigma$$

$$= I + \frac{i}{\varepsilon_0 \omega}(-n_0 e)\hat{O}$$

$$= \begin{pmatrix} 1 & 0 & 0 \\ 0 & 1 & 0 \\ 0 & 0 & 1 \end{pmatrix} + \frac{-in_0 e}{\varepsilon_0 \omega} \cdot \frac{e}{m} \begin{bmatrix} \dfrac{i\omega}{\omega_{ce}^2 - \omega^2} & \dfrac{\omega_{ce}}{\omega_{ce}^2 - \omega^2} & 0 \\[3mm] -\dfrac{\omega_{ce}}{\omega_{ce}^2 - \omega^2} & \dfrac{i\omega}{\omega_{ce}^2 - \omega^2} & 0 \\[3mm] 0 & 0 & \dfrac{-i}{\omega} \end{bmatrix}$$

or

$$\varepsilon = \begin{bmatrix} 1 + \dfrac{\omega_{pe}^2}{\omega_{ce}^2 - \omega^2} & \dfrac{-i\omega_{ce}}{\omega} \cdot \dfrac{\omega_{pe}^2}{\omega_{ce}^2 - \omega^2} & 0 \\[4mm] \dfrac{i\omega_{ce}}{\omega} \dfrac{\omega_{pe}^2}{\omega_{ce}^2 - \omega^2} & 1 + \dfrac{\omega_{pe}^2}{\omega_{ce}^2 - \omega^2} & 0 \\[4mm] 0 & 0 & \dfrac{1 - \omega_{pe}^2}{\omega^2} \end{bmatrix}$$

where $\qquad \omega_{pe}^2 \equiv \dfrac{n_0 e^2}{m\varepsilon_0}$ \hfill (22)

We have thus calculated the cold plasma dielectric tensor ε from considerations of particle dynamics, represented by (14) and (15).

The tensor ε contains a large amount of information. In particular, it governs the propagation of electromagnetic waves in magnetized plasmas.

Now we look for wave fields which oscillate as $\exp\left(i\hat{k}.\hat{r} - i\omega t\right)$ and propagate in cold magnetized plasma.

The wave equation is

$$\nabla^2 \vec{E} - \vec{\nabla}(\vec{\nabla}.\vec{E}) + \frac{\omega^2}{c^2}\bar{\varepsilon}.\vec{E} = 0$$

Replace $\qquad \vec{\nabla}$ by $\imath\vec{k}$, we find

$$-k^2\vec{E} - (ik)(i\vec{k}.\vec{E}) + \frac{\omega^2}{c^2}\bar{\bar{\varepsilon}}.\vec{E} = 0$$

or $\qquad \left(\frac{\omega^2}{c^2}\bar{\varepsilon} + \vec{k}.\vec{k} - k^2 I\right).\vec{E} = 0$

or $\qquad$ M.E.=0 $\hfill (23)$

where $\qquad M = \frac{\omega^2}{c^2}\varepsilon + \vec{k}.\vec{k} - k^2 I \hfill (24)$

Where I is the identity matrix.

The normal modes of the system are accordingly given by

$$\det(M) = 0$$

(25)Equation (25) is the basic dispersion relation.

The wave vector $\vec{k}$ and frequency ω are related to each other by components of the dielectric tensor ε, which themselves contain ω and the parameters ω_{pe} and ω_{ce} which describe the state of plasma. The solutions of equation (25) describe the electromagnetic wave that propagate in a cold magnetized plasma. It is remarkable that all this information can be obtained using only the single-particle dynamics embodied in equation (14) and (15), will be collective behavior described by the simple summation over all electrons in equation (21).

9.4 High- Frequency Waves in a Cold Magnetized Plasma

We now consider the high-frequency normal modes of cold magnetized plasma. The frequency of each normal mode is a root of equation (25); the corresponding polarization is obtained from equation (23). For easy manipulation of the dielectric tensor ε given in (22) we define

$$\varepsilon_1 = 1 + \frac{\omega_{pe}^2}{\omega_{ce}^2 - \omega^2} \hfill (26)$$

$$\varepsilon_2 = \frac{\omega_{ce}}{\omega}.\frac{\omega_{pe}^2}{\omega_{ce}^2 - \omega^2} \hfill (27)$$

$$\varepsilon_3 = 1 - \frac{\omega_{pe}^2}{\omega^2} \tag{28}$$

so that
$$\varepsilon = \begin{bmatrix} \varepsilon_1 & -i\varepsilon_2 & 0 \\ i\varepsilon_2 & \varepsilon_1 & 0 \\ 0 & 0 & \varepsilon_3 \end{bmatrix} \tag{29}$$

Note that $\omega_{ce} \equiv \dfrac{|e|B}{m} \simeq 1.8 \times 10^{11} \left(\dfrac{B}{1\,\text{Telsa}} \right) \text{rad s}^{-1}$ is a positive quantity. It is useful to define a vector $\vec{N}$, as follows, whose magnitude is given by the refractive index and whose direction is given by the wave vector $\vec{k}$.

Thus
$$\vec{N} = \frac{c\vec{k}}{\omega} \tag{30}$$

Then we may write M, defined in equation (24), as

$$M = \frac{\omega^2}{c^2}\varepsilon + \vec{k}\,\vec{k} - k^2\,I$$

$$= \frac{\omega^2}{c^2} \begin{bmatrix} \varepsilon_1 - N_y^2 - N_z^2 & -i\varepsilon_z N_x N_y & N_x N_z \\ i\varepsilon_2 + N_x N_y & \varepsilon_1 - N_x^2 - N_z^2 & N_y N_z \\ 0 & 0 & \varepsilon_3 \end{bmatrix} \tag{31}$$

consider first waves that propagate parallel to the magnetic field. In the case
$$N_x = \quad N_y = 0 \quad \text{and} \quad N_z = N. \text{ by equation (31),}$$

M reduces to

$$M = \frac{\omega^2}{c^2} \begin{bmatrix} \varepsilon_1 - N^2 & -i\varepsilon_2 & 0 \\ i\ \varepsilon_2 & \varepsilon_1 - N^2 & 0 \\ 0 & 0 & \varepsilon_3 \end{bmatrix} \tag{32}$$

• The corresponding normal modes of the plasma have frequencies that are the roots of det (M)=0, that is

$$\varepsilon_3 \left[\left(\varepsilon_1 - N^2 \right)^2 - \varepsilon_2^2 \right] = 0 \tag{33}$$

This equation is satisfied first by

$$\varepsilon_3 = 0 \tag{34}$$

will $\left(\varepsilon_1 - N^2\right)^2 - \varepsilon_2^2 \neq 0$ then by the definition equation (28) of ε_3, equation (34) gives

$$\omega^2 = \omega_{pe}^2 \tag{35}$$

To establish the polarization of the wave, substitute equation (34) into equation (32). In this case, equation (23) will be satisfied by an electric field of the form

$$\vec{E} = (0, 0, E_z).$$

Thus, this normal mode of a cold magnetized plasma is an electrostatic wave, whose wave vector and field amplitude are both directed along the magnetic field. The wave oscillates at the plasma frequency, and is indistinguishable from the Langmuir wave (Plasma wave) in a plasma with no magnetic-field. Physically, each electron is acted on by an electric field that is parallel to the magnetic field, so that there is no perpendicular component field, so that there is no perpendicular component of motion that could be affected by the Lorentz force $\vec{v} \times \vec{B} = 0$.

9.5 Dispersion Relation of Left-Circularly Polarized Wave

The other normal modes that propagate parallel to the magnetic field follow from equation (33) with $\varepsilon_3 \neq 0$ and

$$\varepsilon_1 - N^2 = \pm\varepsilon_2 \tag{36}$$

Let us consider first the case

$$\varepsilon_1 - \varepsilon_2 = N^2 \tag{37}$$

Using equation(26) and (27), this gives the following dispersion relation between $k\left(= \dfrac{\omega}{c}N\right)$ and ω :

$$1 + \frac{\omega_{pe}^2}{\omega_{ce}^2 - \omega^2} - \frac{\omega_{ce}}{\omega}\frac{\omega_{pe}^2}{\omega_{ce}^2 - \omega^2} = \frac{c^2 k^2}{\omega^2}$$

$$\text{or}\qquad k^2 = \frac{\omega^2}{c^2}\left[1 + \frac{\omega_{pe}^2}{\omega_{ce}^2 - \omega^2}\left(1 - \frac{\omega_{ce}}{\omega}\right)\right]$$

$$= \frac{\omega^2}{c^2}\left[1 + \frac{\omega_{pe}^2}{\omega_{ce}^2 - \omega^2}\left(\frac{\omega - \omega_{ce}}{\omega}\right)\right]$$

$$= \frac{\omega^2}{c^2}\left[1 - \frac{\omega_{pe}^2}{\omega\left(\omega_{ce} + \omega\right)}\right]$$

$$k = \frac{\omega}{c}\left[1 - \frac{\omega_{pe}^2}{\omega\left(\omega_{ce} + \omega\right)}\right]^{1/2} \tag{38}$$

The polarization of this wave is obtained by substituting equation (37) into equation (32) and setting

$$M.E = 0$$

or

$$\frac{\omega^2}{c^2}\begin{bmatrix} \varepsilon_2 & -i\varepsilon_2 & 0 \\ i\varepsilon_2 & \varepsilon_2 & 0 \\ 0 & 0 & \varepsilon_3 \end{bmatrix}\begin{bmatrix} E_x \\ E_y \\ E_z \end{bmatrix} = \begin{bmatrix} 0 \\ 0 \\ 0 \end{bmatrix}$$

where we have replaced $\varepsilon_1 - N^2$ by ε_2 this gives

$$\varepsilon_2 E_x - i\varepsilon_2 E_y = 0$$

or

$$\varepsilon_2\left[E_x - i\,E_y\right] = 0$$

Similarly the other equation viz

$$i\varepsilon_2 E_x + \varepsilon_2 E_y = 0 \qquad \text{gives the}$$

result

$$i\varepsilon_2\left[E_x - iE_y\right] = 0$$

and the third equation gives $\varepsilon_3\, E_z = 0$.

This is satisfied by an electric field of the form $\vec{E} = \left(E_x, E_y, 0\right)$, where

$$E_x - i\,E_y = 0 \tag{39}$$

It follows from equation (39) that when $E_x \sim \exp(ikz - i\omega t)$ we must have $E_y \sim \exp\left(ikz - i\omega t + i\frac{\pi}{2}\right)$. Taking real parts, and denoting the phase $kz - \omega t$ by $\phi(t)$, we have

$$E_x \sim \cos\phi(t)$$

and

$$E_y \sim \cos\left(\phi(t) + \frac{\pi}{2}\right)$$

Thus when $\phi = 0$, $\vec{E}$, lies entirely in the positive x-direction.

As ϕ increases from zero, E_x decreases in magnitude and E_y goes in the negative y-direction, that is, $\vec{E}$ rotates in the xy-plane in the direction of the curled fingers of the left hand when the thumb lies along the magnetic field, which as usual defines the z-direction. Therefore equation (38) is the dispersion relation of a left circularly polarized wave.

9.6 Dispersion Relation of Right- Circularly Polarized Wave

Returning to equation (36), we now consider the second case

$$\varepsilon_1 + \varepsilon_2 = N^2 \tag{40}$$

Again using (26) and (27)

$$1 + \frac{\omega_{pe}^2}{\omega_{ce}^2 + \omega^2} + \frac{\omega_{ce}}{\omega} \cdot \frac{\omega_{pe}^2}{\omega_{ce}^2 - \omega^2} = \frac{c^2 k^2}{\omega^2}$$

or

$$k^2 = \frac{\omega^2}{c^2}\left\{1 + \frac{\omega_{pe}^2}{\omega_{ce}^2 - \omega^2}\left(1 + \frac{\omega_{ce}}{\omega}\right)\right\}$$

$$= \frac{\omega^2}{c^2}\left\{1 + \frac{\omega_{pe}^2(\omega + \omega_{ce})}{(\omega_{ce} - \omega)(\omega_{ce} + \omega)\omega}\right\}$$

or

$$k = \frac{\omega}{c}\left\{1 + \frac{\omega_{pe}^2}{\omega(\omega_{ce} - \omega)}\right\}^{\frac{1}{2}} \tag{41}$$

Substituting equation (40) into equation (32), and considering M.E=0, we determine the polarization of this wave.

$$\frac{\omega^2}{c^2} = \begin{bmatrix} -\varepsilon_2 & -i\varepsilon_2 & 0 \\ i\varepsilon_2 & -\varepsilon_2 & 0 \\ 0 & 0 & \varepsilon_3 \end{bmatrix}\begin{bmatrix} E_x \\ E_y \\ E_z \end{bmatrix} = \begin{bmatrix} 0 \\ 0 \\ 0 \end{bmatrix}$$

gives

$$-\varepsilon_2\, E_x - i\varepsilon_2\, E_y = 0$$

and

$$i\varepsilon_2\, E_x - \varepsilon_2 E_y = 0$$

$$\varepsilon_3\, E_z = 0$$

From these equation its follows that

$$E_x + i E_y = 0 \tag{42}$$

By repeating the phase analysis of the preceding paragraph, it can be seen that this is a right circularly polarized wave.

It may be noted that this sense of rotation is the same as that of an electron in the magnetic field. Conversely, the ions, which are oppositely charged, rotate in a left-handed sense about the magnetic field. The ion cyclotron frequency ω_{ce} is given by

$$\omega_{ci} = \frac{z_e B}{M} = 9.6 \times 10^7 Z \left(\frac{B}{1 Telsa}\right)\left(\frac{M}{m_p}\right)^{-1} rad\, s^{-1}$$

where m_p denotes proton mass, Z is the charge number, M is the ion mass. ω_{ci} is much smaller than ω_{ce} by the factor $\sim \frac{m_p}{m_e}$. It is for this reason we could afford to neglect ion dynamics.

Noting that ion dynamics are neglected, we consider the left circularly polarized dispersion relation (38) :

$$k = \frac{\omega}{c}\left\{1 - \frac{\omega_{pe}^2}{\omega(\omega_{ce} + \omega)}\right\}^{\frac{1}{2}}$$

We note first that k is zero, corresponding to infinite wavelength when

$$1 - \frac{\omega_{pe}^2}{\omega(\omega_{ce} + \omega)} = 0$$

Or $\qquad \omega^2 + \omega\,\omega_{ce} - \omega_{pe}^2 = 0$

$$\omega = -\frac{\omega_{ce}}{2} + \frac{\omega_{ce}}{2}\left(1 + \frac{4\,\omega_{pe}^2}{\omega_{ce}^2}\right)^{\frac{1}{2}}$$

or $\qquad \omega \equiv \omega_1 = \frac{\omega_{ce}}{2}\left\{\left(1 + \frac{4\,\omega_{pe}^2}{\omega_{ce}^2}\right)^{\frac{1}{2}} - 1\right\}$ $\qquad\qquad$ (47)

ω_1 is known as the cut-off frequency. If the plasma parameters are such that $\frac{\omega_{pe}^2}{\omega_{ce}^2} < 2$, it follows from equation (47) that $\omega_1 < \omega_{ce}$; and vice versa.

At high frequency $\omega \gg \omega_{pe}$ and ω_{ce} equation (38) gives $\omega \approx ck$, so that the waves propagate with a phase velocity, close to the speed of light in vacuum, which is almost independent of k. These waves are unaffected by the presence of the plasma for two related reasons. First, the plasma particles are too heavy to be able to respond coherently to the rapidly changing wave field. Second, the timescale of electron cyclotron is much longer than that of the wave oscillation. The electrons hardly move on their Larmor orbits while the wave phase reverses, so that the electrons are effectively unmagnetized with respect to the wave.

Now we consider the dispersion relation for right circularly polarized wave (equation 41). We note first that as ω approaches ω_{ce} from below, the magnitude k of the wave vector tends to infinity, corresponding to zero wave length.

This phenomenon occurs in general when the normal mode corresponds to some resonance of the system. In this case, it reflects the equality of the frequency of the right circularly polarized wave with the frequency ω_{ce} of electron gyration in the magnetic field, which is also a right-handed motion.

The analogous resonance for left circularly polarized waves occurs when $\omega = \omega_{ci}$ .Returning to equation (41), we see that if ω exceeds ω_{ce} by a small amount, the quantity $\omega_{pe}^2 / \omega(\omega_{ce} - \omega)$ is negative and has a magnitude much greater than unity, so that no real value of k satisfies the equation. As ω is increased further, the magnitude of $\omega_{pe}^2 / \omega(\omega_{ce} - \omega)$ is reduced, and when ω reaches the value.

$$\omega_2 = \frac{\omega_{ce}}{2} \left\{ \left(1 + \frac{4\,\omega_{pe}^2}{\omega_{ce}^2} \right)^{\frac{1}{2}} + 1 \right\} \tag{48}$$

Equation (41) has the solution k=0. For $\omega > \omega_2$, equation (41) again has real solutions. Thus ω_2 is the low frequency cutoff a second branch of right circularly polarized waves.

By equation (41) when ω is sufficiently small (though still large compared to the characteristic ion frequencies) that $\omega_{ce} >> \omega$ and $\dfrac{\omega_{pe}^2}{\omega_{ce}\,\omega} >> 1$, the Whistler wave frequency is given approximately by

$$\omega = \left(\frac{c\,k}{\omega_{pe}} \right)^2 \omega_{ce} \tag{49}$$

At very high frequencies $\omega \gg \omega_{pe}$ and ω_{ce} , this second branch has $\omega \simeq ck$, which repeats the behavior that we noted for high-frequency left circularly polarized waves.

9.7 Dispersion Relation when Wave Vector K is Perpendicular to The Magnetic Field Direction

We shall now consider the normal modes that have wave vector $\vec{k}$ perpendicular to the magnetic field direction. We choose the x-axis to lie in the direction parallel to the wave vector $\vec{k}$. Then by equation (30), $N_y = N_z = 0$ and equation (31) becomes

$$M = \begin{bmatrix} \varepsilon_1 & -i\varepsilon_2 & 0 \\ i\varepsilon_2 & \varepsilon_1 - N^2 & 0 \\ 0 & 0 & \varepsilon_3 - N^2 \end{bmatrix}. \tag{50}$$

The corresponding normal modes of the plasma have frequencies are the roots of

$$\det(M) = 0, \text{ which now gives}$$

$$(\varepsilon_3 - N^2)\left\{\varepsilon_1(\varepsilon_1 - N^2) - \varepsilon_2^2\right\} = 0 \tag{51}$$

This equation is satisfied first by

$$\varepsilon_3 = N^2 \tag{52}$$

with $\varepsilon_1(\varepsilon_1 - N^2) - \varepsilon_2^2 \neq 0$. By the definition of ε_3 in equation (28) gives

$$1 - \frac{\omega_{pe}^2}{\omega^2} = \frac{c^2 k^2}{\omega^2}$$

or $$\omega^2 = \omega_{pe}^2 + c^2 k^2 \tag{53}$$

which is identical to the dispersion relation equation (11) for electromagnetic waves propagating in arbitrary directions in an unmagnetized plasma. The fact that plasma is magnetized does not enter equation (53) , and this normal mode is accordingly referred to as the ordinary mode(O-made).Physicsally ,the non appearance of the magnetic field indicates that the particle dynamics associated with the O- mode must take place exclusively in the direction parallel to the magnetic field, so that $\vec{v} \times \vec{B} = 0$. Therefore the O- mode must be polarized with an amplitude (0,0,E).

Equation (51) is also satisfied by

$$\varepsilon_1(\varepsilon_1 - N^2) - \varepsilon_2^2 = 0 \qquad (54)$$

with $\varepsilon_3 - N^2 \neq 0$ using equation (26) and (27), this gives

$$\left(1 + \frac{\omega_{pe}^2}{\omega_{ce}^2 - \omega^2}\right)\left(1 + \frac{\omega_{pe}^2}{\omega_{ce}^2 - \omega^2} - \frac{c^2 k^2}{\omega^2}\right) - \left(\frac{\omega_{ce}}{\omega} \cdot \frac{\omega_{pe}^2}{\omega_{ce}^2 - \omega^2}\right)^2 = 0$$

which can be simplified to

$$k = \frac{1}{c}\left\{\frac{(\omega^2 - \omega_1^2)(\omega^2 - \omega_2^2)}{\omega^2 - \omega_{UH}^2}\right\}^{\frac{1}{2}} \qquad (55)$$

Here ω_1 and ω_2 are the frequencies that were defined in equation (47) and (48) respectively, and ω_{UH} denotes the upper hybrid frequency :

$$\omega_{UH} = (\omega_{pe}^2 + \omega_{ce}^2)^{\frac{1}{2}} \qquad (56)$$

The term hybrid is used because this frequency combines two distinct aspects of electron plasma dynamics: space-charge density oscillion (ω_{pe}) and electron-cyclotron motion (ω_{ce}). We first consider the magnitudes of ω_1 and ω_2 relative to ω_{UH}.

Combining equation (47) and (56),

$$\omega_1 = \frac{\omega_{ce}}{2}\left\{\left(\frac{\omega_{ce}^2 + 4\omega_{pe}^2}{\omega_{ce}^2}\right)^{\frac{1}{2}} - 1\right\}$$

$$= \frac{\omega_{ce}}{2}\left\{\frac{(\omega_{UH}^2 + 3\omega_{pe}^2)^{\frac{1}{2}}}{\omega_{ce}^2} - 1\right\}$$

or $$\omega_1 = \left[\omega_{UH}^2 - \frac{3}{4}\omega_{ce}^2\right]^{\frac{1}{2}} - \frac{1}{2}\omega_{ce} \qquad (57)$$

So that ω_1 is always less than ω_{UH} , irrespective of the values of ω_{pe} and ω_{ce}. Similarly combining equation (49) and (56)

$$\omega_2 = \frac{\omega_{UH}}{2}\left\{\left(1+\frac{3\omega_{pe}^2}{\omega_{UH}^2}\right)^{\frac{1}{2}}+\left(1-\frac{\omega_{pe}^2}{\omega_{UH}^2}\right)^{\frac{1}{2}}\right\} \tag{58}$$

The first square root term in equation (58) exceeds unity by more than amount by which the second square-root term is less than unity. Thus the value of the entire term in {} in equation (58) must exceed two, and ω_2 is always greater than ω_{UH}, irrespective of the values of ω_{pe} and ω_{ce}. So quite generally

$$\omega_1 < \omega_{UH} < \omega_2 \tag{59}$$

Now we return to eq. (55). If follows from equation (59) that there are real solutions k only when $\omega_{UH} > \omega > \omega_1$ or $\omega > \omega_2$ in addition, the value of k tends to infinity as ω approaches ω_{UH} from below. It follows that there are two regions of propagation for this wave, which is known as the extraordinary mode (X-mode). The low-frequency branch has a low-frequency cut-off at $\omega = \omega_1$, and a resonance at $\omega = \omega_{UH}$. The high-frequency branch has low-frequency cut off at $\omega = \omega_2$, and by equation (55) satisfies $\omega \simeq ck$ at high frequencies. Figure 2. depicts the frequencies of the ordinary mode and both of the extraordinary mode as a function of k.

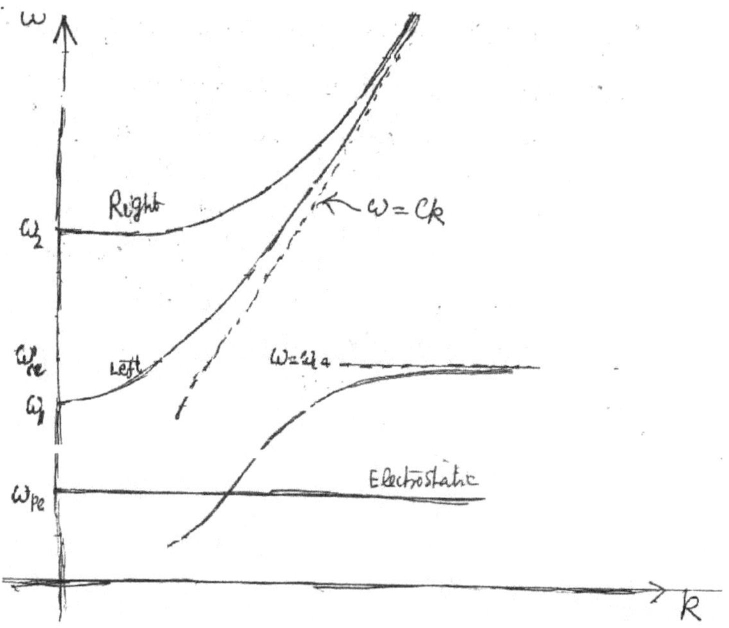

Figure1: Relation between ω and k for wave that propagate parallel to the magnetic field.

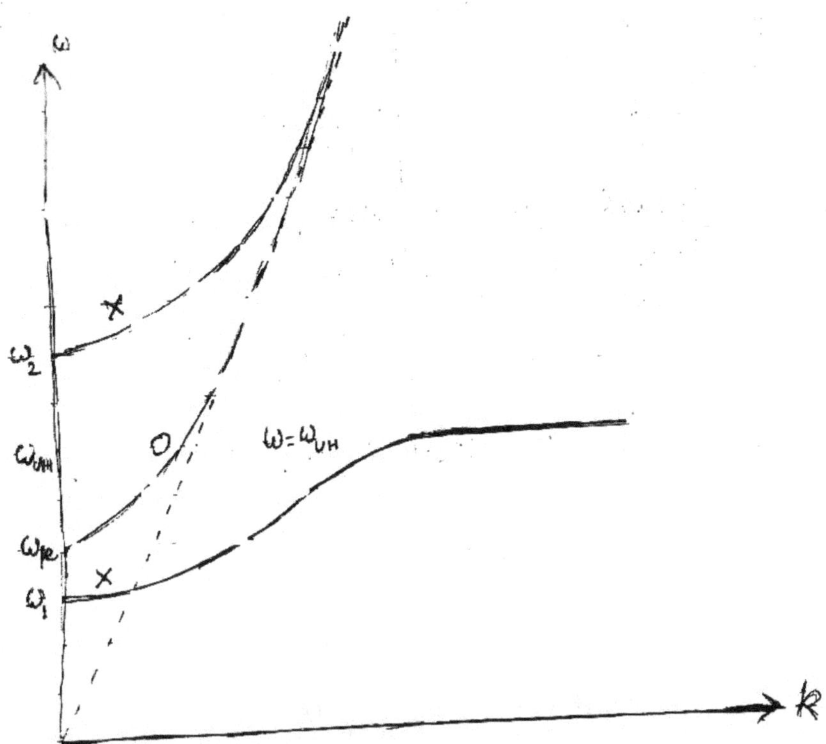

Figure 2: $\omega - k$ dispersion diagram showing frequencies of waves that propagate perpendicular to the magnetic field. O is the ordinary mode and both branches of the extraordinary mode.

9.8 Illustrative Examples

Example 1. Prove that transverse waves with frequencies $\omega < \omega_p$ are evanescent in a cold plasma.

Solution: Dispersion relation of transverse waves is

$$\omega^2 = \omega_{pe}^2 + c^2 k^2 \Rightarrow k^2 = \frac{1}{c^2}(\omega^2 - \omega_{pe}^2)$$

$$k = \frac{1}{c}i(\omega_{pe}^2 - \omega^2)^{\frac{1}{2}} \equiv ik_i$$

Thus k is purely imaginary, hence wave propagating factor $e^{ikz} \equiv e^{-k_i z}$,

when $k_i = \frac{1}{c}(\omega_{pe}^2 - \omega^2)^{\frac{1}{2}}$ the wave will be damped in a distance $z \sim \frac{1}{k_i}$ .That is the wave is evanescent.

Example 2. An electromagnetic plane wave of frequency ω impinges on a half-space filled with plasma frequency $\omega_p > \omega$. What is the characteristic depth of perpetration? prove that for $\omega \ll \omega_p$ this distance is $\dfrac{c}{\omega_p}$, called the collision less skin depth.

Solution: from the solution of Example 1, penetration depth $Z \sim \dfrac{1}{k_t} = \dfrac{c}{(\omega_{pe}^2 - \omega^2)^{\frac{1}{2}}}$

or
$$Z = \frac{c}{\omega_{pe}\left(1 - \dfrac{\omega^2}{\omega_{pe}^2}\right)^{\frac{1}{2}}}$$

For $\qquad \dfrac{\omega}{\omega_{pe}} \ll 1, Z \sim \dfrac{c}{\omega_{pe}}$

9.9 Self Learning Exercise

Q.1 Define electron cyclotron frequency.

Q.2 Derive an expression for the dielectric tensor of an unmagnetized plasma in the from

$$\varepsilon = \left(1 - \frac{\omega_{pe}^2}{\omega^2}\right) I$$

Q.3 Derive dispersion relation of left circularly polarized wave.

9.10 Summary

We now summarize the properties of the normal modes of a cold magnetized plasma that have $\vec{k}$ parallel to the magnetic field direction and frequency ω significantly greater than ω_{ci}. Right circularly polarized waves exist for $\omega < \omega_{ce}$ and $\omega > \omega_2$. Relation between ω and k for waves that propagate parallel to the magnetic field are shown in Fig. 1. The branch with $\omega = \omega_{ce}$, is known as the Whistler wave.

9.11 Glossary

Evanescent wave:A wave which is unable to propagate through the plasma

Whistler Wave: A wave that propagates parallel to the magnetic field and whose frequency $\omega \ll \omega_{ce}$ and is approximately given by $\omega \ll \left(\dfrac{ck}{\omega_{pe}}\right)^2 \omega_{ce}$

9.12 Answers to Self Learning Exercise

Ans.1: $\omega_{ce} = \dfrac{|e|B}{m} \simeq 1.8 \times 10^{11} \left(\dfrac{B}{1Tesla}\right) rad\,s^{-1}$

Ans.2: See derivation of equation (9).

Ans.3: Please see the derivation of eq.(38)

9.13 Exercise

Q.1 What is the frequency of an upper hybrid wave.

Q.2 Define upper hybrid frequency.

Q.3 Using Maxwell's equations derive an expression for the dielectric tensor in the form

$$\varepsilon = 1 + \frac{i}{\varepsilon_0 \omega}\sigma \quad \text{where } \sigma \text{ is the conductivity}$$

Q.4 Show that the frequency ω of an electromagnetic wave and its wave vector k while propagating through a plasma of plasma frequency ωpe satisfy the dispersion relation $\omega^2 = \omega_{pe}^2 + c^2 k^2$ What is the cutoff frequency?

Q.5 Calculate the normal modes of an unmagnetized plasma.

Q.6 Derive an expression of the dielectric tensor of a cold magnetized plasma.

Q.7 Derive the dispersion relation of high frequency normal modes of a cold magnetized plasma. Show that normal mode of a cold magnetized plasma is an electrostatic wave and other normal modes that propagate parallel to the magnetic field are left-circularly, right-circularly polarized electromagnetic wave.

Q.8 Draw $\omega - k$ diagram for waves that propagate parallel to the magnetic field.

Q.9 Draw $\omega - k$ diagram for waves that propagate perpendicular to the magnetic field.

Q.10 What is an ordinary wave? what is an extraordinary wave. Derive their dispersion relation.

9.14 Answers to Exercise

Ans.1: $\omega_{UH} = \sqrt{\omega_{pe}^2 + \omega_{ce}^2}$

Ans.2: $\omega_{UH} = \left(\omega_{pe}^2 + \omega_{ce}^2\right)^{1/2}$

Ans.3: Please see the derivation of eq.(4)

Ans.4: For the relation $\omega^2 = \omega_{pe}^2 + c^2 k^2$,see the derivation of eq.(12)

Cut off frequency ω is given by

$$\omega^2 = \omega_{pe}^2 \text{ so } k = 0$$

Or $\omega = \sqrt{\dfrac{ne^2}{m\varepsilon_0}}$

Ans.5: (i) Transverse mode ($\vec{k}.\vec{E} = 0$)

$$\omega^2 = \omega_{pe}^2 + c^2 k^2$$

See equation (12) of the text.

(ii) Electrostatic mode (i.e. $\vec{k}$ and $\vec{E}$ parallel; $\omega^2 = \omega_{pe}^2$

See equation (13) of the text.

Ans.6: See the derivation of equation (22)

Ans.7: See the derivation of high-frequency waves in a cold magnetized plasma.

References and Suggested Readings

1. Fundamentals of plasma physics by J. A. Bittencourt.

2. The physics of High Temperature plasmas by George Schimidt .

3. Plasma physics by F.F. Chen

4. Plasma Dynamics by Dendy.

UNIT-10
Faraday Rotation and Landau Damping

Structure of the Unit

10.0 Objectives

To learn the

• Dispersion relation, cut-offs and resonances

• Faraday rotation

• Waves in hot plasmas

• Wave-particle interaction

• Landau damping

• Ponderomotive force

10.1 Introduction

The Plasma waves are linear cold plasma waves in an infinite, homogeneous plasma. The plasma is not isotropic, however, since the presence of magnetic field provides one preferred direction. Without the magnetic field, the plasma may be represented by a simple dielectric constant and the only wave solution is a simple electromagnetic wave that propagates above the plasma frequency, ω_{pe}.

By Cold plasma, we mean a collection of charged particles without any net charges, and the particles are at rest except as they are induced to move through the action of the self-consistent electric and magnetic fields of the wave, or in other words, the particles have no kinetic thermal motion of their own. In this chapter we examine the general forms of the dispersion relation in the presence of magnetic field and discuss various cut-offs and resonances of various modes of plasma in the presence of magnetic field. We also discuss an important feature of the magnetoactive media is that they lead to Faraday rotation of $\vec{E}$ as the wave propagates through the media. The electromagnetic field interacts with the particles of the plasma and as a result of this interaction the wave may damp giving energy to the particles (Landau damping) or the wave may gain energy from the particles. An important nonlinear interaction originating from $\vec{v} \times \vec{B}$ force leading to ponderomotive force will be elaborated.

10.2 Cut-offs Resonances

In chapter 3 we have derived the dispersion relation of a cold magnetoactive plasma:

$$An^4 - Bn^2 + c = 0 \tag{1}$$

where $\vec{n} = \dfrac{c\vec{k}}{\omega}$, or $n = \dfrac{ck}{\omega}$ is the index of refraction $\tag{2}$

$$A = S\sin^2\theta + P\cos^2\theta \tag{3}$$

$$B = RL\sin^2\theta + PS(1+\cos^2\theta) \tag{4}$$

$$C = PRL \tag{5}$$

where $P = 1 - \sum_j \dfrac{\omega_{pj}^2}{\omega^2} \tag{6}$

j refers to the species of the plasma particle, $j = e$ for electron, $j = i$ for ion etc.

$$R = 1 - \sum_j \dfrac{\omega_{pj}^2}{\omega(\omega + \epsilon_j \, \omega_{cj})} \tag{7}$$

$$L = 1 - \sum_j \dfrac{\omega_{pj}^2}{\omega(\omega - \epsilon_j \, \omega_{cj})} - \tag{8}$$

ω_{pj} is the plasma frequency for species j, given by

$$\omega_{pj}^2 = \frac{n_j q_j^2}{m_j \epsilon_0} \tag{9}$$

$\epsilon_j \equiv \dfrac{q_j}{|q_j|}$ to denote the sign of the charge for species j.

$\epsilon_j = -1$ for electrons and $\epsilon_j = +1$ for positive ion etc.

$\omega_{cj} \equiv \dfrac{|q_j| B_0}{m_j}$ is the cyclotron frequency for species j.

θ is the angle between $\vec{k}$ and direction of external magnetic field $\vec{B}_0$ to be in the z- direction.

The solutions of Eqn. (1) may be written in terms of the angle:

$$\tan^2 \theta = -\frac{P(n^2 - R)\,(n^2 - L)}{(Sn^2 - RL)\,(n^2 - P)} \tag{11}$$

The general condition for a resonance, where $n^2 \to \infty$, or where $\lambda \to 0$, is given by Equation (11) as

$$\tan^2 \theta = -\frac{P}{S} \quad \text{(general resonance condition)} \tag{12}$$

and the general cutoff condition, where n=0, or $\lambda \to \infty$, is given by Eq. (1) as

$$C = P\,R\,L = 0 \qquad \text{General cutoff condition} \tag{13}$$

We note some special cases at this point :

1. Propagation parallel to $\vec{B}_0$, $\theta = 0$ (The numerator of Equation (11) must vanish)

(a) $\quad P \equiv 1 - \sum_j \dfrac{\omega_{pj}^2}{\omega^2} = 0 \qquad$ (Plasma oscillations)

(b) $\quad n^2 = R \equiv 1 - \sum_j \dfrac{\omega_{pj}^2}{\omega(\omega + \epsilon_j\, \omega_{cj})}$ (wave with right-handed polarization)

(c) $\quad n^2 = L \equiv 1 - \sum_j \dfrac{\omega_{pj}^2}{\omega(\omega - \epsilon_j\, \omega_{cj})}$ (wave with left-handed polarization)

2. Propagation perpendicular to $\vec{B}_0$, $\theta = \dfrac{\pi}{2}$ (the denominator of equation (11)

must vanish.)

    (a) $n^2 = P$ (ordinary wave)

    (b) $n^2 = \dfrac{RL}{S}$ (Extraordinary wave)

Principal solutions- Parallel Propagation:

We first define the principal resonances to be those which occur at $\theta = 0$ and $\theta = \dfrac{\pi}{2}$. The general condition for a resonance $(n^2 \to \infty)$ is from Equation (11)

$$\tan^2 \theta = \dfrac{-P}{S}.$$

Hence, for $\theta \to 0$, we require $S \to \infty$ since P=0 is a cutoff. Since $S = \dfrac{1}{2}(R + L)$, this can be satisfied for

       either $R \to \infty$ (electron cyclotron resonance)

       or $L \to \infty$ (ion cyclotron resonance)

The Right-Handed Wave:

For a simple plasma of electrons and one ion species, the dispersion relation for the right-handed wave, $n_R^2 = R$, which propagates parallel to $\vec{B}_0$ is given by equation (11) as

$$n_R^2 = R = 1 - \frac{\omega_{pi}^2}{\omega(\omega + \omega_{ci})} - \frac{\omega_{pe}^2}{\omega(\omega - \omega_{ce})} \tag{14}$$

So, the resonance is clearly at $\omega = \omega_{ce}$. The cutoff frequency, where $n^2 = R = 0$ is given by

$$n_R^2 = 1 - \frac{\omega_{pi}^2}{\omega\,\omega_{ci}} + \frac{\omega_{pe}^2}{\omega\,\omega_{ce}} \equiv 1$$

$$\omega_R = \frac{\omega_{ce} - \omega_{ci}}{2} + \left[\left(\frac{\omega_{ce} + \omega_{ci}}{2} \right)^2 + \omega_p^2 \right]^{\frac{1}{2}} \tag{15}$$

where $\omega_p^2 = \omega_{pe}^2 + \omega_{pi}^2$.

For low and high density limits, the cutoff frequency may be approximated by

$$\omega_R \simeq \begin{cases} \omega_{ce}\left(1+\dfrac{\omega_{pe}^2}{\omega_{ce}^2}\right) & \text{R-wave cutoff- low density} \\[2em] \omega_{pe}+\dfrac{1}{2}\omega_{ce} & \text{R-wave cutoff- high density} \end{cases} \tag{16}$$

where we have assumed $m_e \ll m_i$.

For high and low frequencies, the index of refraction approaches limits

$$n_R^2 \simeq \begin{cases} 1+\dfrac{\omega_{pi}^2}{\omega_{ci}^2} \equiv \dfrac{c^2}{V_A^2}, & \text{as } \omega \to 0 \\[2em] 1-\dfrac{\omega_{pe}^2}{\omega^2}, & \text{as } \omega \to \infty \end{cases} \tag{17}$$

where $V_A = \dfrac{B}{(\mu_0\rho)^{\frac{1}{2}}}$, $\rho = n_0(m^+ + m^-)$ is the mass density.

Fig.1 displays the dispersion curve for R-wave. Fig.(1a) is for the case $\omega_{pe} \gg \omega_{ce}$ and Fig. (1b) is for the low density limit: $\omega_{pe} \ll \omega_{ce}$

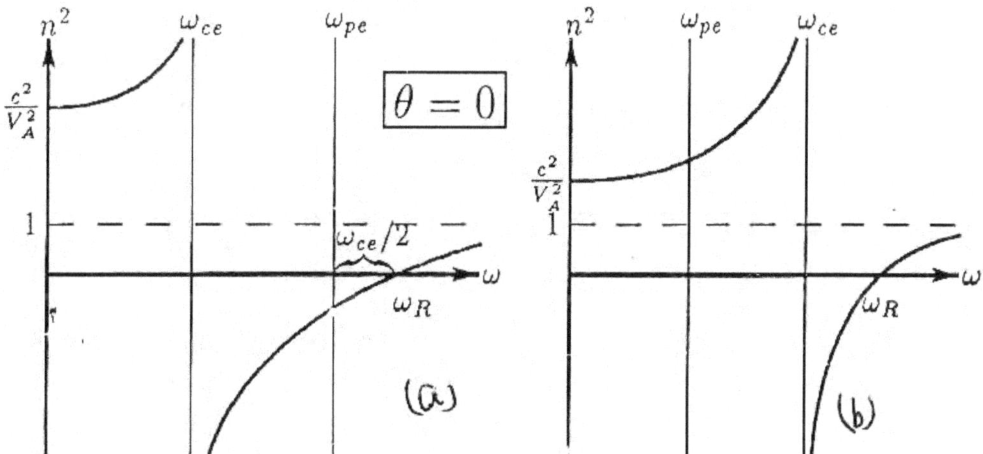

Figure1: *Dispersion relation for R wave (a)high density case $\omega_{pe} \gg \omega_{ce}$ (b) low density case $\omega_{pe} \ll \omega_{ce}$*

The Left- Handed wave:

The dispersion relation for the left - handed wave, $n_L^2 = L$, is also given by equation (11) as

$$n^2 = L = 1-\frac{\omega_{pi}^2}{\omega(\omega-\omega_{ci})}-\frac{\omega_{pe}^2}{\omega(\omega+\omega_{ce})} \tag{18}$$

So the resonance in this case is clearly at $\omega = \omega_{ci}$. The cutoff frequency where $n^2 = L = 0$ is given by

$$\omega_L = \frac{\omega_{ci} - \omega_{ce}}{2} + \left[\left(\frac{\omega_{ci} + \omega_{ce}}{2}\right)^2 + \omega_p^2\right]^{\frac{1}{2}} \tag{19}$$

For low and high densities, the cutoff frequency may be approximated by

$$\omega_L \simeq \begin{cases} \omega_{ci} + \dfrac{\omega_{pi}^2}{\omega_{ci}} & \text{L- WAVE CUTOFF- low density} \\[3ex] \omega_{pe} - \dfrac{1}{2}\omega_{ce} & \text{L- WAVE CUTOFF- high density} \end{cases} \tag{20}$$

where we have again assumed me $m_e \ll m_i$.

For high and low frequencies, the index of refraction approaches the limits

$$n_L^2 \simeq \begin{cases} 1 + \dfrac{\omega_{pi}^2}{\omega_{ci}^2} \equiv \dfrac{c^2}{V_A^2} & \text{as } \omega \to 0 \\[3ex] 1 - \dfrac{\omega_{pe}^2}{\omega^2} & \text{as } \omega \to \infty \end{cases} \tag{21}$$

Fig. 2 displays the dispersion curves for the L-wave.

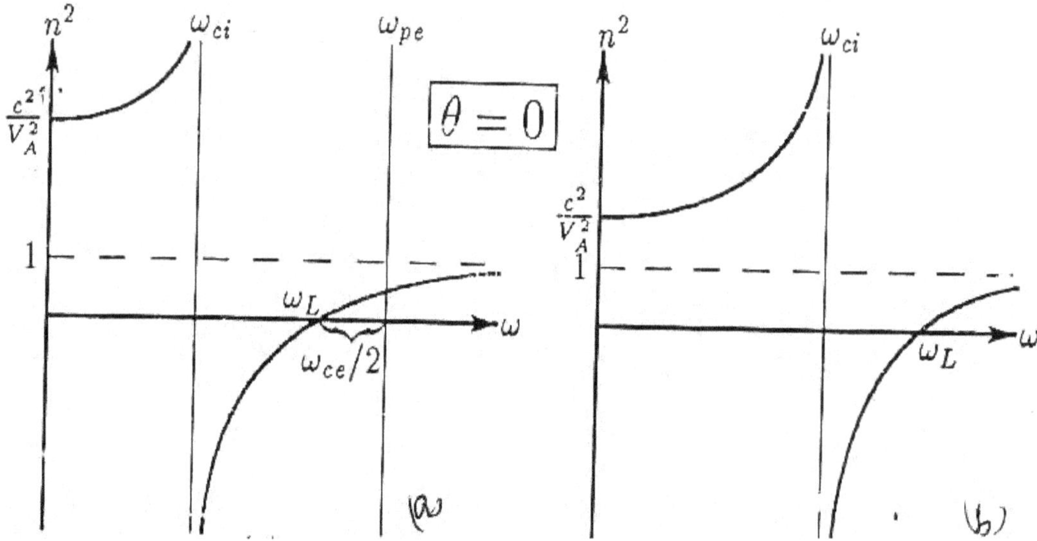

Figure 2: *Dispersion relation for R wave (a)high density case $\omega_{pe} \gg \omega_{ce}$ (b) low density case $\omega_{pe} \ll \omega_{ce}$*

Principal Solutions- Perpendicular Propagation :

As $\theta \to \dfrac{\pi}{2}$, $\dfrac{P}{S} \to \infty$, and since $P \to \infty$ is a trivial solution (either $\omega \to 0$, so no wave at all, or $\omega_p \to \infty$, which is impossible), we require $S \to 0$. These resonance are called hybrid resonance because they generally involve some combination of ω_c and ω_p. The solutions for perpendicular propagation are the ordinary and extraordinary waves.

The ordinary wave:

The dispersion relation for the ordinary wave is the same as in an unmagnetized plasma, and is given simply by

$$n_0^2 = P = 1 - \frac{\omega_{pe}^2}{\omega^2} - \frac{\omega_{pi}^2}{\omega^2} = 1 - \frac{\omega_{pe}^2}{\omega^2} \tag{22}$$

for a single ion species plasma, and it is immediately apparent that this wave has no dependence on the magnetic field at all. This wave has $\vec{E}$ parallel to $\vec{B}_0$, so the particles do not experience any effect of the magnetic field. It is clear that there is no resonance and that the cutoff is at $\omega = \omega_{pe}$, neglecting again terms of order $\dfrac{m_e}{m_i}$ .There is no propagation below ω_{pe} and $n^2 \to 1$ as $\omega \to \infty$.

The Extraordinary Wave:

The dispersion relation for the extraordinary wave, given by

$$n_x^2 = \frac{RL}{S} = \frac{\left[(\omega+\omega_{ci})\,(\omega-\omega_{ce})-\omega_p^2\right]\left[(\omega-\omega_{ci})\,(\omega+\omega_{ce})-\omega_p^2\right]}{\left(\omega^2-\omega_{ci}^2\right)\left(\omega^2-\omega_{ce}^2\right)+\left(\omega_{pe}^2+\omega_{pi}^2\right)\left(\omega_{ce}\omega_{ci}-\omega^2\right)} \tag{23}$$

is the most complicated of these simplified dispersion relation, since neither the resonance nor the cutoffs have simple expressions. The resonance are given by the zeros of the denominator, which lead to the quadratic roots,

$$\omega^2 = \frac{\omega_e^2 + \omega_i^2}{2} \pm \left[\left(\frac{\omega_e^2 - \omega_i^2}{2}\right)^2 + \omega_{pe}^2\,\omega_{pi}^2\right]^{\frac{1}{2}} \tag{24}$$

where $\omega_j^2 = \omega_{pj}^2 + \omega_{cj}^2$, $j = e, i$. One of the roots is simply given by the sum of the electron cyclotron frequency and the electron plasma frequency, namely

$$\omega_{UH}^2 = \omega_{pe}^2 + \omega_{ce}^2 \qquad \text{(Upper Hybrid Resonance)} \tag{25}$$

The other root is more complex, but again neglecting terms of order $\dfrac{m_e}{m_i}$,

$$\omega_{LH}^2 = \omega_{ce}\omega_{ci}\left(\frac{\omega_{pe}^2 + \omega_{ce}\,\omega_{ci}}{\omega_{pe}^2 + \omega_{ce}^2}\right) \qquad \text{[Lower Hybrid Resonance]} \tag{26}$$

10.3 Faraday Rotation

An important feature of magnetoactive media is that they lead to Faraday rotation. In order to analyze the Faraday rotation of $\vec{E}$ as the wave propagates in a magnetoactive plasma, we shall limit the analysis to $\theta = 0$ where the Maxwell equations become

$$\vec{\nabla} \times \vec{E} = -\frac{\partial \vec{B}}{\partial t} \Rightarrow i\vec{k} \times \vec{E} = -(-i\omega)\vec{B} \tag{27}$$

$$\vec{\nabla} \times \vec{B} = \mu_0 \frac{\partial \vec{D}}{\partial t} \Rightarrow i\vec{K} \times \vec{B} = \mu_0(-i\omega)\,\epsilon_0\,\vec{\vec{K}}\vec{E} \tag{28}$$

where $\vec{\vec{K}}$ is the dielectric tensor defined as

$$\vec{\vec{K}} = \begin{pmatrix} S & -iD & 0 \\ iD & S & 0 \\ 0 & 0 & P \end{pmatrix} \tag{29}$$

Now taking $\vec{k} \equiv (0, 0, k_z)$, $\quad \vec{B} = (B_x, B_y,\, 0)$ and $\vec{E} = (E_x, E_y,\, 0)$ we find

$$-k_z\,E_y = \omega\,B_x \tag{30}$$

$$k_z\,E_x = \omega\,B_y \tag{31}$$

$$-k_z\,B_y = -\frac{\omega}{c^2}\left[SE_x + iD\,E_y\right] \tag{32}$$

$$k_z\,B_x = -\frac{\omega}{c^2}\left[iD\,E_x + S\,E_y\right] \tag{33}$$

we then define the rotating coordinate variables

$$E_\pm = E_x \pm iE_y \tag{34}$$

$$B_\pm = B_x \pm iB_y \tag{35}$$

$$S = \frac{1}{2}(R + L) \tag{36}$$

$$D = \frac{1}{2}(R - L) \tag{37}$$

$$S + D = R \tag{38}$$

$$S - D = L \tag{39}$$

We find $k_z E_{\pm} = \pm i\omega\, B_{\pm}$ \hfill (40)

$$k_z B_{\pm} = \pm \frac{i\omega}{c^2}\, E_{\pm}(S \mp D) \tag{41}$$

These may be solved to obtain the result

$$(n^2 - K_{\mp})\, E_{\pm} = 0 \tag{42}$$

where $\quad K_{+} \equiv K_{xx} + iK_{xy}$ \hfill (43)

$$K_{-} \equiv K_{xx} - iK_{xy} \tag{44}$$

Equation (42) has two solutions.

1. Suppose $\quad E_{+} \neq 0$, then $n^2 = L$ and $E_{-} = 0$ So

$E_x = i\, E_y$ which confirms that this is the L-wave.

The E_{+} field may then be represented by

$$E_{+} = \hat{E}_{+}\, \exp\left[i\left(\frac{\omega}{c} n_L z - \omega t \right) \right]$$

where $\hat{E}_{+}$ is the complex amplitude .

2. Suppose $\quad E_{-} \neq 0$, then $n^2 = R$ and $E_{+} = 0$ so $E_x = -i\, E_y$ which confirms our identification of this as the R-wave. The E_{-} field may then the represented by

$$E_{-} = \hat{E}_{-}\, \exp\left[i\left(\frac{\omega}{c} n_R z - \omega t \right) \right]$$

where again $\hat{E}_{+}$ is the complex amplitude.

Constructing the measurable field E_x and E_y from these, we obtain.

$$R_e(E_x) = R_e\left(\frac{E_{+} + E_{-}}{2}\right) = \frac{1}{2} R_e \left\{ \hat{E}_{+} e^{\left[i\left(\frac{\omega}{c} n_L z - \omega t\right) \right]} + \hat{E}_{-} e^{\left[i\left(\frac{\omega}{c} n_R z - \omega t\right) \right]} \right\} \tag{45}$$

If we now take $R_e(E_y(0,t)) = 0$ so that the electric is aligned with the x-axis at z=0, then this demands that $\hat{E}_{+} = \hat{E}_{-} = E_0$ which we take to be real. We may then factor out a common term and represent the result as

$$R_e(E_x) = E_0 \cos\left(\frac{\Delta n}{2}\frac{\omega}{c}x\right)\mathrm{Re}\left[e^{i\left(\frac{\omega}{c}\bar{n}\,z - \omega t\right)}\right] \tag{46}$$

$$\mathrm{Re}(E_y) = E_0 \sin\left(\frac{\Delta n}{2}\frac{\omega}{c}x\right)\mathrm{Re}\left[e^{i\left(\frac{\omega}{c}\bar{n}\,z - \omega t\right)}\right] \tag{47}$$

$$\text{where } \bar{n} = \frac{1}{2}(n_L + n_R) \tag{48}$$

$$\Delta n = n_L - n_R > 0 \tag{49}$$

From this it is apparent that E_x and E_y are being modulated by the sine and cosine terms while the phase velocity of the composite wave is determined from the exponential terms which yields a phase velocity of $v_p = \frac{c}{\bar{n}}$ choosing a point of constant phase, the total electric field rotates in space as the wave propagates, as shown in figure 3. If we take the angle of rotation to be ϕ, then the rate of rotation of the E field vector is given by

$$\frac{d\varphi}{dz} = \frac{d}{dz}\left(\frac{\Delta n}{2}\frac{\omega}{c}z\right) = \frac{1}{2}\frac{\omega}{c}(n_L - n_R) \tag{50}$$

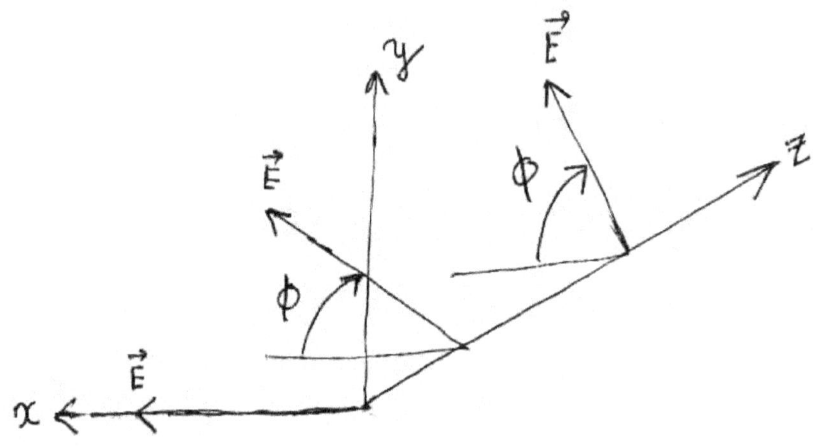

Figure 3. Faraday Rotation of the electric field

High Frequency Limit- Region 1 :

In order to estimate the amount of faraday rotation - for frequencies

206

$\omega \gg \omega_{ce} \gg \omega_{ci}$, we begin with the dispersion relation for the R-wave:

$$n_R^2 = R = 1 - \frac{\omega_{pi}^2}{\omega(\omega + \omega_{ci})} - \frac{\omega_{pe}^2}{\omega(\omega - \omega_{ce})}$$

$$\approx 1 - \frac{\omega_{pe}^2}{\omega^2}\left(1 - \frac{\omega_{ce}}{\omega}\right)^{-1}$$

$$\approx 1 - \frac{\omega_{pe}^2}{\omega^2}\left(1 + \frac{\omega_{ce}}{\omega}\right)$$

$$n_R^2 \approx 1 - \frac{\omega_{pe}^2}{\omega^2} - \frac{\omega_{pe}^2 \cdot \omega_{ce}}{\omega} \tag{51}$$

and similarly for the L-wave

$$n_R^2 \simeq 1 - \frac{\omega_{pe}^2}{\omega^2} + \frac{\omega_{pe}^2}{\omega^2} \cdot \frac{\omega_{ce}}{\omega} \tag{52}$$

These dispersion relation are compared with the definitions

$$n_{R,L}^2 \equiv \left(\bar{n} \mp \frac{1}{2}\Delta n\right)^2 \simeq \bar{n}^2 \mp \bar{n}\,\Delta n \tag{53}$$

but comparison yields the results

$$\bar{n}^2 = 1 - \frac{\omega_{pe}^2}{\omega^2} \approx 1 \text{ (approximately free space propagation)}$$

$$\bar{n}\,\Delta n \simeq \Delta n = \frac{\omega_{pe}^2}{\omega^2} \cdot \frac{\omega_{ce}}{\omega} \tag{54}$$

This result, along with equation (51) then give the rate of rotation as

$$\frac{d\varphi}{dz} = \frac{\Delta n}{2}\frac{\omega}{c} = \frac{1}{2} \cdot \frac{\omega_{pe}^2}{\omega^2} \cdot \frac{\omega_{ce}}{\omega} \cdot \frac{\omega}{c}$$

or $\qquad \dfrac{d\varphi}{dz} = \dfrac{\omega_{pe}^2 \omega_{ce}}{2\omega^2 c} \propto \lambda^2\, n_e\, B_0 \tag{55}$

where λ is the free space wavelength. The total rotation angle then is given by

$$\varphi = \lambda^2 \int_0^L n_e\,(z)\, B_0(z)dz \tag{56}$$

If the Faraday rotation is used as a diagnostic for estimating the magnetic field in a plasma, it is preferable to use long wavelengths (since $\phi \propto \lambda^2$), but one must still keep $\omega >> \omega_{ce} \sim \omega_{pe}$ in order to get a significant rotation.

Since the rotation is proportional to both density and magnetic field strength, one must be known to determine the other, and some idea of the variation along the path is necessary. Simultaneous measurement of the phase can give some additional information, since $\bar{n}$ differs some from unity. The phase is given by

$$\varphi = \frac{\omega}{c}\int_0^L (1-\bar{n})dz \simeq \frac{\omega}{c}\int_0^L \frac{\omega_{pe}^2}{2\omega^2}dz \propto \lambda \int_0^L n_e(z)dz \tag{57}$$

Since $\bar{n} \simeq 1 - \dfrac{\omega_{pe}^2}{2\omega^2}$. Thus measurement of both the phase and rotation can give estimates of the mean density and mean magnetic field along the path of integration.

10.4 Landau Damping (The Wave-Particle Interaction)

Intense laser light couples with the plasma either into electrostatic waves (due to resonance absorption and ion acoustic decay $2\omega_{pe}$ instability) or into both electrostatic and scattered light waves (Raman and Brillion scattering). We need to understand how electrostatic waves are damped. Since electrostatic wave are simply charge density fluctuations and their associated electric fields, these waves do not readily escape from the plasma. Their energy is ultimately transferred to the particles via either linear or nonlinear damping mechanisms.

(A) Collisional damping (electron-ion collisions)

The coherent motion of oscillations of electrons in the electric field of the wave is converted to random (or thermal) motion at the rate at which electron-ion collisions occur. To balance the energy dissipated, the energy of the wave then damps at the rate υ i.e.

$$\upsilon \frac{E^2}{8\pi} = \upsilon_{ei} \frac{n\, m\, v_\omega^2}{2}$$

where $v_\omega = \dfrac{e\, E}{m\omega}$.

Thus $\quad \upsilon = \dfrac{\omega_{pe}^2}{\omega^2}\upsilon_{ei};$ for $\omega = \omega_{pe}$ $\quad \upsilon \approx \upsilon_{ei}$ \hfill (58)

(B) Landau Damping :

Landau damping is also known as "collision-less damping".To understand the mechanism of Landau damping, consider an electrostatic wave :

E sin(kx- ωt).

Most particles of the plasma are non-resonant i.e have velocity u much different than $\dfrac{\omega}{k}$. These particles simply oscillate in the field and experience no gain or loss in energy. In contrast resonant particles with $v \approx \dfrac{\omega}{k}$ experience a nearly constant field and so can be efficiently accelerated or decelerated. These particles do exchange energy with the wave. The particle dynamics are determined by

$$\ddot{x} = \frac{qE}{m}\sin(kx - \omega t) \tag{59}$$

We assume $\qquad x = x_0 + v_0 t + x_1 + x_2 \tag{60}$

where $x_0 + vt$ represents free streaming of the particle in the absence of the electric field.

$\qquad x_1$ represents first - order correction $(\sim E)$

$\qquad x_2$ represents second -order correction $(\sim E^2)$

In a similar way we represent the velocity v of the particle as

$$v = v_0 + v_1 + v_2 \tag{61}$$

The expansion parameter is $k\delta x$, where δx is the charge between the free streaming position and its actual position.

The equation (59) can be written as

$$\dot{v} = \frac{qE}{m}\sin\left(kx - \omega t\right) \tag{62}$$

Now substituting (61) in the left hand side and (60) in the right hand side of eq.(62) and equating the terms of the same order of smallness ,we obtain

$$\dot{v}_1 = \frac{qE}{m}\sin\left(kx_0 - \Omega t\right) \tag{63}$$

where $\Omega = \omega - kv_0$

$$\dot{v}_2 = \frac{qE}{m} kx_1 \cos(kx_0 - \Omega t) \tag{64}$$

Now integrating(63),we find

$$v_1 = \frac{qE}{m\Omega}\left[\cos(kx_0 - \Omega t) - \cos(kx_0)\right] \tag{65}$$

Further integrating (65)

$$x_1 = -\frac{qE}{m\Omega^2}\left[\sin(kx_0 - \Omega t) - \sin(kx_0) + \Omega t \cos(kx_0)\right] \tag{66}$$

Now substituting (66) into (64) we obtain

$$\dot{v}_2 = -\frac{kq^2E^2}{m^2\Omega^2}\cos(kx_0 - \Omega t)\left[\sin(kx_0 - \Omega t) - \sin(kx_0) + \Omega t \cos(kx_0)\right] \tag{67}$$

We next compute the rate of the change of energy $(\delta\dot{\varepsilon})$ of a set of particles with random initial positions.

$$<\delta\dot{\varepsilon}_1> = <mv_0\dot{v}_1> \tag{68}$$

where $<\,>$ represents average over initial positions.

To second order, we obtain

$$<\delta\dot{\varepsilon}_2> = mv_0 <\dot{v}_2> + m<v_1\dot{v}_1> \tag{69}$$

Substituting for $\dot{v}_2$ from (67) and $\dot{v}_1$ from (63) and carrying out the algebra, we obtain

$$<\delta\dot{\varepsilon}_2> = \frac{q^2E^2}{2m}\left[\frac{\sin\Omega t}{\Omega} + \frac{kv_0}{\Omega^2}(\sin\Omega t - \Omega t\cos\Omega t)\right] \tag{70}$$

Taking the long time limit and using the well known formula regarding the Dirac delta function viz:

$$\delta(\Omega) = \lim_{t\to\infty}\frac{\sin\Omega t}{\pi\Omega} \tag{71}$$

We obtain

$$<\delta\dot{\varepsilon}_2> = \frac{\pi q^2E^2}{2m}\left[\delta(\Omega) - kv_0\frac{\partial}{\partial\Omega}\delta(\Omega)\right] \tag{72}$$

Now using the formula

$$\delta(\Omega) \equiv \delta(\omega - kv_0) = k^{-1}\delta(v_0 - \frac{\omega}{k})$$

we obtain

$$<\delta\dot{\varepsilon}_2> = \frac{\pi q^2 E^2}{2mk} \frac{\partial}{\partial v_0}\left[v_0 \delta\left(v_0 - \frac{\omega}{k} \right) \right] \tag{73}$$

Lastly, we average the rate of the energy change over a distribution of initial velocities, $f(v_0)$, we find:

$$\overline{<\delta\dot{\varepsilon}_2>} = \int dv_0 f(v_0) <\delta\dot{\varepsilon}_2> \tag{74}$$

where $f(v_0)$ is the velocity distribution function .Substituting for $<\delta\dot{\varepsilon}_2>$ from (73), we obtain

$$\overline{<\delta\dot{\varepsilon}_2>} = \frac{\pi q^2 E^2}{2m|k|} \cdot \frac{\omega}{k} \frac{\partial f_0}{\partial v}\bigg|_{v=\frac{\omega}{k}} \tag{75}$$

By energy conservation, the rate of change in the energy of the particles must be balanced by a growth or damping of the wave:

$$2\gamma \frac{E^2}{8\pi} + \overline{<\delta\dot{\varepsilon}_2>} = 0 \tag{76}$$

where γ is the rate at which the electric field grows or damps. Equation (76) gives

$$2\gamma \frac{E^2}{8\pi} = \frac{\pi q^2 E^2}{2m|k|} \cdot \frac{\omega}{k} \frac{\partial f}{\partial v}\bigg|_{v=\frac{\omega}{k}}$$

Specializing to the case of an electron plasma wave

$$\frac{\gamma}{\omega} = \frac{\pi}{2} \frac{\omega_{pe}^2}{k^2} \frac{\partial}{\partial v} \overline{f}\left(\frac{\omega}{k} \right) \tag{77}$$

where $f = n\overline{f}$.

For Maxwellian plasma

$$\frac{\gamma}{\omega} = -\sqrt{\frac{\pi}{8}} \frac{\omega_{pe}^2}{|k^3|v_{th}^3} \omega \exp\left(-\frac{\omega^2}{2k^2 V_{th}^2} \right) \tag{78}$$

Note that the Landau damping of an electron plasma wave is a strong function of its phase velocity. The damping becomes sizable whenever $\frac{\omega}{k} \lesssim 3\, v_{th}$,   where v_{th} is the electron thermal velocity .

or $\dfrac{\omega^2}{k^2} \lesssim 9\dfrac{k_B T}{m}$

or $\quad \dfrac{1}{k^2} \lesssim 9 \dfrac{v_{th}^2}{\omega_p{}^2} \qquad\qquad (\omega = \omega_p)$

$$\dfrac{1}{k^2} \lesssim 9\,\lambda_D^2$$

here λ_D is Debye length $\equiv \dfrac{v_{th}}{\omega_p}$. Thus Landau- damping becomes important

$$k\lambda_D \gtrsim \dfrac{1}{3} \tag{79}$$

Physical significance of Landau Damping :

We conclude our discussion of linear Landau damping with a simple mechanical analogy. Consider a group of boxes translating along at a velocity equal to $\dfrac{\omega}{k}$. Inside the boxes are uniformly-distributed particles, some moving slightly slower than $\dfrac{\omega}{k}$, some moving slightly faster,

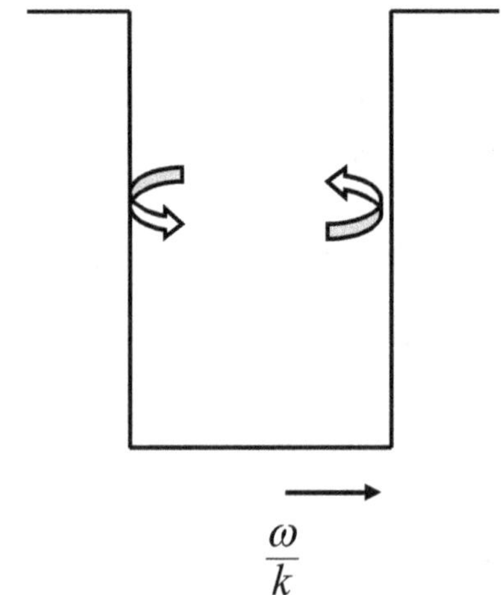

$$\dfrac{\omega}{k}$$

Figure :4

Those particle moving slower than $\dfrac{\omega}{k}$ are overtaken by the wall to their left and gain energy as they are bounced off. Likewise, those particles moving faster than $\dfrac{\omega}{k}$ overtake the right wall and lose energy as they are reflected. For a time less than the transit time of a particle through the box, the net energy change simply depends on whether more particles are initially moving faster or slower than $\dfrac{\omega}{k}$.

Ponderomotive force:

• When waves travel in a plasma, particularly when the wave amplitude is not homogeneous, the wave amplitude itself can effectively modify the plasma density profile through the "Ponderomotive force".

• One of the consequences of this ponderomotive force is that a localized wave can effectively expel plasma from the vicinity of the wave amplitude maximum, and this may tend to make the wave even more localized, leading to a through in the plasma that is called a "Cavtion".

• We give below a brief derivation of the ponderomotive force and its associated potential.

• In an inhomogeneous, high frequency field, the motion of an electron may be divided into two parts which describe the high frequency oscillation about an effective center of oscillation, and relatively slow motion of the oscillation center.

• If we take the electric field to be given by $E(x,t) = E_0(x)\cos\omega t$, and imagine that the amplitude is increasing in the positive x-direction, then as the electron moves into the stronger field, it will be accelerated more strongly back toward the oscillation center.

• On the other hand, as it moves toward the weaker field as x goes negative, it receives a weaker restoring force. On the average, then, it will experience a slow drift toward the weaker field as if under the influence of a steady or slowly varying force, while at the same time experiencing the rapid oscillation at the high frequency.

• In order to make this more quantitative, we examine the equation of motion of a charged particle in an inhomogeneous electric field:

$$m\ddot{\vec{r}} = qE_0(\vec{r})\cos\omega t \tag{80}$$

• We will separate this motion into a slow motion and a fast motion such that

$$\vec{r} = \vec{r}_0 + \vec{r}_1 \qquad \text{where} \quad <\vec{r}> = \vec{r}_0 \text{ is the average over the fast time scale, or}$$

over the period, $T = \dfrac{\partial \pi}{\omega}$. $\vec{r}_0$ hence describes the oscillation center.

$\vec{r}_1$ describes the rapidly oscillating motion, and is determined by

$$m\ddot{\vec{r}}_1 = q\vec{E}_0 \cos\omega t \tag{81}$$

where $\vec{E}_0 = \vec{E}_0(\vec{r}_0)$. The solution is simply

$$\vec{r}_1 = -\left(\frac{q\vec{E}_0}{m\omega^2}\right)\cos\omega t . \qquad (82)$$

• For the slow variation, we will expand $\vec{E}_0(\vec{r})$ about $(\vec{r}_0)$ such that the equation of motion becomes

$$m(\ddot{\vec{r}}_0 + \ddot{\vec{r}}_1) = q\left[\vec{E}_0 + (\vec{r}_1.\vec{\nabla})\vec{E}_0\right]\cos\omega t \qquad (83)$$

and we wish to average this over a period such that

$$m\ddot{\vec{r}}_0 = q <\vec{r}_1\cos\omega t >.\vec{\nabla}\ \vec{E}_0 \qquad (84)$$

Using (82) for $\vec{r}_1$, the average is simply $-\dfrac{q\vec{E}_0}{2m\omega^2}$, so (5)becomes

$$m\ddot{\vec{r}}_0 = -\frac{q^2}{2m\omega^2}\vec{E}_0.\vec{\nabla}\ \vec{E}_0 = -\frac{q^2}{4m\omega^2}\nabla(E_0^2) \qquad (85)$$

The ponderomotive force and its associate ponderomotive potential, then are given by

$$F_p = -\frac{q^2}{4m\omega^2}\nabla(E_0^2) = -\nabla\psi_p \qquad (86)$$

$$\psi_p = \frac{q^2}{4m\omega^2}E_0^2 \qquad (87)$$

If inhomogeneous magnetic fields are included, there is a drift of guiding center in addition to the motion from the ponderomotive force, but the ponderomotive force is unchanged.

10.5 Self Learning Exercise

Q.1. What is collisional damping? show that a wave damps at a rate $v = \dfrac{\omega_{pe}^2}{\omega^2}v_{ei}$, where v_{ei} represents electron-ion collision frequency.

Q.2. What is Landau damping? Explain its physical significance. Show that a plasma wave will get damp is the condition $k\lambda_D \gtrsim 1$ is satisfied.

10.6 Summary

In this chapter we have learnt that in the presence of a magnetic field the plasma becomes anisotropic. This is because of the presence of a magnetic field a preferred direction is provided to the plasma. Without the magnetic field, the plasma may be represented by a simple dielectric constant and the only wave solution is a simple electromagnetic wave that propagates above the plasma frequency ω_{pe} .In the presence of the magnetic field the direction of the plasma becomes a tensor. We have discussed E.M. wave propagation parallel to $\vec{B}_0$, $\theta = 0$ and also propagation perpendicular to $\vec{B}_0$ .Finally we have discussed Faraday rotation of the electric field.

10.7 Glossary

Ponderomotive force: A force experienced by a charged particle which is proportional to the gradient of the amplitude of the wave field .The ponderomotive force is independent of the sign of the charge.

Faraday Rotation: In case of parallel propagation of electromagnetic wave ,parallel to magnetic field , in a plasma the R wave has a greater phase velocity than the L-wave. As a result ,the plane of polarization is rotated after these waves emerge out of the plasma and combine.

Evanescent Wave: The wave unable to propagate through the plasma. The wave becomes damped or in other words evanescent.

10.8 Exercise

Q.1 What is ponderomotive force? Derive on expression for the ponderomotive force and ponderomotive potential. Give the physical significance of the ponderomotive force. Justify that the ponderomotive force leads to the expulsion of plasma from the high intensity region.

Q.2 What is a "caviton"?

Q.3 Derive an expression for the Landau damping rate of plasma wave:

$$\frac{\gamma}{\omega} = -\sqrt{\frac{\pi}{8}} \frac{\omega_{pe}^2 \, \omega}{|k| V_{th}^3} \, \exp\left(-\frac{\omega^2}{2k^2 V_{th}^2}\right)$$

Q.4 Explain the phenomenon of Faraday rotation of an electromagnetic wave where it passes through a magnetoactive media. How does it help in knowing the plasma density of a plasma medium.

Q.5 Explain the term "Cut-off" and "resonance". Derive the dispersion relation in the form

$$\tan^2\theta = -\frac{P(n^2 - R)\,(n^2 - L)}{(S\,n^2 - RL)\,(n^2 - P)}$$

where the symbols have their own meaning.

References and Suggested Readings

1. Physics of High temperature Plasmas by George Schmidt (Academic Press)

2. Fundamentals of Plasma Physics by J.A. Bittencourt (Springer)

3. Plasma Waves by D. G. Swanson (Academic Press)

4. Plasma Physics by F.F. Chen

UNIT-11
Beam- Plasma Interaction and
Parametric Instabilities

Structure of the Unit

11.0 Objectives

To learn

(a) The two-stream instability.

(b) Growth rate of the instability of the beam- plasma system.

(c) Common characteristics of parametric instabilities

(d) Modulated harmonic oscillator model

(e) Grow Rates and thresholds

(f) Excitation of Coupled –mode oscillations

11.1 Introduction

In this chapter we shall discuss a typical instance of instability that is afforded by a directed beam of electrons passing through a plasma at rest (A.I. Abhiezer and Ya.B. Fainberg 1949, D. Bohm and E. P. Gross 1949). In plasmas near thermodynamic equilibria, the collective modes are stable elementary excitations. Those excitations suffer damping through resonant interactions with individual particles; their lifetimes are finite.

However, When a plasma is away from thermodynamic equilibrium, collective modes may become unstable; amplitude of such an excitation tends to grow exponentially. Most of the plasmas in nature are significantly far from thermodynamic equilibrium. A current carrying plasma, for example, involves a relative drift motion between the electrons and ions. A beam-plasma system is unstable. In this chapter, we investigate the conditions under which the beam-plasma system becomes unstable. The physical mechanism responsible for the onset of plasma instability may be thought in the excess of free energy due to a departure from thermodynamic equilibrium. Beam-plasma system is an example of a situation in which wave-particle interaction leads to a growing wave amplitude, at the expense of the kinetic energy of the plasma particles, we consider here the so- called two- stream instability.

Parametric instabilities are not unique to plasmas, as they relate to any oscillatory system where one of the parameters (hence the name "parametric") is modulated at an appropriate frequency.

11.2 What is a beam ?

Here we consider plasmas whose velocity distributors deviate strongly from a Maxwellian. In particular, we consider beam-plasma systems, where the distribution in the velocity of one of the plasma species- for example, the electrons- has two district components. These are a background, Maxwellian component, and an energetic component whose velocity in a particular direction greatly exceeds that of the background. The energetic component is referred to as "beam", and it represents a source of free energy.

11.3 Dispersion Relation of Beam Plasma System

In order to examine how the free energy of streaming electrons (i.e. beam of electrons) may be released, we investigate the stability of the beam-plasma system. This requires the calculation of its response to small perturbation.

For simplicity, we assume that the difference in velocity between the beam electrons and the background electrons greatly exceeds the spread in velocity of either population. That is, we consider a beam-plasma system which, to leading order, is cold. In this approximation, the initial electron velocity distribution function $f_0(v)$ is Zero everywhere except at $\vec{v} = 0$ or at $v_z = v_o$, with v_x and v_y negligibly where v_o is the velocity of the beam, whose direction we choose to define the z- axis.

- We shall consider electrostatic perturbations, and choose the wave $\vec{k}$ vector to lie along the z-axis, parallel to the beam.

- Physically the suppression of k_x and k_y does not matter. The propagation of electrostatic waves in these directions is governed by the distribution of electron velocities in the x- and y directions. These distributions have no beam component, so that the cold plasma treatment will apply. Thus, we restrict attention to the only component of $\vec{k}$ for which the presence of the beam is significant.

- We denote the fraction of electrons in the beam by ξ, So that the remaining fraction $(1 - \xi)$ forms the background plasma.

- The beam is assumed to be electrically compensated: the sum of the electron charge densities in the plasma and the beam is equal to the ion charge density in the plasma. The system is homogeneous and unbounded, i.e. both the beam and the plasma extend throughout space, and the directed velocity v_o of the beam is everywhere the same. We shall assume that v_0 is non-relativistic.

- In the electron oscillation frequency range, the longitudinal permittivity of the plasma-beam system has the form

$$\varepsilon(k,\omega) - 1 = -(1-\xi)\frac{\omega_{pe}^2}{\omega^2} - \frac{\xi\,\omega_{pe}^2}{\left(\omega - kv_0\right)^2} \tag{1}$$

The first term on the right corresponds to the plasma at rest; the second

Term is due to the beam electrons. In a frame of reference K' moving with the

beam, the contribution of the beam electrons to $\xi - 1$ is $-\left(\dfrac{\omega_{pe}'}{\omega'}\right)^2$ where ω' is the

oscillation frequency in that frame, and $\omega_{pe}'^2 = \xi\omega_{pe}^2$.

On return to the original frame K, the frequency ω' is replaced by

$$\therefore \omega' = \omega - kv_0 \tag{2}$$

- The spectrum of the longitudinal oscillations of the cold beam-plasma system is given by $\varepsilon(k,\omega) = 0$, i.e.

$$\varepsilon(k,\omega) = 1 - (1-\xi)\frac{\omega_{pe}^2}{\omega^2} - \frac{\xi\omega_{pe}^2}{\left(\omega - kv_0\right)^2} = 0 \tag{3}$$

- In the limit where the number of electrons in the beam tends to Zero, ξ vanishes

and eqn.(3) reduces to $1 - \dfrac{\omega_{pe}^2}{\omega^2} = 0$. We note also that $\omega - kv_0$ is the Doppler –

shifted wave frequency that is experienced in the rest frame of the beam electrons. It is clear from (3) that we may regard the background and beam populations as distinct plasmas, both with their own plasma frequency proportional to their density. We define

$$\omega_{po}^2 = (1-\xi)\omega_{pe}^2 \tag{4}$$

$$\omega_{pb}^2 = \xi\omega_{pe}^2 \tag{5}$$

Then eqn. (3) can be written as

$$\varepsilon(k,\omega) = 1 - \frac{\omega_{po}^2}{\omega^2} - \frac{\omega_{pb}^2}{\omega^2} = 0 \tag{6}$$

- This dispersion relation is approxianately satisfied under two conditions.

$$\text{Either } \omega^2 \approx \omega_{po}^2 \tag{7}$$

So that $\dfrac{\omega_{po}^2}{\omega^2} \approx 1$, cancelling the first term in the expression for $\varepsilon(k,\omega)$:

$$\text{or } (\omega - kv_0)^2 \approx \omega_{pb}^2 \tag{8}$$

and eqn. (6) is again approximately satisfied. Thus, both the background and the beam components of the plasma support families of electrostatic waves, given by eqn. (7) and (8) respectively.

- We note that we may write eqn. (6) in the form

$$(\omega^2 - \omega_{po}^2)[(\omega - k\,v_0)^2 - \omega_{pb}^2] = \omega_{pb}^2 \omega_{po}^2 \tag{9}$$

Using the binominal theorem, this becomes

$$(\omega - \omega_{po})(\omega + \omega_{po})(\omega - kv_o - \omega_{pb})(\omega - kv_o + \omega_{pb}) = \omega_{pb}^2 \omega_{po}^2 \tag{10}$$

This is an alternative way of displaying the approximate roots of eqn. (6).

- In general, the beam population is small, so that

$$\xi << 1 \tag{11}$$

Then the right hand side of eqn. (10) is small compared to ω_{po}^4, This restricts the magnitude of the left-hand ride, so that at least one among the four factors must also be small.

$$\text{Either} \qquad \omega \approx \pm \omega_{pb} \tag{12}$$

Corresponding to eqn. (7); or

$$\omega \approx kv_o \pm \omega_{pe} \tag{13}$$

corresponding to eqn. (8).

11.4 Growth Rate of Beam- Plasma Instability

- Now let us examine what happens when the two families of waves (described by eqns. (12) and (13)) overlap in frequency.

- The condition for frequency overlap can be expressed in terms of the critical wave number k_c, defined by

$$k_c = \frac{\omega_{po}}{v_o} \tag{14}$$

- When $k = k_c$, eqn (13) gives $\omega \simeq \omega_{po} \pm \omega_{pb}$ for the beam –supported waves. But $\omega_{pb}^2 = \xi\omega_{po}^2$ and $\xi \ll 1$ thus, ω_{pb} is negligibly small in comparison to ω_{po}, hence when $k = k_c$, $\omega \approx \omega_{po}$, which is the characteristics frequency of background- supported waves.

Thus, $\quad k \simeq k_c,$ \tag{15}

is the condition for frequency resonances to be possible between waves form the two families.

- We now seek solutions of eqn. (9) which have the form

$$\omega = \omega_{po} + \eta \ , \ |\eta| << \omega_{po} \tag{16}$$

For the case $k = k_c$, substitution of eqns (16) and (14) into eqn. (9) yields:

$$(\omega^2 - \omega_{po}^2)\{(\omega - kv_o)^2 - \omega_{pb}^2\} = \omega_{pb}^2\omega_{po}^2$$

or $\quad (\omega - \omega_{po})\{\omega + \omega_{po}\}\{(\omega_{po} + \eta - kv_o)^2 - \omega_{pb}^2\} = \omega_{pb}^2\omega_{po}^2$

or $\quad \eta(2\omega_{po})\{(\omega_{po} + \eta - kv_o + \omega_{pb})(\omega_{po} + \eta - kv_o - \omega_{pb})\} = \omega_{pb}^2\omega_{pb}^2$

or $\quad \eta(2\omega_{po})\{(\eta + \omega_{pb})(\eta - \omega_{pb})\} = \omega_{pb}^2\omega_{po}^2$

where we have used $\omega_{po} - kv_o = 0$

finally we obtain

$$\eta^3 - \omega_{pb}^2\eta = \frac{\omega_{pb}^2\omega_{po}}{2} \tag{17}$$

If we assume $|\eta| >> \omega_{pb}$ \tag{18}

- Which we shall check subsequently for consistency, we may neglect $\omega_{pb}^2\eta$ compared to η^3 in eqn (17) leaving

$$\eta^3 = \frac{\omega_{pb}^2\omega_{po}}{2} = \frac{\xi\omega_{pe}^3}{2} \tag{19}$$

- Now unity has three complex cube roots $\left\{1, \exp\left(\frac{2i\pi}{3}\right), \exp\left(\frac{4i\pi}{3}\right)\right\}$, and eqn (19) has three corresponding roots. These are

$$\eta_o = \left(\frac{\xi}{2}\right)^{\frac{1}{3}} \omega_{pe} \tag{20}$$

$$\eta_1 = \eta_0 \left(-\frac{1}{2} + \frac{i\sqrt{3}}{2}\right) \tag{21}$$

$$\eta_2 = \eta_0 \left(-\frac{1}{2} - \frac{i\sqrt{3}}{2}\right) \tag{22}$$

- As we know that the amplitude of a perturbation with wave vector $\vec{k}$ in a homogeneous unbounded medium has the asymptotic from (as $t \to \infty$) $e^{-i\omega(\vec{k})t}$, where $\omega(\vec{k})$ is the frequency of waves propagating in the medium. $\omega(\vec{k})$ are the roots of the equation $\varepsilon(\omega, \vec{k}) = 0$.

  The frequency $\omega(\vec{k})$ are in general complex. If the imaginary part in $\omega < 0$, the perturbation is damped in the course of the time. If, however in $\omega > 0$, in some range of $\vec{k}$, such perturbations grow: the medium is unstable an $|Im\,\omega|$ is then called the 'instability growth rate'.

- Now combining eqns (20) to (22) with (16), the first root has a real frequency. The effect of the linear frequency for this root is to introduce a small frequency shift η_0 with respect to the background plasma frequency ω_{po}.

  The second and third roots include an imaginary term, which we write as $\pm i\gamma$ where

$$\gamma = \left(\frac{\sqrt{3}}{2}\right)\eta_0 = \frac{\sqrt{3}}{2^{\frac{4}{3}}} \xi^{\frac{1}{3}} \omega_{pe} \tag{23}$$

- Following the earlier discussion, the root $\omega - \omega_{po} + \eta_2$ grows exponentially at the rate γ.

  It may be noted that at resonance the waves supported by the background and beam plasmas are indistinguishable, because their wave number k_c and frequency $\omega_{po} + \eta_2$ are identical.

- We have shown that this collective beam-plasma electrostatic oscillation can grow exponentially.

- Let us briefly confirm the consistency of our results, eqns (20) to (22) . First, using eqn (11) it is clear that $|\eta| << \omega_{po}$ as required at eqn (16) . Second, using (11) and (5), we are indeed consisted with (18).

- The phenomenon that we have identified is known as the "two-stream instability", the two streams being the beam $(v_z = v_o)$ and the background plasma $(v_z = o)$.

   It is a mechanism by which the beam and background components of the plasma may interact, mediated by resonance between the collective oscillations that they support.

- The instability rests on linear frequency resonance which occurs when

   $$\text{or} \quad k = k_c \ \text{or} \ k = \frac{\omega_{po}}{v_o} \ \text{or} \ \omega_{po} = k\,v_o .$$

   It results in the growth of the average electrostatic field energy that is associated with the resonant oscillation.

   There is only one possible source for this energy: the free kinetic associated with the directed motion of the beam electrons.

11.5 Illustrative Examples

Example 1. Show that the oscillation frequency of the background plasma ω' as seen by the beam electrons is given by

$\omega' = \omega - \vec{k}.\vec{V}$, where $\vec{V}$ is the velocity of the beam electrons?

Solution: The law of transformation for the frequency is easily found by transforming the phase factor of the wave. The phase factor of the wave is

$$\vec{k}.\vec{r}. - \omega t \tag{i}$$

Now the position vector $\vec{r}$ in the beam frame is given by

$$\vec{r}' = \vec{r} - \vec{V}t \tag{ii}$$

Substituting (ii) into (i), we obtain phase as seen by beam:

$$Phase = \vec{k}.(\vec{r}' + \vec{V}t) - \omega t$$
$$= \vec{k}.\vec{r}' - (\omega - \vec{k}.\vec{V})t$$

$$= \vec{k}.\vec{r}\,' - \omega' t$$

Thus $\omega' = \omega - \vec{k}.\vec{V}$

11.6 Introduction to Parametric Instabilities

1. As an example, consider a simple system of a child's swing, whose period depends on the length of the swing from its support, and the parametric instability occurs when the child lengthens or shortens the effective length twice each period.

2. This modulation at twice the natural frequency leads to growth of the fundamental oscillation, through what it called "pumping".

3. Quite generally, parametric excitations require a minimal set of common characteristics.

11.7 Common Characteristics of Parametric Instabilities

Parametric excitations satisfy the following characteristics in general.

(A) Matching condition: The modulation and the natural oscillation should satisfy a phase matching condition, such as $\omega T = n\lambda$ $n = 1,2.................$ where ω is the natural frequency and T is the period of the modulation. The example above has n=1.

(B) Threshold: Instability or amplification occurs only when the amplitude of the modulation exceeds a critical value.

(C) Frequency Locking: The frequencies of the amplified oscillations are determined by the modulation frequency rather than the natural frequency. For the examples listed above for n=1, amplification is at frequency $\dfrac{\pi}{T}$ (the natural frequency), but for n=2, amplification occurs at $\dfrac{2\pi}{T}$ (the natural frequency again) and at zero frequency.

- The matching and frequency locking conditions follow form the intrinsic nonlinearity of the multiple frequency system, and can be viewed either as coming from.

Conservation of energy and momentum or as coming from resonance conditions such that

$$\omega_o = \omega_i + \omega_s$$
$$k_o = k_i + k_s$$

where the subscripts o, i ,s stand for pump, idler and signal, respectively.

- The instability occurs when the pump exceeds a certain threshold and the idler and signal waves grow.

- From the nonlinear nature of the coupling, it is apparent that energy can be drawn from the pump wave and diverted to the idler and signal waves, or daughter waves.

  This process can become so efficient that as a pump wave propagates, it loses energy to the daughter waves. Until it is depleted to the extent, it falls below threshold.

- Among the various effects which limit this efficiency of coupling are: finite wavelength effects, where the phase matching conditions cannot be satisfied everywhere.

  In the following, we examine a few basic models that illustrate the fundamental phenomena.

11.8 Modulated Harmonic Oscillator Model

We consider first a damped oscillator described by

$$\frac{d^2x}{dt^2} + 2\gamma_o \frac{dx}{dt} + (\Omega^2 + \gamma_o^2)x = 0 \tag{24}$$

which has the simple solution

$x = A\exp(-i\Omega t - \gamma_0 t)$ if Ω and γ_0 are constants and represent the frequency and damping rate of the oscillator.

- If, however, we take the frequency to be modulated at frequency ω_o, such that

$$\Omega^2 = \Omega_o^2(1 - 2\varepsilon \cos\omega_o t) \tag{25}$$

where Ω_0 is the natural frequency and ω_o and ε the pump frequency and amplitude of the modulation, respectively then the transformation $x(t) = e^{-\gamma_o t} y(t)$ brings (24) into the form

$$\frac{d^2 y}{dt^2} + \Omega_0{}^2 (1 - 2\varepsilon \cos \omega_o t) y = 0 \tag{26}$$

This equation is known as the "Mathieu equation".

- All of the characteristics of the parametric instability are contained in the properties of Mathieu functions, but since these are neither trivial nor commonly known, it is more instructive to examine the properties of Eq (26) by a perturbational analysis assuming that the damping decrement $\dfrac{\gamma_0}{\Omega_0}$ and modulational amplitude ε are both small.

Taking the Fourier transform of (24) leads to

$$D(\omega)\tilde{x}(\omega) = \varepsilon \Omega_0^2 [\tilde{x}(\omega + \omega_0) + \tilde{x}(\omega - \omega_0) \tag{27}$$

where $D(\omega) \equiv -\omega^2 - 2i\omega\gamma_0 + \Omega_0^2 + \gamma_0^2$

- We examine two special cases:

Case I: $\omega_0 \simeq 2\Omega_0$

If we choose $\omega_0 \simeq \Omega_0$, then first term on the right of (27) represents the response at $\simeq 3\Omega_0$, so is far form resonant. The second term on the right gives a response at frequency $\omega - \omega_0 \simeq -\Omega_0$, so this term is nearly resonant. Keeping only the resonant term, we need $\tilde{x}(\omega - \omega_0)$ which we obtain form (27) to be

$$D(\omega - \omega_0)\tilde{x}(\omega - \omega_0) \simeq \varepsilon \Omega_0^2 \, \tilde{x}(\omega) \tag{28}$$

Then the dispersion relation becomes

$$D(\omega)D(\omega - \omega_0) = \varepsilon^2 \Omega^4 \tag{29}$$

Making simple resonant approximations for the $D(\omega)$,

$$D(\omega) = -(\omega + \Omega_0 + i\gamma_0)(\omega - \Omega_0 + i\gamma_0)$$
$$D(\omega) \simeq -2\Omega_0(\omega - \Omega_0 + i\gamma_0)$$
$$D(\omega - \omega_0) \simeq 2\Omega_0.(\omega - \omega_0 + \Omega_0 + i\gamma_0)$$

- The dispersion relation may be written

$$(\omega - \Omega_0 + i\gamma_0)(\omega - \Omega_0 - \Delta + i\gamma_0) + \frac{1}{4}\varepsilon\Omega_0^2 = 0 \tag{30}$$

where $\Delta \equiv \omega_0 - 2\Omega_0$ is the frequency mismatch.

Then using the definition

$\omega \equiv \Omega_\circ + \delta + i\gamma$, where δ is the real frequency shift and γ is the growth rate, separating the real and imaginary parts of (30) results in

$$\delta(\delta - \Delta) - (\gamma + \gamma_0)^2 + \frac{1}{4}\varepsilon\Omega_0^2 = 0 \tag{31}$$

$$(2\delta - \Delta)(\gamma + \gamma_0) = 0, \tag{32}$$

So there are two types of solutions form(32) :

(i) Damped solution: $\gamma = -\gamma_0$,this describes damped oscillations with frequencies given by

$$\delta = \frac{1}{2}(\Delta \pm \sqrt{\Delta^2 - \varepsilon^2\Omega_0^2})$$

which requires $\Delta^2 > \varepsilon^2\Omega_0^2$ or that the frequency mismatch be sufficiently large or the modulation amplitude be sufficiently small.

(ii) Locked solution: $\delta = \dfrac{\Delta}{2}$, this describes frequency locked oscillations

since $\mathrm{Re}\ \ \mathrm{Re}\,\omega = \delta + \omega_0 = \dfrac{\omega_0}{2}$, so that the frequency is independent of the natural frequency. Then from (31), this root gives

$$\gamma = -\gamma_0 \pm \frac{1}{2}\sqrt{\varepsilon^2\Omega_0^2 - \Delta^2} \tag{33}$$

which is complementary to the other case in the sense that this case requires

$$\Delta^2 \leq \varepsilon^2\Omega_0^2,$$

The more weakly damped root of Eq. (33) becomes unstable when

$$\varepsilon^2 > \frac{\left(\Delta^2 + 4\gamma_0^2\right)}{\Omega_0^2} \tag{34}$$

• So this is the parametric instability and (34) gives the threshold as a function of the mismatch.

The maximum growth rate occurs when $\Delta = 0$, where the minimum threshold and maximum growth rate are given by

$$\varepsilon_{\min} = \frac{2\gamma_0}{\Omega_0} \tag{35}$$

$$\gamma_{\min} = -\gamma_0 + \frac{\varepsilon}{2\Omega_0} \tag{36}$$

We note that in the limit as $\gamma_0 \to 0$, there is no threshold, so an infinitesimal excitation can drive the instability .

Case II: $\qquad \omega_0 \approx \Omega_0$

For this case, we expect the coupling near $\omega \sim 0$, which is the difference frequency. The coupling from the symmetric forms of Eq. (28),

$$D(\omega \pm \omega_0)\tilde{x}(\omega \pm \omega_0) \simeq \varepsilon \Omega_0^2 \, \tilde{x}(\omega)$$

leads to the dispersion relation

$$1 = \frac{\varepsilon^2 \Omega_0^2}{D(\omega)}\left[\frac{1}{D(\omega + \omega_0)} + \frac{1}{D(\omega - \omega_0)} \right] \tag{37}$$

Using the approximation $D(\omega \pm \omega_0) \simeq 2\Omega_\circ(\omega \pm \delta + i\gamma_0)$,

where the frequency mismatch is given in this case by $\Delta = \omega_0 - \Omega_0$, and approximating $D(\omega)$ by D(0), Eq. (37) simplifies to

$$1 = \frac{\varepsilon^2 \Omega_0^2}{2}\left[\frac{1}{\omega - \Delta + i\gamma_o} + \frac{1}{\omega + \Delta + i\gamma_o} \right] \tag{38}$$

Separating this into real and imaginary parts, with $\omega \equiv \omega_r + i\gamma$

$$\omega_r^2 - \Delta^2 - (\gamma + \gamma_0)^2 = \varepsilon^2 \Omega_0 \Delta \tag{39}$$

$$\omega_r(\gamma + \gamma_0) = 0 \tag{40}$$

Thus, comparing these several results, we find that each case has a damped solution and a locked solution, and there is a threshold for the instability. In case I, the threshold is lower and the growth rate is higher than the corresponding case II.

11.9 Stimulated Brillouin Scattering (SBS) and Stimulated Raman Scattering (SRS)

• We have seen that two large waves can drive other waves at the beat frequency and wave number. If the beat wave is an eigenmode of the system, it will be driven to large amplitude.

In this case it is not necessary that both driving waves be large to obtain coupling.

- The matching conditions (say $\omega_0 = \omega_1 + \omega_2$, $k_0 = k_1 + k_2$) require that the three waves, represented a vectors in (ω,k) space, satisfy the rules of vector addition.

- The requirement that the three waves be eigenmodes implies that the end points of these vectors be situated on dispersion curves.

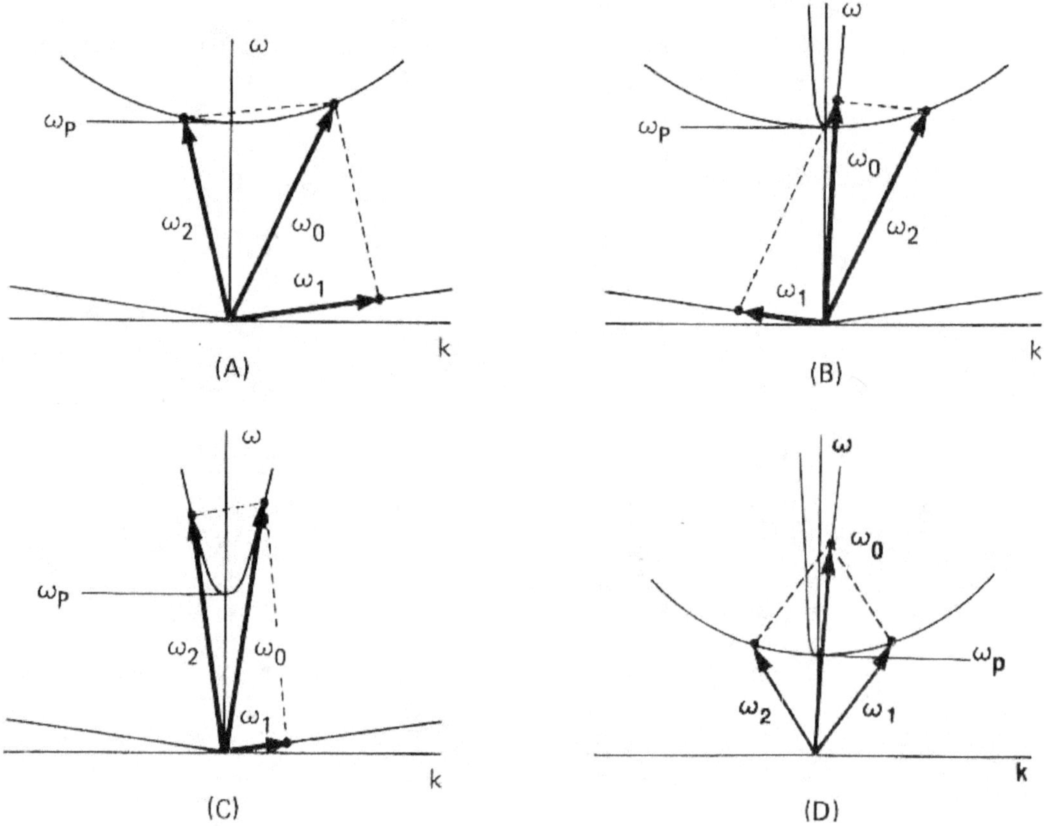

Fig.1(A) *Parallelogram constructions showing the $\omega-$ and $k-matching$ conditions for electron decay instability.*

(B) Parametric decay instability of an electromagnetic wave into a plasma wave and ion sound wave.

(C)Decay of an electromagnetic wave into an ion-acoustic wave and a back scattered electromagnetic wave(stimulated Brillouin back scattering).

(D)Two Plasmon decay instability.

- An example for collinear waves in an unmagnetized plasma is shown is Fig.1.

  Here the pump wave is an electromagnetic wave, which decays into an ion acoustic wave and an other backward propagating electromagnetic wave.

  This process is called "stimulated Brillouin backscatter".

- A similar constructive can be made for an electromagnetic wave decaying into an electron plasma wave and a backward propagating electromagnetic wave, called " Stimulated Raman backscatter".

  From the constructions it is easy to see that an electromagnetic wave can not decay into two other electromagnetic waves, nor can an electron wave decay into two other electron waves.

- An ion wave, however, if it is on the linear portion of the dispersion curve $\left(k << \dfrac{\omega_{pi}}{c} \right)$ may decay into two ion waves.

11.10 Physical Interpretation of the Backscatter Instabilities

- The backscatter instabilities have a simple physical interpretation. Consider a light with wavelength λ_0 impinging on plasma with an electron density varying with a wavelength $\lambda = \dfrac{1}{2}\lambda_0$.

- This density perturbation forms a one-dimensional lattice, resulting in a partial reflection of the electromagnetic wave.

  The reflected wave and incoming wave produce a standing wave component, whose intensity maxima are located at the minima of the electron density wave.

- We know that the electrons will be expelled by the poderomotive force from regions where the field intensity is large. This enhances the density perturbation, leading to more backscattering and larger standing wave amplitudes and so on.

- The effect is most pronounced when density perturbation corresponds to an eigenmode, viz, an electron or ion wave.

  This is in agreement with Fig. 1, where the ion wave has roughly half the wavelength (twice the wave number) of the pump wave and the frequency of the backscattered wave $\omega_1 = \omega_0 - c_s k_2$ due to the Doppler shift.An electron plasma wave can backscatter on ion waves in the same manner.

11.11 Self Learning Exercise

Q.1. Derive dispersion relation of a beam-plasma system. Derive an expression for the growth rate of the beam- plasma instability.

Q.2. Show that the beam-plasma dispersion relation can be written in the form.

$$(\omega^2 - \omega^2_{po}) \; [(\omega - kv_0)^2 - \omega^2_{pb}] = \omega^2_{pb}\omega^2_{po}$$

Show that the roots of the dispersion relation are

$$\omega \approx \pm\omega_{pe} \text{ and } \omega \approx k\, v_0 \pm \omega_{pe}$$

Q.3. What are parametric instabilities?

11.12 Summary

Beam-plasma instability is the main mechanism of energy transfer of the beam into electromagnetic modes of the plasma. Almost all sources of electromagnetic radiations employ this mechanism. For the generation of high frequency radiations. We use an electron beam whose electrons are moving at relativistic speed. Important sources of electromagnetic waves e.g. Free electron lasers, gyrations, backward wave oscillators etc. use relativistic electron beams propagating through plasma filled cavities.

In this chapter we have seen that two large waves can derive other waves at the beat frequency and wavenumber .If the beat wave (i.e. a wave at different frequency) is an eigenmode of the system ,it will be driven to large amplitude.

The matching conditions (say $\omega_0 = \omega_i + \omega_S, k_0 = k_L + k_S$),require that the three waves ,when represented as vectors in (ω, k) space ,satisfy the rules of vector addition. The requirement that the three waves be eigenmodes implies that the end points of these vectors be situated on dispersion curves.An example for collinear waves in an unmagnetized plasma is shown in figure1.Here the pump wave is an electromagnetic wave,which decays into an ion acoustic wave and an other backward propagating electromagnetic wave.This process is called "stimulated Brillouin backscatter".A similar construction can be made for an electromagnetic wave decaying into an electron plasma wave and a backward propagating electromagnetic wave,called "stimulated Raman Backscatter".It may be noted that in the presence of damping , a finite threshold pump intensity is required for the onset of these instabilities.

When combination frequency is an eigenmode(or close to it),excitation is strong.If a strong pump wave interacts with two small eigenmodes such that (ω,k) matching conditions are satisfied ,exponential growth of the small modes can result.We have explained these fundamental phenomena on the basis of modulated harmonic oscillator model.

11.13 Glossary

Parametric instabilities : they are not unique to plasmas, as they relate to any oscillatory system where one of the parameters (hence the name "parametric") is modulated at an appropriate frequency.

F.E.L. : Free electron lasers

Stimulated Brillouin Scattering (SBS) : A pump wave ,which is an electromagnetic wave decays into an ion acoustic wave and an other backward propagating electromagnetic wave.

Stimulated Raman Scattering (SRS) : An electromagnetic wave decays into an electron plasma wave and a backward propagating electromagnetic wave.

Theshold of instability : Instability occurs only when the amplitude of the pump wave exceeds a critical value.

11.14 Answers to Self Learning Exercise

Ans.3: Parametric instabilities are related to any oscillatory system where one of the parameters of the system is modulated at an appropriate frequency. In such a case the system becomes unstable.

11.15 Exercise

Q.1. What are the common characteristics of parametric instabilities?

Q.2 Explain stimulated Brillouin scattering.

Q.3 Explain stimulated Raman scattering

Q.4 What is the source of energy for the growth of the beam –plasma instability. Give physical explanation.

11.16 Answers to Exercise

Ans.1: Common characteristics are:

(i) Matching condition: the modulation and the natural oscillation should satisfy a phase matching condition: $\omega T = n\lambda, \quad n = 1, 2, \ldots\ldots\ldots,$

(ii) Threshold: The instability or amplification occurs only when the amplitude of modulation exceeds a critical value.

(iii) Frequency Locking :

$$\omega_0 = \omega_2 + \omega_s$$
$$\vec{k}_0 = \vec{k}_i + \vec{k}_s$$

Ans.2: Decay of a photon into an scattered photon and an ion acoustic wave.

Photon $\rightarrow$ photon + ion acoustic wave (phonon)

Ans.3: Decay of a photon into an scattered photon + plasmon :

Photon $\rightarrow$ photon + plasmon.

References and Suggested Readings

1. Fundamentals of Plasma Physics by J.A. Bettencourt

2. Theory of Plasma waves by T.H. Stix

UNIT-12
Introduction to Laser Operation

Structure of the Unit

12.0 Objectives

We know that the natural radioactive delay process is inherent in all excited states of all materials and is referred to as spontaneous emission. However, such emission is not always the dominant decay process. Excitation or de-excitation can also occur by the photons (light Particles) that have specific energies. The phenomenon of light producing excitation is called absorption. This process is also known as stimulated absorption since it requires electromagnetic energy to stimulate the electron and thereby produce the excitation. The inverse of this process was never considered by anyone until Einstein suggested the concept of stimulated emission in 1917. All the lasers are based on the amplification of light by means of stimulated radiation of atoms or molecules. In this unit, we will present an introduction to laser operation, lasers and laser light. We will also discuss light emission and absorption in quantum theory, Einstein Theory of light-matter interaction, stimulated absorption and emission rates.

12.1 Introduction

The word LASER is stands for Light Amplification by Stimulated Emission of Radiation. Each laser contains material capable of amplifying radiation. This material is called the Gain or active medium because radiation gains energy passing through it. The physical principle responsible for this amplification is called stimulated emission and it was discovered by Albert Einstein in 1917. Lasers produce intense beams of light which are monochromatic, coherent and highly collimated. In 1952, Charles Townes, J. Gordon and H. Zeiger in U.S.A. and Nikolai Basov and Aleksander Prokhorov in USSR, independently suggested the principle of generating and amplifying microwave oscillation based on the concept of stimulated radiation. It leads to the invention of MASER (Microwave Amplification by Stimulated Emission of Radiation) in 1954. MASERS used a two level system. In 1955, Basov and Prokhonov suggested use of three level system. In 1958 C. Townes and A.L. Schawlow, and N. Basov and A. Prokhorov independently expressed their ideas about extending the maser concept to optical frequencies. They developed the concept of an optical amplifier surrounded by an optical mirror resonant cavity to allow for growth of the light beam. The 1964 Nobel Prize in Physics was awarded to N. Basov and A. Prokhorov with C. Townes for their pioneering work the field of lasers and masers. A.L. Shawlow was also awarded a share of the 1981 Noble Prize in Physics for his research work on lasers.

The first successful laser device was built by T.H. Maiman of Hughes Research laboratories in 1960 using a synthetic ruby rod. The helical flash lamp surrounded a rod shaped ruby crystal and the optical cavity was formed by coating the flattened ends of the ruby rod with a highly reflecting material. An intense red beam was observed to emerge from the end of the rod when the flash lamp was fired. In 1961, the first gas laser was developed by A. Javan, W. Bennett and D. Harriott of Bell Laboratories using a mixture o helium and neon gases. At the same laboratories, I.F. Johnson and K. Nassau demonstrated the first neodymium laser, which has become one of the most reliable lasers available. In 1962 R. Hall demonstrated first semiconductor laser at the General Electric Research Laboratories.

12.2 Introduction to Laser Operation

There are three basic elements of laser operation.

(i) Lasing or amplifying material amplifies a light signal directed through (Crystal, gas, Semi conductor, dye etc.

(ii) Pump source to add energy to the lasing material.

(iii) Optical or resonator cavity consisting of reflectors to act as the feedback mechanism for light amplification.

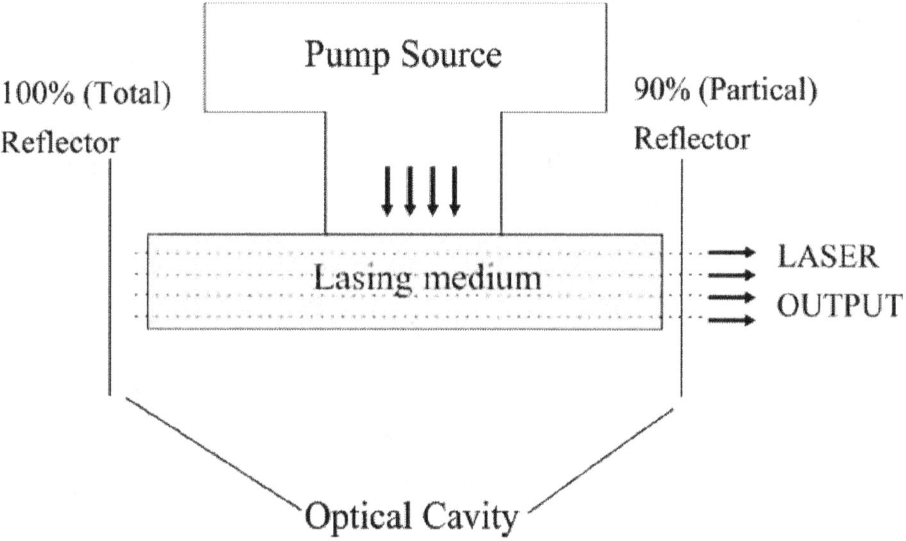

Fig. 12.1 *Schematic diagram of a basic laser*

The basic theory of laser action is as follows: The lasing medium is pumped continuous through pumping source to create a population inversion *i.e.* large number of atoms in the higher energy level than the lower energy level. As the excited atoms start to decay, they emit photons spontaneously in all directions. Some of the photons travel along the axis of the lasing medium, but most of the photons are directed out the sides. The photons traveling along the axis have an opportunity to simulate atoms when they encounter to them and emit photons. These simulated photons travel in phase at the same wavelength and in the same direction as incident photon. If the direction is parallel to the optical axis, the emitted photons travel back and forth in the optical cavity through the lasing material between the totally reflecting mirror and the partially reflecting mirror. The light energy is amplified in this manner until sufficient energy is built up for a burst of laser light to be transmitted through the partially reflecting mirror. The important requirement to the laser action is that there should be more atoms in the excited state than in the lower (ground) state. This phenomenon is known as population inversion. To occur population inversion, a lasing medium must have a longer lived state known as meta stable state. The average lifetime before spontaneous emission occurs for a meta-stable state is in the order of a 10^{-3} second, which is quite lengthy period of time on the atomic timescale (10^{-9} sec) as shown in fig. 12.2.

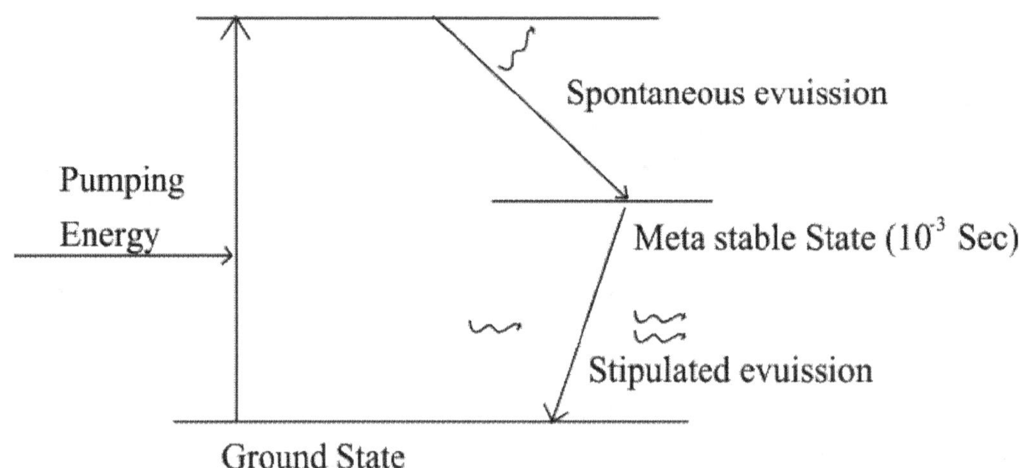

Fig. 12.2 *Three level laser energy diagram*

In this case, excited atoms can produce significant amounts of stimulated emission. The most common approach for producing a population inversion in a lasing medium is to add energy to the system in order to excite atoms into higher energy levels. The ratio of the number of atoms at two energy levels (1 and 2) under thermodynamic equilibrium is given by the following equation:

$$N_2/N_1 = \exp\left[-\left(E_2 - E_1\right)/kT\right]$$

where N_1 and N_2 are the number of atoms in level 1 and 2, E_1 and E_2 are the energies of the two levels, k is Boltzmann constant and T is the temperature in Kelvin.

Resonator Cavity and Shaping of a Beam (Operation of Laser)

In a laser, the active system is enclosed in an optical cavity, usually made in the form of a rod or a tube. The ends of the laser material are closed by accurately parallel reflectors, one of which is perfect reflector and the other has 90% reflectivity (or 10% transmission). The cavity makes the energy density large, when stimulated emission dominates over the spontaneous emission. The multiple reflection also makes the stimulated emission more coherent. This process is called shaping of the beam. Fig 12.3 schematically shows how the beam shapes with time. The hollow circles here represent the unexcited atoms and black dots represent the excited atoms. The left end mirror represents the full reflector and the right end mirror the partial reflector. The description stage wise is as follows

(1) Almost all atoms are initially in the unexcited state. An outside agent (such as an electrical discharge for example) excites atoms inside the laser as shown in figure 12.3 (a).

(2) When pumping radiation falls on the active material, the atoms start reaching the excited state. (Population inversion not yet reached, black dots less than hollow circles)as shown in figure 12.3 (b).

(3) Pumping radiation continues. The majority of certain atoms in the active material area excited to the metastable states in which the excited atoms remain for an appreciable time. Population inversion is reached at this stage. The dominant radiations initially are the spontaneous ones. Two cases of coherent (stimulated) emission are shown (multiple-line arrows) one along the axis, another at slant angle as shown in figure 12.3 (c).

(4) One excited atom falls to a lower state and emits a photon. This photon wave travels past the other excited atoms and triggers them to fall to the lower state. In doing so, they emit photons with waves in step with the original photon wave. The beam travels back and forth again and again due to multiple reflection on the end faces. On each trip, it causes more excited atoms to emit waves in phase with it. Spontaneous component of emissions is still not negligible as shown in figure 12.3 (d).

(5) Only those waves that travel accurately along the axis of the tube will remain in the tube after many trips up and down it. Therefore, the beam is almost entirely composed of waves going in exactly the same direction. Its rays are

accurately parallel. Thus radiations that are mutually coherent go on becoming dominant at successive stages. Spontaneous component of emission is now very weak as shown in figure 12.3 (e).

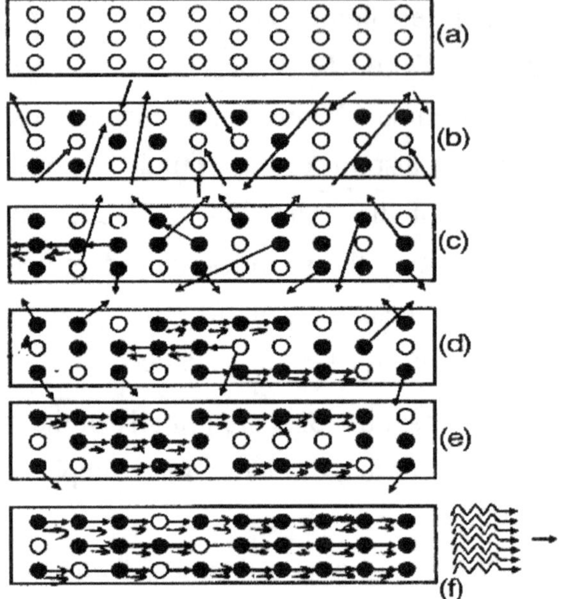

Fig 12.3 *Shaping of laser beam*

(6)   Almost the entire emission becomes coherent. The wave-shapes shown at the exit on the right show phase agreement of waves. This is called the laser beam as shown in figure 12.3 (f).

12.3 Lasers and Laser Light

Lasers are devices that produce intense beams of light which are monochromatic, coherent and highly collimated. Laser light is available in all colors from red to violet and also for outside these conventional limits of the optical spectrum. Laser light produced by dye lasers has the property of emitting light of any wavelength chosen within a range of wavelength, i.e. laser light is also tunable. Laser light has very low divergence. It can travel over great distances or can be focused to a very small spot with a brightness which exceeds that of the sun. The wavelength and the area A of the laser output aperture determine the order of magnitude of the beam's solid angle $\Delta\Omega$ and vertex angle $\Delta\theta$ of divergence through the following relation

$$\Delta\Omega \approx \frac{\lambda^2}{A} \approx (\Delta\theta)^2$$

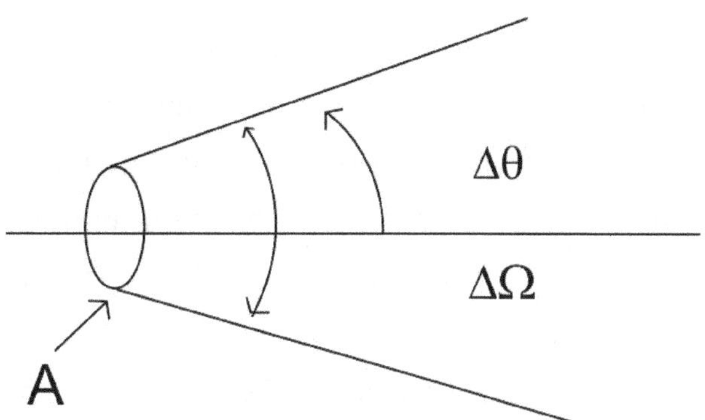

Fig. 12.4 *Sketch at a laser cavity showing angular beam divergence $\Delta\theta$ at the output*

It is well known that lasers produce very pure colours. The laser light would be fully monochromatic if a laser produces exactly one wavelength. This is not possible in principle as well as for practical reasons. Since lasers have definite bandwidth around 100 Hz. The bandwidth of sun is very broad~$10^{14}Hz$. Therefore, the relative spectral purity of a laser beam is quite impressive. Hence the wavelength achieved by the laser light is much closer approach to monochromatic light when compared to other sources of light. The existence of a finite bandwidth(Δv) means that the different frequencies present in a laser beam can eventually get out of phase with each other. The time required for two oscillations differing in frequency by(Δv) to get out of phase by a full cycle is $\frac{1}{\Delta v}$. After this amount of time, the different frequency components in the beam can begin to interfere destructively and the beam loses coherence. Thus $\Delta\tau = \frac{1}{\Delta v}$ is called the beam's coherence time. The extremely small values possible for Δv in laser light make the coherence times of laser light extraordinarily long. The $\Delta L = C\Delta\tau$ distance is called the beam's coherence length. Only portions of the same beam that are separated by less than ΔL are capable of interfering constructively with each other.

12.4 Light in Cavities

The optical cavity or resonators is an important component of laser. Here we will consider only a simplify theory of resonators. This simplification allows us to introduce the concept of cavity modes and to infer certain features of cavity modes that remain valid in more general circumstances. If we consider three dimensional reflecting cavity, then the number of available modes grows extremely rapidly as a function of frequency. For example, a cubical three dimensionally reflecting cavity of 1cm on a side has about 400 million resonant frequencies within the useful gain band of a He-Ne laser. This will eliminate any possibility of achieving the important narrow-band (nearly monochromatic character) of laser light since lasing action occurs across the whole band. The solution of this multimode dilemma was suggested independently in 1958 by Townes and Schawlow, Dicke and Prokhorov. They recognized that a one dimensional rather than a three dimensional cavity was desirable and this could be achieved with open resonator consisting of two parallel mirrors as shown in fig. 12.5.

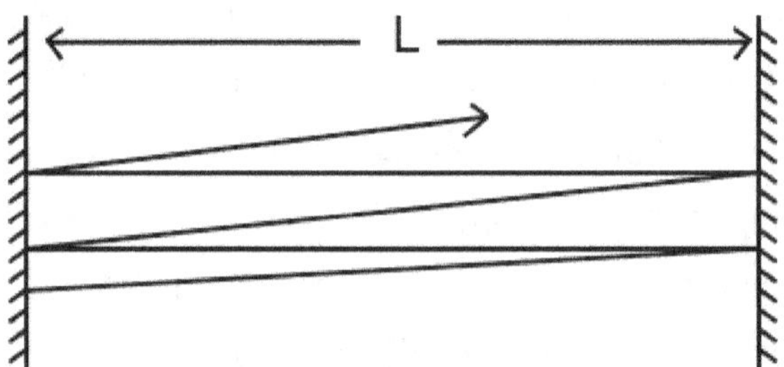

Fig. 12.5 *One dimensional cavity*

If we consider one dimensional (X-axis) cavity of length L, then the cavity electric field component can be written as

$$E = E_0 \, Sinkx$$

The electric field should vanish at both ends of the cavity. It will do so if we fit exactly an integer number of half wavelength into cavity along its axis. This means that along the x axis is determined by the relation

$$L = n \left(\frac{\lambda}{2}\right)$$

where n = 1,2,3....... is a positive known as the mode order and L is the cavity length. Hence,

$$\lambda = \frac{2L}{n} = \frac{c}{v} \qquad \therefore v = \frac{cn}{2L}$$

The electric field should vanish as both ends of the cavity. It will do so if we fit exactly an integer number of half wavelength into cavity along its axis.

Therefore, frequency separation between any two adjacent modes n and n+1 is given by $\Delta v = v_{n+1} - v_n = \dfrac{C(n+1)}{22} - \dfrac{cn}{2L} = \dfrac{c}{2L}$. This gives the separation in frequency of adjacent resonator modes for 10 cm long cavity the separation in frequency of adjacent modes is $\Delta = \dfrac{3\times10^8}{2\times20\times10^{-2}} = 750\ MHz$ .If laser has 1500 MHz gain curve, then the number of possible modes that can lase is 2 as shown in fig. 12.6.

$$\text{Since} \quad \frac{1500\ MHz}{750\ MHz} = 2$$

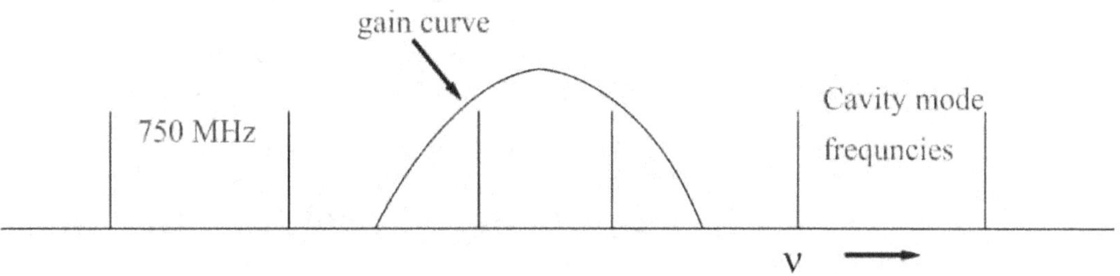

Fig. 12.6 *Mode frequencies separated by 1500 MHz corresponding to a 20 cm long one dimensional cavity*

12.5 Light Emission and Absorption in Quantum Theory

In 1900, Max Planck proposed that light consists of discrete bundles of energy. The amount of energy of bundle is hʋ. These bundles of radiant energy are called Quanta. In 1905, Einstein refined the quantum hypothesis of Planck and provided theoretical justification for the features of photoelectric effect. He gave the name photon to the quanta of light energy. Einstein assumed the difference in energy of the electron before and after its photo ejection to be equal to the energy hʋ of the photon absorbed in the process. This picture of light absorption was extended in two ways by Bohr in his model (1) an electron can go from its ground state (lowest energy orbit) to a higher (excited) state or (2) it can decay from a higher state of lover state, but it cannot remain between these states. These allowed energy states are called Quantum jump and the amount of energy involved in a quantum jump depends on the quantum system. Atoms usually absorb and emit photons in or near the optical region of the spectrum when they have quantum jumps whose energies are typically in the range of 1-6 eV. Quantum jumps by inner shell atomic electrons usually require much more energy and are associated with X-Ray photons. On the other hands, quantum jumps among the so-called Rydberg energy level, those outer level electrons lying far from the ground level and near to the ionization limit involve only a small amount of energy and are associated with far-infrared or even microwave photons. Molecules have vibrational and rotational degrees of freedom whose quantum jumps are much smaller than the quantum jumps in free atoms. Many crystals are transparent in the optical region which means that they do not absorb or emit optical photons because they do not have quantum energy levels that permit jumps in the optical range. However, colored crystals like ruby have impurities that do absorb and emit optical photons. These impurities are frequently atomic ions and they have both discrete energy levels and broad bands of levels that allow optical quantum jumps. Ruby crystal is a good absorber of green photons and so appears red.

12.6 Einstein Theory of Light-Matter Interactions

Einstein predicted the phenomenon of stimulated emission in 1917. In a Laser, each atom of gain medium jump to a higher orbit when the electron receives an amount of energy equal to the difference of energy of the ground state and one of the excited state (higher orbit) as a result of pumping process and converts it into light energy (photon) when it jumps to a lower orbit. At the same time, each atom must deal with the photons that have been emitted earlier and reflected back by the mirrors. The photons already channeled along the cavity axis are responsible for stimulated emission of subsequent photons.

Therefore, three quantum mechanical processes may occur when radiation is incident on a medium.

(a) **Stimulated absorption** - The atom or molecule absorbs a photon of energy $H\nu = E_2 - E_1$ and goes from a Lower energy level 1 to the higher level 2.

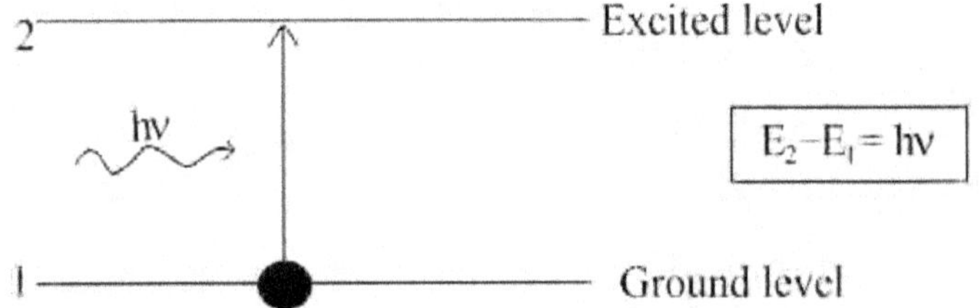

(b) **Spontaneous emission** - The atom or molecule jumps down from higher level 2 to the lower level 1 and emit a photon of energy $h\nu = E_2 - E_1$. The process occurs spontaneously without any external influence.

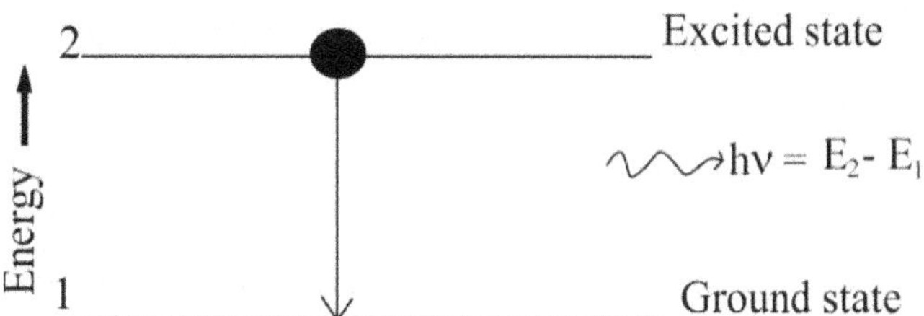

(c) **Stimulated Emission** - When the incoming photon energy matches the energy level difference, the photon can stimulate a transition to a lower energy level. This energy difference is released as another photon. Hence the process is induced or stimulated by the incident photon. The emitted photon has the same characteristics (frequency, polarization and phase) as the incoming one.

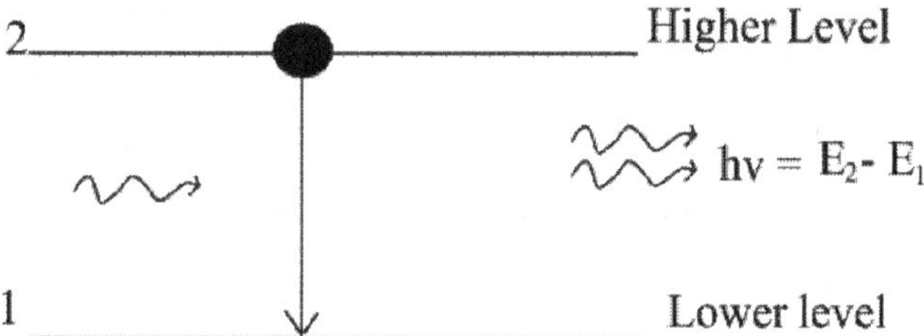

All these processes occur in the gain medium of a laser. Lasers are often classified according to the nature of the pumping process, which is the source of energy for the output laser beam. In electric discharge lasers, the pumping occurs as a result of collisions of electrons in a gaseous discharge with the atoms or molecules of the gain medium. In an optically pumped laser, the pumping photons are supplied by a lamp or another laser. In a diode laser, an electric current at the junction of two different semiconductors produces electrons in excited energy states from which they can jump into lower energy states and emit photons.

12.7 Stimulated Absorptions and Emission Rates

An atom residing in the lower energy level E_1 may absorb the incident photon and jump to the excited state E_2. This transition is known as stimulated absorption or simply as absorption. This process may be represented as

$$A + h\nu \rightarrow A^*$$

where A denotes an atom in the lower state and A* an excited atom.

The number of atoms per unit volume that makes upward transitions from the lower level to the upper level per second is called the rate of absorption. It is represented by

$$R_{abs} = -dN_1/dt$$

where - ve sign stands for the rate of decrease of population at the lower level.

The rate of absorption will be proportional to the population in the lower level (N_1) and the number of photons per unit volume (energy density) in the incident beam i.e.

$$\frac{dN_1}{dt} = -B_{12}\rho(\nu)N_1$$

where B_{12} is a constant of proportionality and $\rho(\nu)$ is the energy density of incident light.

The process of emission of photons by an excited atom through a forced transition occurring under the influence of an external photon is called stimulated emission. The process may be represented as $A^* + h\nu \rightarrow A + 2h\nu$ In this case, the number of atoms per unit volume that makes downward transitions from the higher level to the lower level per second is called the rate of stimulated emission. It is represented by

$$R_{st} = -\frac{dN_2}{dt}$$

The rate is negative since the population is decreasing in higher level. The rate of stimulated emission is proportion to the population of the higher level N_2 and the energy density of incident light *i.e.*

$$\frac{dN_2}{dt} = -B_{21}\,\rho(v)N_2$$

where B_{21} is a constant of proportionality

12.8 Illustrative Examples

Example 12.1 Determine the coherence time and coherence length for white light ($\lambda = 4000\text{Å}$ to 7000 Å).

Sol. Frequency band width $\Delta v = v_{\text{violet}} - v_{\text{red}}$

$$= \frac{c}{\lambda v} - \frac{c}{\lambda r}$$

$$= \frac{3 \times 10^8}{4 \times 10^{-7}} - \frac{3 \times 10^8}{7 \times 10^{-7}}$$

$$= 7.5 \times 10^4 - 4.28 \times 10^{14}$$

$$\Delta v = 3.22 \times 10^{14} \qquad \text{Hz}$$

$\therefore$ Coherence time $\quad \tau = \dfrac{1}{\Delta v}$

$$= \frac{1}{3.22 \times 10^{14}} = \frac{10}{3.22} \times 10^{-15}$$

$$= 3.10 \times 10^{-15} \text{ Sec} \qquad\qquad \textbf{Ans.}$$

Coherence length $L = C\tau$

$$= 3 \times 10^8 \times 3.10 \times 10^{-15}$$

$$L = 9.30 \times 10^{-7} m \qquad\qquad \textbf{Ans.}$$

Example 12.2 Calculate the order of magnitude of the vertex angle of divergence when wavelength (λ) is 500 nm and the area (A) of the laser output aperture is 5X5 mm^2.

Sol. $\qquad (\Delta\theta)^2 = \dfrac{\lambda^2}{A}$

where d = 500 mm $= 5 \times 10^{-7} m$ $\textit{and}$ $A = 25 \times 10^{-6} m$

$\therefore \qquad (\Delta\theta)^2 = \dfrac{25 \times 10^{-14}}{25 \times 10^{-6}}$

$$= 10^{-8}$$

$$\Delta\theta = 10^{-4} = \frac{10^{-3}}{10} rad$$

$$\Delta\theta = \frac{1}{10} \text{ mrad.} \qquad\qquad \textbf{Ans.}$$

Example 12.3 A pulsed laser is constructed with a ruby crystal as the active element. The ruby rod contains typically a total of $3 \times 10^{19} cr^{+3}$ ions. If the laser emits light at 6943 Å wavelength find (a) the energy of one emitted photon (in eV) and (b) the total energy available per laser pulse(assuming total population inversion).

Sol. (a) The photon energy in eV is given by

$$E = \frac{12400}{\lambda(\text{Å})} eV$$

$$= \frac{12400}{6943}$$

$$E = 1.78 \, eV \text{Ans.}$$

(b) Energy per pulse = Energy of one photon $\times$ Total number of photons

$$= \text{Energy of one photon} \times \text{Total number of atoms in the excited state}$$

$$= 1.75 \times 1.6 \times 10^{-19} \times 3 \times 10^{19}$$

$$E_t = 8.54 \, Joule \qquad \textbf{Ans.}$$

Example 12.4 Find the intensity of laser beam of power 30 mw and diameter 2 mm. Assume that intensity is uniform throughout the beam.

Sol. Given $P = 30 \, mW = 30 \times 10^{-3} W$

and $d = 2 \, mm$

$\therefore r = 1mm = 1 \times 10^{-3} m$

Intensity is given by $I = \dfrac{P}{A}$

$$= \frac{30 \times 10^{-3}}{3.14 \times (1 \times 10^{-3})^2}$$

$$= \frac{30}{3.14} \times 10^3$$

$$= 9.55 \times 10^3 \text{W}/m^2$$

$$I = 9.55 \text{ KW}/m^2 \qquad \textbf{Ans.}$$

12.9 Self Learning Exercise

Q.1 LASER stands for..

Q. 2 What are the characteristics of a laser beam.

Q.3 What is pumping.

Q.4 The photons emitted by stimulated emission are................. .

Q. 5 Why a Laser requires optical cavity.

Q. 6 What is the order of ratio of the life times of meta-stable state and the excited state.

Q. 7 The higher energy state having a longer mean life is called..............................

Q. 8 The active or gain medium in a laser can be..............................

Q. 9 The life time of the atom in excited state is the order of...........................

Q. 10 Why the number of laser photons changes in the cavity.

Q.11 Write the formula for the mode frequency spacing.

12.10 Summary

This chapter undertakes a superficial introduction to lasers. It presents an overview of the properties of laser light with the goal of understanding what a laser is, in the simplest terms. The chapter introduces the theory of light in cavities, cavity modes and describes an elementary theory of laser action. Every stage of laser operation from the injection of energy into the amplifying medium to the extraction of light from the cavity is an opportunity for energy loss and energy gain. This chapter provides a brief overview of these properties. This chapter also discusses Einstein theory of light-matter interaction. In the end of unit, some examples on above topics are given.

12.11 Glossary

Coherence : If the phase difference between the waves from two sources reaching any point in space does not change with time, then these sources of light are known as 'coherent sources of light'. This property of the waves is known as coherence.

Laser : It stands for Light Amplification by Stimulated Emission of Radiation. It produces intense beam of light which is monochromatic, coherent and highly directional.

Metastable states : There are certain energy levels in an atom from which transitions to the ground state are forbidden by the selections rules. Such states re called 'Metastable states. The mean life time of metastable state $\left(\sim 10^{-3}\ \text{sec}\right)$ is much longer than the mean life time of excited states $\left(\sim 10^{-8}\ \text{sec}\right)$.

Population inversion : If the number of atoms in excited states is greater than the number of atoms in ground state, then it is called population inversion.

Pumping : The methods used to achieve population inversion are called pumping.

Active Medium : A medium in which population inversion is to be achieved is called an active medium.

Quanta : In 1900, Max Planck proposed that light consists of discrete bundles of energy. The amount of energy of bundle is $h\nu$. These bundles of radiant energy are called Quanta.

Stimulated Emission : The process of emission of photons by an excited atom through a forced transition occurring under the influence of an external photon is called stimulated emission.

12.12 Answers of Self Learning Exercise

Ans.1: Light Amplification by stimulated Emission of Radiation

Ans.2: There are four main characteristics of a laser beam.

    (i) Coherence

    (ii) Monochromaticity

    (iii) Directionality and

    (iv) Higher Intensity.

Ans.3: It is mechanism which can be used to create population inversion in active medium.

Ans.4: Coherent

Ans.5: To provide the optical feedback.

Ans.6: 10^{6}

Ans.7: Meta stable state with 10^{-3} sec.

Ans.8: Solids, gases and liquids.

Ans.9: 10^{-9} sec.

Ans.10: There are two main reasons for the same.

(i) Laser photons are continually being added because of stimulated emission.

(ii) Laser photons are continually being lost because of mirror transmission, scattering or absorption at the mirrors.

Ans.11: $\Delta v = \dfrac{c}{2L}$

12.13 Exercise

Section A : Very Short Answer Type Questions

Q.1 Define the terms "active medium for laser".

Q.2 What is the order of life time of meta-stable state?

Q.3 What are the basic components of a laser?

Q.4 What is the principle of laser action?

Q.5 Why the selection of laser medium is important in the lasers.

Section B : Short Answer Type Questions

Q.6 Define the phenomenon of stimulated emission and absorption.

Q.7 What is population inversion how can write the relative population between two energy states with energy values E_1 and E_2.

{Hint: - $N_2 = N_1\, e^{-(E2-E1)/KT}$}

Q.8 A laser beam of 10 mw has a wavelength 650 mm. Calculate the number of photons emitted per second. [3.27×10^{16}]

Q.9 Define coherence time and coherence length.

Q.10 How can get the pumping process in electric discharge lasers.

Section C : Long Answer Type Questions

Q.11 Describe the key elements of laser operation.

Q.12 Explain the role of optical resonator and meta-stable state in laser action.

Q.13 Explain the Einstein theory of light-matter interactions.

Q.14 What are the essential requirements of a laser? Explain how these requirements are achieved.

Q.15 Give the reasons for the following basic properties of a laser

(i) Directionality

(ii) Monochromaticity

(iii) Coherency and

(iv) High intensity.

References and Suggested Readings

1. W.T. Silfvast, Laser Fundamentals, Cambridge University Press (1998).

2. A.K. Ghatak and K. Thyagarajan, Lasers Theory and Applications, MacMillan India Ltd. New Delhi, 1981.

3. M.J. Beesley, Lasers and their Applications, Tylor and Francis Ltd. London, 1976.

4. B. A. Lengyel, Introduction to Laser Physics, John Wiley and Sons, New York, 1967.

5. P. W. Milonni and J.H. Eberly, Laser Physics, John Wiley & Sons, New York, 2010.

6. A. Maitland and M. H. Dunn, Laser Physics, North-Holland Publication co., Amsterdam, 1969.

7. P. A. M. Dirac, Quantum theory of emission and absorption in quantum electrodynamics, Dover Publication, New York, 1958.

$$\boxed{\begin{array}{c} \textbf{UNIT-13} \\[4pt] \textbf{Laser Oscillation: Population} \\[4pt] \textbf{Rate Equations, Gain and Threshold Gain} \end{array}}$$

Structure of the Unit

13.0 Objectives

In the previous chapter, we learnt that light energy cannot take arbitrary values but must be multiples of Photon energy $h\nu$. It was similarly established that the electrons in an atom cannot have arbitrary amounts of energy but they take only discrete energies. This is a consequence of the Bohr's postulate of permitted orbits the passing of an electron from one energy level to another energy level within the atom occurs in a jump which is called a quantum transition. The electron transitions may be induced by a variety of ways. Interaction with light photons is one of the possible way of supplying energy to orbital electrons which causes upward electron transitions and sends the atom into its excited state.

In this chapter, we will discuss about the population rate equation Einstein coefficients, relation among these coefficients and laser oscillation with gain and threshold gain.

13.1 Introduction

Einstein predicated in 1917 that the photons in the light field induce or stimulate the excited atoms to fall to lover energy state and give up their excess energy in the form of photons. He called this type of emission as stimulated emission. When photons are incident on a medium, three quantum mechanical processes occur in the medium namely (I)stimulated absorption (ii) spontaneous emission, and (iii)stimulated emission.

In the process of stimulated emission, photons are multiplied and their characteristics are related to each other.

Hence, amplification of light take place. This process is made to dominate in the laser light source. A medium amplifies light only when the following three conditions are fulfilled

1. The population at excited level should be greater than that at the lower energy level.

2. The ratio $\dfrac{B_{21}}{A_{21}}$ should be large and

3. A very high density of radiation should be present in the medium.

13.2 Population Rate Equations

The population of energy levels of the lasing medium change under the action of radiation. These changes can be described conveniently by means of population rate equations. The upper and lower level population densities N_2 and N_1 change in time due to absorption, spontaneous emission, stimulated emission and collisions. These collisions are inelastic collisions. The effects of the these processes on the rate equations for N_2 and N_1 are given by the following equations

$$\frac{dN_2}{dt} = -\Gamma_2 N_2 - \frac{\sigma(v)}{hv} I_v \left(N_2 - N_1 \right) - A_{21} N_2 \qquad (13.1a)$$

and
$$\frac{dN_1}{dt} = -\Gamma_1 N_1 + \frac{\sigma(v)}{hv} I_v \left(N_2 - N_1 \right) + A_{21} N_2 \qquad (13.1b)$$

The first term on the right of these equations represents populations.

Changing due to inelastic collision and rest terms represent population changing due to three processes(absorption, spontaneous and simulated emission).

Since stimulated emission rate

$$= \frac{\lambda^2 A_{21}}{8\pi} S(v) \, number \text{ of incident photons/ area-time} \tag{13.2}$$

Therefore quality $\sigma(v) = \frac{\lambda^2 A_{21}}{8\pi} S(v)$ has the dimension of an area. It is the cross section for stimulated emission(and absorption). The relation (13.2) identifies the cross section as an effective area associated with the atomic transition, such that every photon intercepted by this area would induce an atom to undergo stimulated emission or absorption. There is no actual geometric object associated with this area. The cross-section is nothing more than a conventional measure of the absorption strength of a transition. It depends not only on the transition wavelength and spontaneous emission rate, but also on the line shape function $S(v)$.We must also account for the pumping process that produces the positive population inversion $\left(N_2 - N_1 \right)$.

For this, we add a term K to the population equations and call it the pumping rate into the upper level. Therefore, we have the following set of coupled equations for the light and the atoms in the laser cavity.

$$\frac{dN_1}{dt} = -\Gamma_1 N_1 + A_{21} N_2 + g(v)\phi_v \tag{13.3a}$$

$$\frac{dN_2}{dt} = -(\Gamma_1 + A_{21})N_2 - g(v)\phi_v + K \tag{13.3b}$$

$$\frac{d\phi_v}{dt} = -\frac{Cl}{L} g(v)\phi_v - \frac{C}{2L}(1 - r_1 r_2)\phi_v \tag{13.3c}$$

Here we have used $I_v = hv\phi_v$ (where ϕ_v is the Photon flux)and gain coefficient $g(v) = \sigma(v)\,(N_2 - \frac{g_2}{g_1} N_1)$ (where $g_2 \& g_1$ are degeneracies of levels 2 and 1, respectively).Here, we have assumed that degeneracies of levels are equal *i.e.*

$g_1 = g_2$. The term $-\frac{C}{2L}(1 - r_1 r_2)$ represents the rate at which intensity is lost to the imperfect reflectivity of the mirrors of a laser resonator.

Eqs. (13.3) rewrite in terms to absolute numbers rather than densities of atoms and photons. The total number of atoms in level 2 is $n_2 = N_2 V_g$, where V_g is the volume of the gain medium is. Similarly the total number of atoms in the lower level of the laser transition is $n_1 = N_1 V_g$. The electromagnetic energy density. (U_v) in the cavity is related to intensity I_v and photon flux ϕ_v by

$$u_v = \frac{I_v}{C} = \left(\frac{hv}{C}\right)\phi_v \tag{13.4}$$

But u_v is related to photon number by

$$u_v = \frac{hvq_v}{V} \tag{13.5}$$

Where V is the cavity volume and q_v is the number of photons in the laser cavity. These relations assume a uniform distribution of intensity within the cavity and a refractive index $n \approx 1$. Thus from eqs. (13.4) and (13.5), we get

$$\phi_v = \frac{cq_v}{V} \tag{13.6}$$

Hence Eqs. (13.3) may be written as

$$\frac{dn_1}{dt} = -\Gamma_1 n_1 + A_{21} n_2 + \frac{Cl}{L} g(v)q_v \tag{13.7a}$$

$$\frac{dn_2}{dt} = -(\Gamma_2 + A_{21})n_2 - \frac{Cl}{L} g(v)q_v + p \tag{13.7b}$$

$$\frac{dq_v}{dt} = \frac{Cl}{L} g(v)q_v - \frac{C}{2L}(1 - r_1 r_2)q_v \tag{13.7c}$$

where we have used the valuations

$$\frac{V_g}{V} = \frac{l}{L} \text{ and } KV_g = p \tag{13.8}$$

Eqs. (13.7) imply that

$$\frac{d}{dt}(n_2 + q_v) = -(\Gamma_2 + A_{21})n_2 + p - \frac{C}{2L}(1 - v_1 v_2)q_v \tag{13.9}$$

The interpretation of this equation is as follows:-

The left hand side is the rate of change of the total number of excitation *i.e.,* the number of atoms in the upper level 2 of the lasing transition plus the number of photons in the cavity. The first term on the right is the rate of decrease in the number of these excitations as a result of inelastic collisions and spontaneous emission from level 2. The second term is the rate of change associated with pumping of level 2.The last term is the rate at which excitation in the form of photons is lost from the cavity. In this equation, contributions from stimulated emissions or absorption do not appear because they have canceled out. Since an increase in q_v is always accompanied by an equal decrease in n_2.

13.3 Einstein A and B Coefficients

Let us consider two energy levels in a particular atom. The lower level is designated as 1 and upper level as 2 as shown in Fig. 13.1. If the atom is initially in the state 1, it can be raised to state2 by absorbing a photon of light whose frequency is

$$v = \frac{E_2 - E_1}{h} \tag{13.10}$$

This process of absorption is called induced or stimulated absorption.

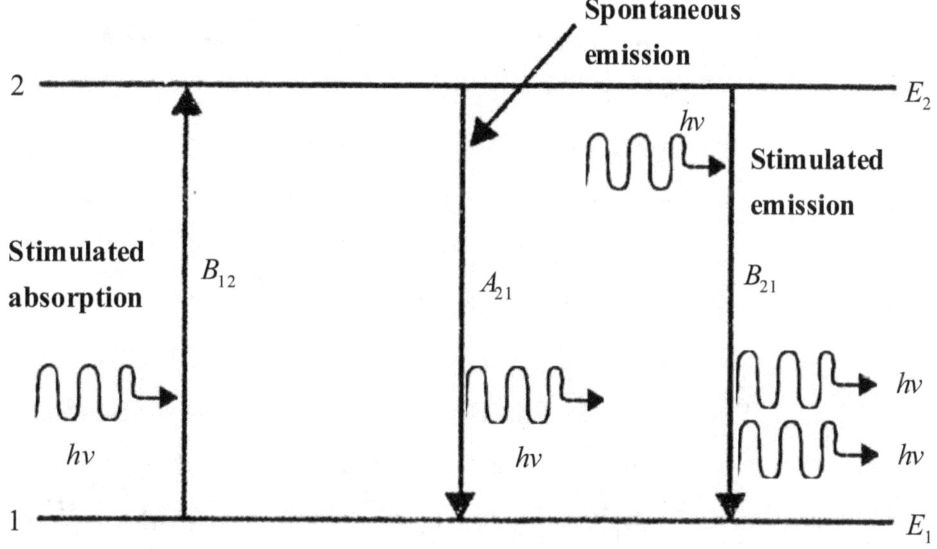

Fig 13.1 *Transitions between two energy levels in an atom can occur by stimulated absorption, spontaneous emission and stimulated emission*

The possibility that the atom will actually undergo this transition is proportional to the rate at which the photons of frequency ν fall on it and therefore to the spectral energy density $u(\nu)$. The transition probability also depends on the properties of state 1 and 2, but we can include this dependence in some constant of proportionality B_{12} and it is known as the Einstein's Coefficient for Stimulated absorption. Hence if we shine light of frequency υ and energy density $u(\nu)$ on the atom when it is in the lower state 1, and the probability for it to go to higher state 2 is

$$P_{1 \to 2} = B_{12} u(v) \qquad \text{(Stimulated Absorption)} \qquad (13.11)$$

If the atom is initially in the upper state 2 two processes can occur:

1. The atom has a small life time (mean lifetime $\approx 10^{-8}$ sec) in the excited state so that it can spontaneously drop to the state 1 by emitting a photon of frequency ν. This is called spontaneous emission as mentioned in fig. 13.1. The probability of such a process will, of course, depend on the two energy level but it will be independent of the presence of any photon incident on the atom. Let us express this probability by the coefficient A_{21} and it is known as the Einstein's Coefficient for spontaneous emission.

$$P_{2 \to 1} = A_{21} \qquad \text{(Spontaneous emission)} \qquad (13.12)$$

2. While discussing the interaction of radiation with matter, Einstein proposed that if a photon of right frequency is present, it is not essential that it may always be absorbed, it can also interact with an atom in excited state and induce that atom to emit a new photon as shown in fig. 13.1. Such a process of emission is known as *induced* or *simulated emission*. The probability of induced emission will thus be proportional to the spectral energy density $u(\nu)$ and we can write

$$P_{2 \to 1} = B_{21} u(v) \qquad \text{(Stimulated emission)} \qquad (13.13)$$

Where B_{21} is the Einstein's Coefficient for stimulated emission. The total probability for an atom in state 2 to fall to the lower state 1 is therefore

$$P_{2 \to 1} = A_{21} + B_{21} u(v) \qquad \text{(Emission)} \qquad (13.14)$$

13.4 Relation among A and B Coefficients

Now we consider an assembly of N_1 atoms in the state 1 and N_2 atoms in the state 2, all in thermal equilibrium at the temperature T with light of frequency ν and energy density $u(\nu)$. The number of atoms in state 1 that absorb a photon and go the state 2 per second, *i.e.*, the rate of transition is given by

$$R_{12} = N_1 P_{1 \to 2} = N_1 B_{12} u(\nu) \tag{13.15}$$

The number of atoms that go from 2 to 1 per second will have two components

(a)   Due to spontaneous emission

$$\left(R_{21}\right)_{\text{spentaneous}} = N_2 P_{2 \to 1} = N_2 A_{21}$$

(b)   Due to stimulated emission

$$\left(R_{21}\right)_{\text{stimulated}} = N_2 P_{2 \to 1} = N_2 B_{21} u(\nu)$$

Thus, the total rate of transition from state 2 to state 1 will be

$$R_{21} = N_2 P_{2 \to 1} = N_2 \left[A_{21} + B_{21} u(\nu)\right] \tag{13.16}$$

In thermal equilibrium, the number of atoms going from state 1 to state 2 per second must be equal to the number going from state 2 to 1. This is in accordance with statistical principle of detailed balancing. Hence from Eqs. (13.15) and (13.16).

$$N_1 P_{1 \to 2} = N_2 P_{2 \to 1}$$
$$\Rightarrow \quad N_1 B_{12} u(\nu) = N_2 \left[A_{21} + B_{21} u(\nu)\right] \tag{13.17}$$

Dividing both sides of this equation by $N_2 B_{21}$, we get

$$\left(\frac{N_1}{N_2}\right)\left(\frac{B_{12}}{B_{21}}\right) u(\nu) = \frac{A_{21}}{B_{21}} + u(\nu)$$

Solving for $u(\nu)$, we get

$$u(\nu) = \frac{A_{21} / B_{21}}{\left(\dfrac{N_1}{N_2}\right)\left(\dfrac{B_{12}}{B_{21}}\right) - 1} \tag{13.18}$$

Now, in thermal equilibrium, by Maxwell-Boltzmann distribution law $\left(N_1 = N_0 e^{-E_1 / k_B t}\right)$, the ratio between the populations of states 1 and 2 is given by

$$\frac{N_1}{N_2} = e^{\frac{(E_2 - E_1)}{kT}} = e^{h\nu / kT} \tag{13.19}$$

With this result, we have

$$u(v) = \frac{A_{21}/B_{21}}{\left(\dfrac{B_{12}}{B_{21}}\right)e^{hv/kT} - 1} \tag{13.20}$$

Eq. (13.20) is a formula for the energy density of photons of frequency v in equilibrium at the temperature T with atoms whose possible energies are E_1 and E_2. This formula can be compared with the Planck's radiation law, according to which

$$u(v)\,dv = \frac{8\pi hv^3/c^3}{e^{hv/kT} - 1}\,dv \tag{13.21}$$

The comparison yields

$$B_{12} = B_{21} \tag{13.22}$$

and $\quad \dfrac{A_{21}}{B_{21}} = \dfrac{8\pi hv^3}{c^3} \tag{13.23}$

Equations (13.22) and (13.23) are known as the Einstein's relations. Equation (13.23) gives the relationship between the A and B coefficients. The first relation show that the probability of induced or stimulated emission per incident photon is equal to the probability of absorption per incident photon. *i.e.,* in the presence of an incident photon absorption and induced emission are equally probable.

The second relation shows that the ratio of coefficients of spontaneous emission to stimulated emission is proportional to the third power of frequency of the radiation. That is why it is difficult to achieve laser action in higher frequency ranges such as X-rays.

The net absorption rate R_{12} is usually much more than the net stimulated emission rate $\left(R_{21}\right)_{\text{stimulated}}$ because of large difference in the population in the two states. Let us now compare the probability of spontaneous emission to that of stimulated emission.

$$\frac{\left(P_{2\to1}\right)_{\text{spontaneous}}}{\left(P_{2\to1}\right)_{\text{stimulated}}} = \frac{A_{21}}{B_{21}u(v)} = \frac{u(v)\left[\left(\dfrac{B_{12}}{B_{21}}\right)e^{hv/kT} - 1\right]}{u(v)}$$

$$= \left(e^{hv/kT} - 1 \right) \tag{13.23 a}$$

(i) If $hv \gg kT$, the spontaneous emission will be much more probable than induced (stimulated) emission (which can even be considered negligible). This is ordinarily true for electronic transition in atoms and molecules and in radiative transitions in nuclei.

(ii) If $hv \ll kT$, then

$$\left(e^{hv/kT} - 1 \right) = \frac{hv}{kT}$$

So that $\quad \dfrac{\left(P_{21} \right)_{spontaneuos}}{\left(P_{21} \right)_{stimulated}} = \dfrac{hv}{kT}$

In such a case the stimulated emission is no more negligible and in even predominates over the spontaneous emission. Such a situation can be achieved for radiation in microwave region.

Principle of Laser

The population of atoms in different energy states is governed by Maxwell-Boltzmann distribution law. Further, according to the process of detailed balancing the total emission and absorption rates, in thermal equilibrium, remain equal. In the presence of incident radiation the equilibrium is distributed and the ratio of emission rates is given by-

$$\frac{\text{Rate of emission}}{\text{Rate of absorption}} = \frac{R_{21}}{R_{12}} = \frac{A_{21} + B_{21} u(v) N_2}{B_{12} u(v) N_1}$$

$$= \left[1 + \frac{A_{21}}{B_{21} u(v)} \right] \frac{N_2}{N_1} \tag{13.23 b}$$

For Laser Actions:

(a) The emission rate must be greater than the absorption rate and

(b) The probability of spontaneous emission (which produces incoherent radiations) must much smaller than the probability of stimulated emission.

Thus, with the second condition *i.e.*, $A_{21} \ll B_{21} u(v)$, we have

$$\frac{\text{Rate of emission}}{\text{Rate of absorption}} = \frac{N_2}{N_1}$$

Under normal conditions the number N_2 in the upper energy state 2 is always smaller than the number N_1 in the lower energy state. If, by some means, N_2 is made greater than N_1, the emission rate will become larger than the absorption rate and consequently amplification will take place.

To achieve the basic requirement, *i.e.*, $A_{21} << B_{21}u(v)$, the following possibilities can be used.

(i) $hv << kT$, such a situation exists at microwave frequencies. Ammonia maser is based on this principle. MASER stands for Microwave Amplification by Stimulated Emission of Radiation.

(ii)   **Metastable States**: There are certain energy levels in an atom from which transitions to the ground state are forbidden by the selection rules. Such states are called 'Metastable States'. The atom can, however, go from the metastable state to the ground state either by giving up the appropriate amount of energy to another atom during a collision process or it may absorb radiation and go to a higher energy state (to which transitions are not forbidden) and from there it may return to normal state a metastable state $\left(\sim 10^{-3} \sec\right)$ to be much longer than the mean life time of other excited states $\left(\sim 10^{-8} \sec\right)$. Thus, if certain atoms are excited to the metastable state, the probability of spontaneous emission will be quite negligible. However, induced transitions, being a sort of forced oscillations can take place.

There are several mechanisms which can be used to invert the normal population density of the energy levels. This process of population inversion is called pumping. The commonly used methods are

(i)   Optical pumping
(ii)   Electron impact
(iii)   Inelastic atom-atom collision
(iv)   Chemical pumping
(v)   Gas dynamic pumping

13.5 Laser Oscillation: Gain and Threshold Gain

Consider the propagation of narrowband radiation in a medium of atoms that have a transition frequency equal or nearly equal to the frequency of radiation as shown in fig 13.2.

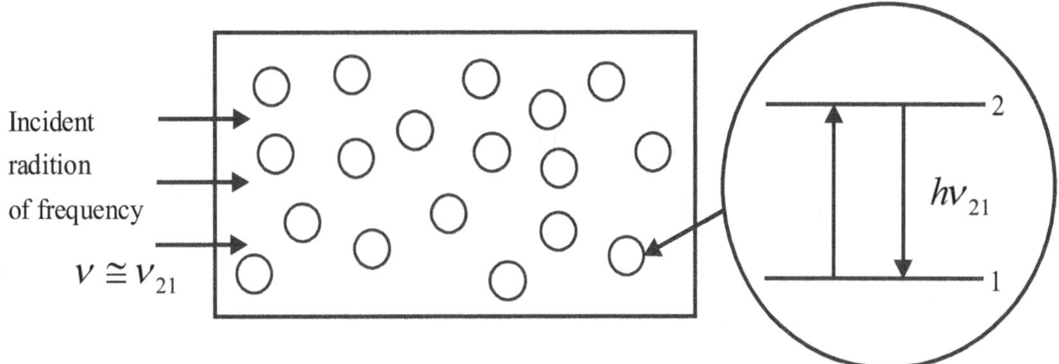

Fig 13.2 *Propagation of radiation of frequency V in a medium of atoms with a transition frequency V_{21}*

If there are more atoms in the upper level of the transition than the lower, stimulated emission will be more than absorption and the radiation can be amplified as it propagates. In such a case, we can say that there is a gain at the resonant frequency.

The growth rate of the number of laser photons in the cavity is described by the gain coefficient.

The gain coefficient is given by

$$g(v) = \sigma(v)\left(N_2 - \frac{g_2}{g_1}N_1\right)$$

$$g(v) = \frac{\lambda^2 A_{21}}{8\pi}\left(N_2 - \frac{g_2}{g_1}N_1\right)S(v) \tag{13.24}$$

The sign of population invention $\left(N_2 - \frac{g_2}{g_1}N_1\right)$ should be positive for amplifying media. Here, g_2 and g_1 are the degeneracies for upper and lower level, respectively. This expression for the gain coefficient may be generalized to include the refractive index of the host medium *i.e.*,

$$g(v) = \frac{\lambda^2 A_{21}}{8\pi n^2}\left(N_2 - \frac{g_2}{g_1}N_1\right)S(v) \tag{13.25}$$

where n is the refractive index at the frequency v. This modification is signified in solid state lasers, where n may differ appreciably from unity.

Here λ = Wavelength of radiation

$$A_{21} = \text{Einstein's coefficient for spontaneous on the } 2 \rightarrow 1 \text{ transition}$$

$$N_2, N_1 = \text{Number of atoms per unit volume in levels 2 and 1}$$

and $\quad S(v) = \text{Line-shape function}$

Threshold Gain

In a laser, there is not only an increase in the number of cavity photons due to stimulated emission but also a decrease in the number of photons due to scattering, absorption of radiation at the mirrors and output coupling of radiation in the form of the usable laser beam. To sustain laser oscillation, the stimulated amplification must be sufficient to overcome these losses. This sets a lower limit on the gain coefficient $g(v)$, below which laser oscillation does not occur. The condition that the gain coefficient is greater than or equal to this lower limit is called the threshold condition for laser oscillation.

In general, the scattering and absorption of radiation within the gain medium of active atoms is quite small compared to the loss occurring at the mirrors of the laser. Therefore, we will consider in detail only the losses associated with the mirrors.

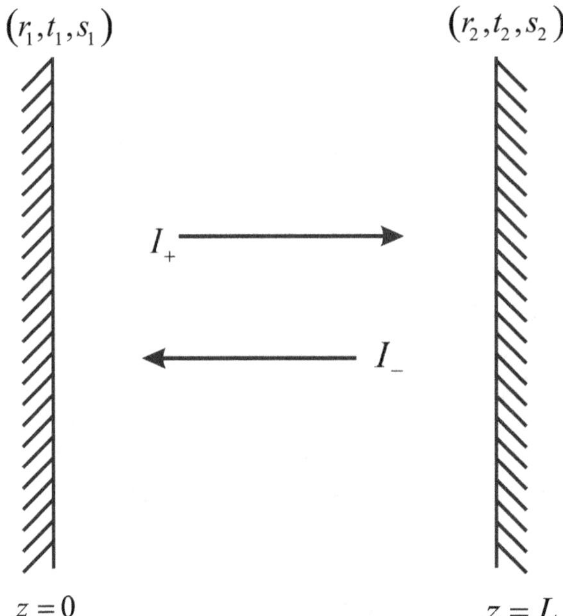

Fig. 13.3 *Two oppositely propagating beams in a laser resonator*

This figure shows an empty space bounded on two sides by highly reflecting mirrors, *i.e.*, laser resonator. A beam of intensity I incident upon one of these mirrors is transformed into a reflected beam of intensity rI, where r is the reflection coefficient of the mirror. A beam of intensity tI (where t is the transmission coefficient) passes through the mirror. Therefore, from the law of conservation of energy

$$r + t = 1 \tag{13.26}$$

Actually, some of the incident beam may be absorbed by the mirror or scattered away from the mirror surface because it is not perfectly smooth. Thus, the law of conservation of energy takes the form

$$r + t + s = 1 \tag{13.27}$$

where r = the fraction of power reflected

t = the fraction of power transmitted

and s = the fraction of power that is absorbed or scattered by the mirror.

Each of the mirrors of laser resonator is characterized by a set of coefficients r, t and s. For the mirror at $z = L$, we have

$$I_v^-(L) = r_2 I_v^+(L) \tag{13.28 a}$$

And for the mirror at $z = 0$

$$I_v^+(0) = r_1 I_v^-(0) \tag{13.28 b}$$

Equations (13.28) are boundary conditions that must be satisfied by the solution of the equations describing the propagation of intensity inside the laser cavity.

Now we will write the equations describing the propagation of intensity inside the laser cavity. Here we are interested only in steady state or continuous wave laser oscillations. The intra-cavity intensity is very small near the threshold of laser oscillations. Therefore, the equation for light propagating in the positive z-direction may be written as

$$\frac{dI_v^+}{dz} = g(v) I_v^+ \tag{13.29 a}$$

where g may be taken to be constant. Similarly, the equation for light propagating in the negative z direction in the same gain medium is

$$\frac{dI_v^-}{dz} = -g(v) I_v^- \tag{13.29 b}$$

The solutions of these equations are

$$I_\nu^+(z) = I_\nu^+(0)e^{g(\nu)z} \qquad (13.30\ a)$$

and $\quad I_\nu^-(z) = I_\nu^-(L)\exp\left[g(\nu)(L-z)\right] \qquad (13.30\ b)$

Thus, from eq. (13.30a) the right going beam intensity at the right mirror (Z=L) is

$$I_\nu^+(L) = I_\nu^+(0)e^{g(\nu)L} \qquad (13.31)$$

and the left going beam intensity at the left mirror(z=0) is

$$I_\nu^-(0) = I_\nu^-(L)e^{g(\nu)L} \qquad (13.32)$$

In steady state, the left going beam has a fraction r_1 of itself reflected at the left mirror (z=0) and this fraction is just the right x going beam at z=0. A similar consideration applies at the right mirror. Thus, we get by using eqs. (13.28), (13.31) and (13.32)

$$I_\nu^+(0) = r_1 I_\nu^-(0) = r_1\left[e^{g(\nu)L}I_\nu^-(L)\right] = r_1 e^{g(\nu)L}\left[r_2 I_\nu^+(L)\right]$$

$$I_\nu^+(0) = r_1 r_2 e^{g(\nu)L}\left[I_\nu^+(0)e^{g(\nu)L}\right] = \left[r_1 r_2 e^{2g(\nu)L}\right]I_\nu^+(0) \qquad (13.33)$$

Similar manipulations applied to any of the quantities $I_\nu^+(L)$, $I_\nu^-(L)$ and $I_\nu^-(0)$ lead to the same result. Therefore If $I_\nu^+(0)$ is not zero, then at steady state we must have

$$r_1 r_2 e^{2gL} = 1 \qquad (13.34)$$

The steady state value of gain that allows equation (13.34) to be satisfied is also the value at which laser action begins. There is net attenuation of I_ν in the cavity for smaller values. Thus the value of g that satisfies equation (13.34) is marked g_t and called the threshold gain, $i.e.$,

$$\ln\left(r_1 r_2 e^{2g_t L}\right) = 0$$

$$\Rightarrow \quad \ln\left(r_1 r_2\right) + 2g_t L = 0$$

$$\Rightarrow \quad g_t = -\frac{1}{2L}\ln\left(r_1 r_2\right) \qquad (13.35)$$

For high reflectivities $r_1 r_2 \approx 1$, then we define $r_1 r_2 = 1-x$ or $x = 1-r_1 r_2$ and use the first term in the Taylor series expansion $\ln(1-x) \approx -x$ when $x \ll 1$. Thus equations (13.35) becomes

$$g_t \approx \frac{1}{2L}\left(1 - r_1 r_2\right)$$ (13.36)

The derivation of equation (13.34) assumes that the gain medium fills the entire distance L between the mirrors. This assumption is valid for many solid state lasers in which the ends of the gain medium are polished and coated with reflecting material. However, in gas and liquid lasers, the gain medium is usually contained in a cell of length $l < L$ as shown in fig. 13.4.

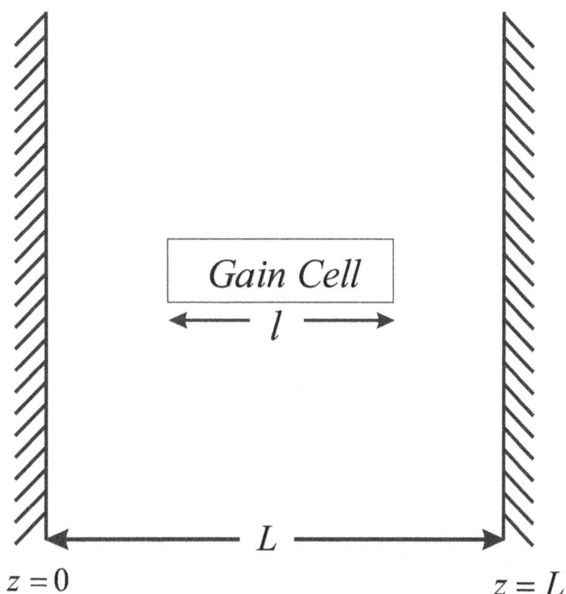

Fig 13.4 *Gain medium between the mirrors*

In this case, the threshold conditions is

$$g_t = -\frac{1}{2l}\ln\left(r_1 r_2\right) \approx \frac{1}{2l}\left(1 - r_1 r_2\right) \text{ for high reflectivities}$$ (13.37)

This threshold condition assumes that losses occur only at the mirrors. These losses are associated with transmission through mirrors, absorption by the mirrors and scattering off the mirrors into non-lasing modes. Absorption and scattering are minimized as much as possible by using mirrors of high optical quality.

If we count other losses arise due to scattering and absorption within gain medium, then the threshold condition is modified as follows:-

$$g_t = -\frac{1}{2l}\ln\left(r_1 r_2\right) + a$$ (13.38)

where a is the effective loss per unit length associated with these additional losses. Such losses are usually small, but they are not difficult to account for in the threshold condition.

13.6 Illustrative Examples

Example 13.1 The wavelength of emission is 6000 A^0 and the lifetime $\left(\tau_{sp}\right)$ is 10^{-6} sec. Determine the coefficient for the stimulated emission.

Sol. The coefficient for stimulated emission is given by

$$B_{21} = \frac{C^3 A_{21}}{8\pi h v^3}$$

Given $A_{21} = \dfrac{1}{\tau_{sp}} = 10^6 \sec^{-1}$ and $\lambda = 6\times10^{-7}\,m$

$$\therefore \quad B_{21} = \frac{\lambda^3 A_{21}}{8\pi h} = \frac{\left(6\times10^{-7}\right)^3 \times 10^6}{8\times3.14\times6.626\times10^{-34}}$$

$$= \frac{216\times10^{-19}}{166.45}$$

$$B_{21} = 1.29\times10^{19}\,\frac{m^3}{J.S^2} \qquad\qquad \textbf{Ans.}$$

Example 13.2 The length of a laser tube is 15 cm and the gain factor of the laser material is 0.0005 /cm. If one of the cavity mirrors reflects 100%light that is incident on it, what is the required reflectance of the other cavity mirror?

Sol. The gain factor

$$g_t = \frac{1}{2L}\ln\frac{1}{r_1 r_2}$$

$$\therefore \quad r_2 = \frac{1}{r_1 e^{2Lg_t}} = \frac{1}{1\times e^{2\times15\times0.0005}}$$

$$r_2 = 0.958$$

It means the second mirror should have a reflectance of 98.5%. **Ans.**

Example 13.3 A typical He-Ne laser of wavelength 632.8 nm have a gain cell of length l=50 cm and mirrors with reflectivities $r_1 = 0.998$ and $r_2 = 0.980$. Calculate the value of threshold gain.

Sol. Given $\lambda = 632.8\ nm$, $l = 50 cm$, $r_1 = 0.998$ and $r_2 = 0.980$

The threshold gain is

$$g_t = -\frac{1}{2l}\ln\left(r_1\, r_2\right)$$

$$= -\frac{1}{2 \times 50}\ln\left(0.998\right)\left(0.980\right)$$

$$gt = 2.2 \times 10^{-4}\, cm^{-1} \qquad\qquad \textbf{Ans.}$$

13.7 Self-Learning Exercise

Q.1 The upper and lower level population densities change in time due to ……………

Q.2 Write the Einstein's coefficient for stimulated absorption.

Q.3 If the atom is initially in the upper state, how many processes can occur.

Q.4 The probability of stimulated emission is proportional to……………….

Q.5 How can distinguish amplifying media from absorbing media?

Q.6 What is the threshold gain for high reflectivities?

Q.7 Why not laser action can achieve in higher frequency ranges?

Q.8 The intra-cavity field is spatially uniform when……………………

Q.9 How can minimized the absorption and scattering losses as much as possible.

Q.10 What is relation between electromagnetic energy density $\left(u_v\right)$ and photon flux $\left(\phi_v\right)$.

13.8 Summary

In this chapter, we have introduced some fundamental concepts such as gain and threshold. By giving the these concepts, the population rate equations for our understanding of lasers are discussed in detail. These equations generally include stimulated emission, absorption spontaneous emission and collisions. Einstein's coefficients and relations among these coefficients have also been derived in this unit. In the end of unit, some examples on above topics are given.

Non Radiative Decay : When the energy difference between two energy level $\left(E_2 - E_1\right)$ is delivered in some form of energy other than EM radiation (photon) *i.e.,* it may go into the kinetic or internal energy of the surrounding atoms or molecules, then the phenomena is called non radiative decay.

Population of the Level : The number of atoms or molecules per unit volume occupy a given energy level.

Amplifier Material – When population inversion is positive, then the material behaves as an amplifier.

Laser Amplifier: If the transition frequency falls in the optical region, the amplifier is called a laser amplifier.

Laser Oscillation : Laser oscillation begins when the gain of the active material compensates the losses in the laser.

Population Rate Equations : The populations of energy levels of the lasing medium change under the action of radiation. These changes can be described by population rate equations.

Threshold Gain : It is a lower limit of gain, below which laser oscillations does not occur.

Inelastic Collision : An inelastic collision is a collision in which kinetic energy is not conserved due to the action of internal friction. In collisions of macroscopic bodies, all kinetic energy is turned into vibrational energy of the atoms, causing a heating effect, and the bodies are deformed.

Degeneracy : Two or more different states of a quantum mechanical system are said to be degenerate if they give the same value of energy upon measurement.

Spectral Energy Density: It is the amount of energy (Photons) stored in a given system or region of space per unit volume.

Laser (Optical) Cavity : It is an arrangement of mirrors that forms a standing wave cavity resonator for light waves. Optical cavities are a major component of lasers, surrounding the gain medium and providing feedback of the laser light.

Ans.1: Stimulated emission, absorption, spontaneous emission and collisions.

Ans.2: B_{12}

Ans.3: Two (i) Spontaneous emission and (ii) Stimulated emission.

Ans.4: The spectral energy density $u(v)$.

Ans.5: From the sign of population inversion.

Ans.6: $g_t = \dfrac{1}{2l}\left(1 - r_1 r_2\right)$

Ans.7: The ratio of coefficients of spontaneous emission of stimulated emission is proportional to the third power of frequency of the radiation. That's why it is difficult to achieve laser action in higher frequency ranges.

Ans.8: $\left(1 - r_1 r_2\right)$ is small, *i.e.*, mirrors are highly reflecting.

Ans.9: Using mirrors of high optical quantity.

Ans.10: $u_v = \dfrac{hv}{c}\,\phi_v$

13.11 Exercise

Section A: Very Short Answer Type Questions

Q.1 What do you mean by the population rate equations?

Q.2 What is the probability of stimulated emission?

Q.3 What will be threshold condition for gain in the case of distributed losses?

Q.4 Write the formula of gain coefficients.

Q.5 Write the rate at which intensity is lost due to the imperfect reflectivity of the mirrors.

Section B: Short Answer Type Questions

Q.6 Write short note on the gain at the resonant frequency.

Q.7 Define the Einstein's coefficients and write relation among these coefficients.

Q.8 What do you mean by the threshold condition for laser?

Q.9 Give the brief explanation about the relationship between the Einstein's coefficients A and B.

Q.10 Give the interpretation of following equation.

$$\frac{d}{dt}\left(n_2 + q_v\right) = -\left(\Gamma_2 + A_{21}\right)n_2 + p - \frac{c}{2L}\left(1 - r_1 r_2\right)q_v$$

Section C: Long Answer Type Questions

Q.11 Derive the relation between Einstein's coefficients and discuss the result.

Q.12 Derive the population rate equations for lasers.

Q.13 What is the threshold condition of gain for laser oscillations and derive it's formula.

References and Suggested Readings

1. A.K. Ghatak and K. Thyagarajan, Optical Electronics, Cambridge press, New York, 1989.

2. O. Svelto and D.C. Hanna, Principles of Lasers, Plenum Press, New York, 1976.

3. P.W. Milonni and J.H. Eberly, Laser Physics, John Wiley & sons, New York, 2010.

4. W.T. Silfast, Laser Fundamentals, Cambridge press, New York, 1996.

5. K. Thyagarajan and A.K. Ghatak, Lasers-Theory & Applications, Macmillan, 1981.

UNIT-14

Laser Oscillation: Three & Four Level Laser Schemes Spatial Hole Burning

Structure of the Unit

14.0 Objectives

In the 13 unit, we introduced certain concepts related to laser oscillation, gain and population rate equations, and indicated their importance in our understanding of lasers. We have also derived the relations between Einsten's Coefficients. These relations state that the probability of stimulated emission is equal to the probability of the absorption and the ratio of coefficients of spontaneous emission to stimulated emission is proportional to the third power of frequency of the incident radiation. These relations help to achieve the basic conditions for laser action.

In this unit, we will briefly discuss the concept of feedback and threshold for a laser action. We will also obtain an expression for rate equations for photons and populations. This is followed by a discussion of three-level-laser scheme and four-level-laser schemes using the rate equation approach. Finally, we will explain about the small signal gain and gain saturetion, and spatial hole buring.

14.1 Introduction

Laser is a physical system of atoms or molecules (lasing or active medium) between two mirrors. Some of these atoms are promoted to excited states due to pumping process. The excited atoms begin radiating spontaneously like an ordinary fluorescent lamp. A spontaneously emitted photon can induce an excited atom to emit another photon of the same frequency and direction as the first. The more such photons are produced by stimulated emission, because the stimulated emission rate is proportional to the flux of photons already in the stimulating field. The mirrors of laser keep photons from escaping completely, so that they can be redirected into the active laser medium to stimulate the emission of more photons. By making one of the mirror partially reflector (90% reflection or 10% transmission) bunch of photons are allowed to escape to constitute the output laser beam. The intensity of the output laser beam is determined by the rate of production of excited atoms, the reflectivities of the mirrors and certain properties of the active medium.

14.2 Feedback

Consider two arbitrary energy levels 1 and 2 of a given material with g_1-Fold and g_2-Fold degenerate, respectively. Also let N_1 and N_2 be their respective population. The material behaves as an amplifier if $N_2 > \dfrac{g_2 N_1}{g_1}$. In this case we say that there exists a population inversion in the material. A material in which this population inversion is produced is referred to as an active medium. If the transition frequency falls in the optical region, the amplifier is called a laser amplifier.

To make an oscillator from an amplifier, it is necessary to introduce suitable positive feedback. In the case of a laser, feedback is often obtained by placing the active material between two highly reflecting mirrors. In this case, a plane EM wave (photon) traveling in a direction perpendicular to the mirrors bounced back and forth between the two mirrors, and is amplified on each passage through the active material. If one of the two mirrors is partially transparent, a useful output beam is obtained from that mirror.

14.3 Threshold

In order to generate radiation or photon, the amplifying medium is placed in an optical resonator, which consists of a pair of mirrors facing each other. Radiation, which bounced back and forth between the mirrors is amplified by the amplifying medium and also suffers losses due to the finite reflectivity of the mirrors, output coupling and other scattering and diffraction losses. If the oscillations have to be sustained in the laser cavity, then the losses must be exactly compensated by the gain. This condition is known as the threshold condition for laser oscillation.

The gain at threshold value is given by

$$\gamma = \alpha_1 - \frac{1}{2d}\ln r_1 r_2$$

where γ = gain coefficient

α_1 = Attenuation coefficient due to all loss mechanism other than the finite reflectivity.

d= Length of resonator cavity

r_1 & r_2 = Reflectivities of the mirror of resonator cavity.

14.4 Rate Equations For Photons and Populations

The equations describing the propagation of the light inside the laser cavity for continuous wave laser oscillations in positive and negative z directions are as follows:-

$$\frac{dI_v^+}{dz} = g(v)I_v^+ \tag{14.1 a}$$

and $\quad -\dfrac{dI_v^{\,-}}{dz} = g(v)I_v^{\,-}$ $\hspace{4cm}$ (14.1 b)

We must include the time derivative $\dfrac{\partial I_v}{\partial t}$ in the propagation equation to describe time-dependent phenomena. Therefore equation (14.1) can be written as

$$\dfrac{\partial I_v^{\,+}}{\partial z} + \dfrac{1}{c}\dfrac{\partial I_v^{\,+}}{\partial t} = g(v)I_v^{\,+} \hspace{3cm} (14.2\ a)$$

And $\quad -\dfrac{\partial I_v^{\,-}}{\partial z} + \dfrac{1}{c}\dfrac{\partial I_v^{\,-}}{\partial t} = g(v)I_v^{\,-}$ $\hspace{2.5cm}$ (14.2 b)

Addition of equations (14.2 a) and (14.2 b) gives

$$\dfrac{\partial}{\partial z}\left(I_v^{\,+} - I_v^{\,-}\right) + \dfrac{1}{c}\dfrac{\partial}{\partial t}\left[I_v^{\,+} + I_v^{\,-}\right] = g(v)\left[I_v^{\,+} + I_v^{\,-}\right] \hspace{2cm} (14.3)$$

Sine in many lasers, there is very small variation of $\left(I_v^{\,+} - I_v^{\,-}\right)$ with z. Therefore equation (14.3) may be written as

$$\dfrac{d}{dt}\left[I_v^{\,+} + I_v^{\,-}\right] = cg(v)\left[I_v^{\,+} + I_v^{\,-}\right] \hspace{3cm} (14.4)$$

If the gain medium does not completely fill the resonator (i.e. l<L) as shown in figure 14.1, then $g(v) = 0$ outside it.

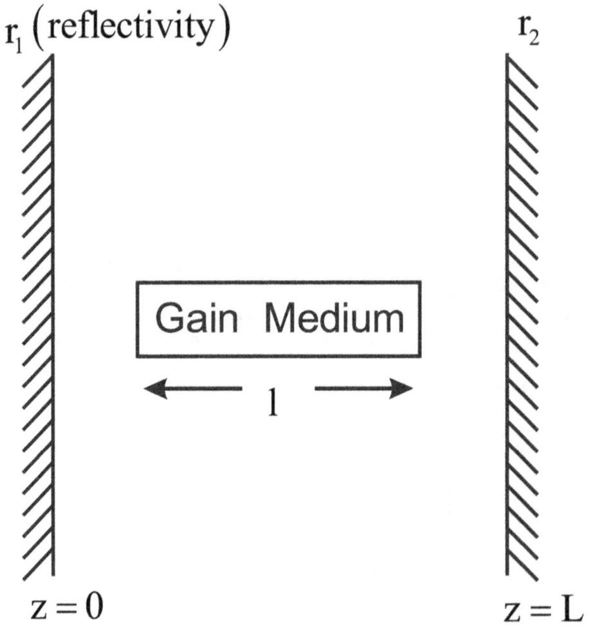

Fig. 14.1 *A laser in which gain medium does not completely fill the resonator.*

In this case, equations (14.4) can be written as

$$\frac{d}{dt}\left[I_\nu^{+}+I_\nu^{-}\right]=\frac{cl}{L}g(\nu)\left[I_\nu^{+}+I_\nu^{-}\right] \tag{14.5}$$

Since the number of photons inside the cavity is proportional due to the total intensity. Therefore equation (14.5) may write as

$$\frac{dq_\nu}{dt}=\frac{cl}{L}g(\nu)q_\nu \tag{14.6}$$

Where q_ν is the number of cavity photon associated with the frequency ν.

Equation (14.6) describes the growth in time of the number of cavity photons as a result of the absorption and stimulated emission of photons by the gain medium. The factor $\dfrac{cg(\nu)l}{L}$ is the growth rate.

Now we consider the loss associated with the output coupling of laser radiation from the cavity. In this case, radiation reflected from the mirror z=L has an intensity that is r_2 times the incident intensity. After it is reflected from the mirror at z=0, it has an intensity r_1r_2 times its intensity before the round trip inside the resonator. In other words, a fraction $\left(1-r_1r_2\right)$ of intensity is lost. Since the time it takes to make a round trip is $\dfrac{2L}{c}$, the rate at which intensity is lost due to the imperfect reflectivity of the mirrors is $c\left(1-r_1r_2\right)/2L$. In terms of photons, this loss rate is

$$\left(\frac{dq_\nu}{dt}\right)_{output\ coupling}=-\frac{c}{2L}\left(1-r_1r_2\right)q_\nu \tag{14.7}$$

Therefore, the total rate at which the number of cavity photons changes is

$$\frac{dq_\nu}{dt}=\left(\frac{dq_\nu}{dt}\right)_{gain}+\left(\frac{dq_\nu}{dt}\right)_{output\ coupling}$$

$$=\frac{cl}{L}g(\nu)q_\nu-\frac{c}{2L}\left(1-r_1r_2\right)q_\nu$$

$$\frac{dq_v}{dt} = \frac{cl}{L}g(v)q_v - \frac{cl}{L}g_t q_v \tag{14.8}$$

where $g(v)$ = gain coefficient = $\dfrac{\lambda^2 A}{8\pi}\left(N_2 - \dfrac{g_2}{g_1}N_1\right)S(v)$

and g_t = threshold gain = $\dfrac{1}{2l}(1 - r_1 r_2)$

Equation (14.8) gives the rate of change with time of the number of cavity photons. This equation may write in term of the total intensity inside the cavity.

$$\frac{dIv}{dt} = \frac{cl}{L}g(v)I_v - \frac{cl}{L}g_t I_v \tag{14.9}$$

If we assume equal upper and lower level degeneracies are equal, *i.e.*, $g_1 = g_2$, then we may write equation (14.9) as

$$\frac{dI_v}{dt} = \frac{cl}{L}\frac{\lambda^2 A}{8\pi}(N_2 - N_1)S(v)I_v - \frac{c}{2L}(1 - r_2 r_2)I_v \tag{14.10}$$

14.5 Three-Level Laser Scheme

It is not possible to achieve steady state population inversion in two-level-laser scheme. The mechanism we use in two level laser, scheme to excite atoms to level 2 can also de-excite them, i.e., if we try to pump atoms from level 1 (ground level) to level 2 (excited level) by irradiating the medium, the radiation will induce both upward transitions $1 \rightarrow 2$ (due to absorption) and downward transitions $2 \rightarrow 1$ (due to stimulated emission). This optical pumping process produces nearby equal number of atoms in level 2 as in level 1. Therefore, we cannot obtain a positive steady-state population inversion using only two atomic levels in the pumping process. Thus, in order to produce a steady state population inversion, one makes use of either a three-level-level or a four-level laser scheme. In this section, we shall discuss a three level laser scheme.

We consider a three-level scheme consisting of energy levels E_1, E_2 and E_3 all of which are assumed to be nondegenerate. Let N_1, N_2 and N_3 represent the population densities of the three levels as shown in Figure 14.2.

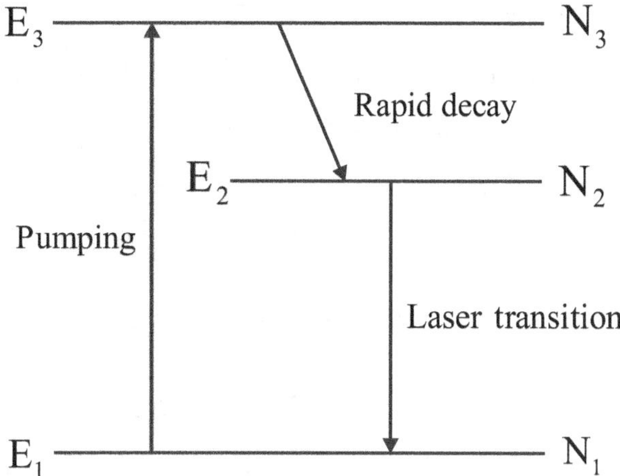

Fig 14.2 *A three-level laser scheme*

In such a laser, pumping process acts between level 1 and level 3. Due to this pumping, atoms lift from level 1 to level 3, from which they decay rapidly to level 2 through some non-radiative process. Thus, the key to the three-level inversion scheme is to have atoms in the pumping level 3 drop very rapidly to the upper laser level 2. This accomplishes two purposes:

(i) The pumping from level 1 is directly from level 1 to the upper laser level 2. Because every atom finding itself in level 3 converts quickly to an atom in level 2 and

(ii) The rapid depletion of level 3 does not give the pumping process much chance to act in reverse and repopulate the ground level 1.

Now we will write the rate equations describing the rate of change of N_1, N_2 and N_3.

The rate of the population N_1 of atoms per cubic centimeter in level 1 due to pumping is

$$\left(\frac{dN_1}{dt}\right)_{pumping} = -PN_1 \tag{14.11}$$

Since the pumping takes atoms from level 1 to level 3 and level 3 is assumed to decay very rapidly to level 2, we may also write the rate of change of population of level 2 due to pumping

$$\left(\frac{dN_2}{dt}\right)_{pumping} \approx \left(\frac{dN_3}{dt}\right)_{pumping} = -\left(\frac{dN_1}{dt}\right)_{pumping} = PN_1 \tag{14.12}$$

Now we assume that level 2 decays only into level 1 by spontaneous emission of via collisions and we will denote the rate by Γ_{21}. Then the population changes associated with the decay of level 2 is

$$\left(\frac{dN_2}{dt}\right)_{decay} = -\Gamma_{21}N_2, \left(\frac{dN_1}{dt}\right)_{decay} = \Gamma_{21}N_2 \tag{14.13}$$

Therefore total rates of changes of the population of levels 1 and 2 are

$$\frac{dN_1}{dt} = -PN_1 + \Gamma_{21}N_2 + \sigma\phi_v\left(N_2 - N_1\right) \tag{14.14 a}$$

$$\frac{dN_2}{dt} = PN_1 - \Gamma_{21}N_2 - \sigma\phi_v\left(N_2 - N_1\right) \tag{14.14 b}$$

The last terms in eq. (14.14) is due to stimulated emission.

$$\Rightarrow \quad \frac{d}{dt}\left(N_1 + N_2\right) = 0$$

$$\Rightarrow \quad N_1 + N_2 = N_T = \text{constant} \tag{14.15}$$

Here we are assuming that each active atom of the gain medium must be either in level 1 or level 2 since level 3 decays practically instantaneously into level 2. Therefore the conserved quantity N_T is simply the total number of active atoms per unit volume.

Using equation (14.14), we can determine the threshold pumping rate necessary to achieve a population inversion, together with the threshold power expended in the process.

In the steady state, N_1 and N_2 are not changing in time, i.e., $\dfrac{dN_1}{dt} = \dfrac{dN_2}{dt} = 0$ and denoted steady state values by $\overline{N_1}$ and $\overline{N_2}$. The number of cavity photon $\left(\phi_v\right)$ near threshold is small enough that stimulated emission may be omitted from eq. (14.14). Therefore for steady state eq. (14.14) becomes

$$\overline{N_2} = \frac{P}{\Gamma_{21}}\overline{N_1} \tag{14.16}$$

Since (14.15) must hold for all possible values of N_1 and N_2 including the steady state values $\overline{N_1}$ and $\overline{N_2}$, we also have

$$\overline{N_1} + \overline{N_2} = N_T \tag{14.17}$$

From eqs. (14.16) and (14.17), we get

$$\overline{N_1} = \frac{\Gamma_{21}}{P + \Gamma_{21}} N_T \tag{14.18 a}$$

and $\quad \overline{N_2} = \dfrac{P}{P + \Gamma_{21}} N_T \tag{14.18 b}$

Therefore steady state threshold region population inversion is

$$\overline{N_2} - \overline{N_1} = \frac{P - \Gamma_{21}}{P + \Gamma_{21}} N_T \tag{14.19}$$

To have a positive steady-state population inversion, i.e., a positive gain, we must have

$$P > \Gamma_{21} \tag{14.20}$$

Which simply says that the pumping rate into the upper laser level must exceed the decay rate. The greater the pumping rate with respect to the decay rate, the greater the population inversion and gain.

The pumping of an atom from level 1 to level 3 requires an energy

$$E_3 - E_1 = h\nu_{31} \tag{14.21}$$

Therefore, the power per unit volume delivered to the active atoms in the pumping process is

$$\frac{\text{Power}}{V} = h\nu_{31} P \overline{N_1} \tag{14.22}$$

Put the value of $\overline{N_1}$ from eq. (14.18 a) to (14.22), we get

$$\frac{\text{Power}}{V} = \frac{h\nu_{31} P \Gamma_{21}}{P + \Gamma_{21}} N_T \tag{14.23}$$

where V is the volume of the active medium.

For the minimum pumping rate necessary to reach positive gain, eq. (14.20) may write

$$P_{min} = \Gamma_{21} \tag{14.24}$$

Substituting P_{min} for P in equation (14.23), we obtain

$$\left(\frac{\text{Power}}{V}\right)_{\min} = \frac{1}{2}\Gamma_{21}N_T h\nu_{31} \tag{14.25}$$

It is minimum power per unit volume that must be exceeded to produce a positive gain. With this amount of pumping power delivered to the active medium, half the active atoms are in the lower level of the laser transition and half are in the upper level [see from eq. 14.18(a) and (b)]. A pumping power density greater than (14.25) makes $\overline{N_2} > \overline{N_1}$.

14.6 Four-Level Laser Scheme

This scheme is also useful model for achieving population inversion. In three-level laser scheme, the lower level has the inversion. In three-level laser scheme, the lower level was the ground level and one has to lift more than 50% of the atoms in the ground level in order to obtain population inversion. This problem can be overcome by using another level of the atomic system and having the lower laser level also as an excited level. This type scheme is four-level laser scheme as shown in figure 14.3.

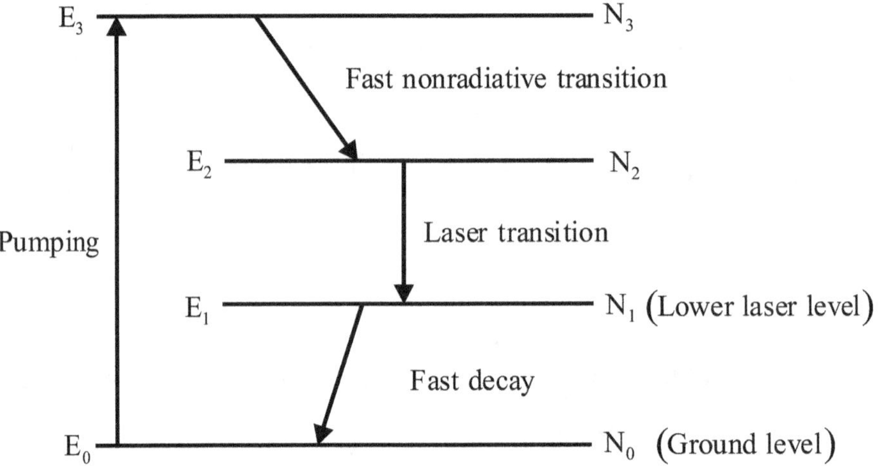

Fig. 14.3 *A four-level laser scheme*

Level 0 is the ground level and levels 1, 2 and 3 are excited levels of scheme. Atoms from level 0 are pumped to level 3 from where they make a fast non-radiative relaxation to level 2. Level 2, which corresponds to the upper laser level is usually a metastable level having a long lifetime. The transmission from level 2 to level 1 forms the laser transition. In order that atoms do not accumulate in level 1 and hence destroy the population inversion between levels 2 and 1, level 1 must have a very small lifetime so that atoms from level 1 are quickly removed to level 0 ready for pumping to level 3. If the relaxation rate of atoms from level 1 to level 0 is faster than the rate of arrival of atoms to level 1, then one can obtain population inversion between levels 2 and 1 even for very small pump powers. Level 3 is not a laser level, it can be a collection of a large number of levels or a broad level. In such a case, an optical pump source emitting over a broad range of frequencies can be used to pump atoms from level 0 to level 3 effectively. In addition, level 1 required to be sufficiently above the ground level so that level 1 is almost unpopulated at ordinary temperatures.

Now we shall write the rate equations corresponding to the populations of the four levels. The decay from level 3 to level 2 is extremely rapid or instantaneous. Thus we may take $N_3 \approx 0$ and the population rate equations for the four-level laser take the form

$$\frac{dN_0}{dt} = -PN_0 + \Gamma_{10}N_1 \qquad (14.26\ a)$$

$$\frac{dN_1}{dt} = -\Gamma_{10}N_1 + \Gamma_{21}N_2 + \sigma(\nu)(N_2 - N_1)\phi_\nu \qquad (14.26\ b)$$

$$\frac{dN_2}{dt} = PN_0 - \Gamma_{21}N_2 - \sigma(\nu)(N_2 - N_1)\phi_\nu \qquad (14.26\ c)$$

Where P is the pumping rate out of the ground Level 0 and PN_0 is the upper level pumping rate. Γ_{21} and Γ_{10} are the rates for the decay processes $2 \rightarrow 1$ and $1 \rightarrow 0$, respectively.

$$\Rightarrow \quad \frac{d}{dt}(N_0 + N_1 + N_2) = 0$$

$$\Rightarrow \quad N_0 + N_1 + N_2 = \text{constant.} \qquad (14.27)$$

If the stimulated emission rate is very small compared to the pumping and decay rates then it may be omitted from eq. (14.26). In the steady state condition

$$\frac{dN_0}{dt} = \frac{dN_1}{dt} = \frac{dN_2}{dt} = 0 \text{ denoted by } \overline{N_0}, \overline{N_1} \text{ and } \overline{N_2}. \text{ Thus steady state}$$

populations from equations (14.26) and (14.27).

$$\overline{N_0} = \frac{\Gamma_{10}\Gamma_{21}}{\Gamma_{10}\Gamma_{21} + \Gamma_{10}P + \Gamma_{21}P}N_T \qquad (14.28\ a)$$

$$\overline{N}_1 = \frac{\Gamma_{21}P}{\Gamma_{10}\Gamma_{21} + \Gamma_{10}P + \Gamma_{21}P}N_T \qquad (14.28\ b)$$

$$\overline{N}_2 = \frac{\Gamma_{10}P}{\Gamma_{10}\Gamma_{21} + \Gamma_{10}P + \Gamma_{21}P}N_T \qquad (14.28\ c)$$

Therefore, the steady state population inversion of the laser transition is

$$\overline{N}_2 - \overline{N}_1 = \frac{P(\Gamma_{10} - \Gamma_{21})N_T}{\Gamma_{10}\Gamma_{21} + \Gamma_{10}P + \Gamma_{21}P} \qquad (14.29)$$

Thus, the pumped $(P \neq 0)$ four level scheme will always have a steady state population inversion when

$$\Gamma_{10} > \Gamma_{21} \qquad (14.30)$$

i.e., when the lower laser level decays more rapidly than the upper laser level.

14.7 Small Signal Gain and Gain Saturation

Equation (14.19) gives the steady-state population inversion for a three-level laser, when the stimulated emission rate is negligible. In general, the stimulated emission rate is not negligible. In this case, the steady-state population inversion [using eqs. (14.14 a) & (14.14 b)] is

$$\overline{N}_2 - \overline{N}_1 = \frac{(P - \Gamma_{21})N_T}{P + \Gamma_{21} + 2\sigma\phi_v} \qquad (14.31)$$

Here ϕ_v is the steady state (time independent) cavity photon flux. Therefore steady state gain coefficient for a three level laser is

$$g(v) = \frac{\sigma(v)(P - \Gamma_{21})N_T}{P + \Gamma_{21} + 2\sigma(v)\phi_v}$$

(Assuming $g_1 = g_2$)

$$= \frac{\sigma(v)(P - \Gamma_{21})N_T}{P + \Gamma_{21}} \cdot \frac{1}{1 + \left[\dfrac{2\sigma(v)\phi_v}{P + \Gamma_{21}}\right]}$$

$$g(v) = \frac{g_0(v)}{1 + \dfrac{\phi_v}{\phi_v\,\text{sat}}} = \frac{g_0(v)}{1 + \dfrac{I_v}{I_v\,\text{sat}}} \qquad (14.32)$$

where $g_0(\nu)$ = small signal gain = $= \dfrac{\sigma(\nu)(P-\Gamma_{21})N_T}{P+\Gamma_{21}}$

$$\text{(14.33)}$$

and $\quad \phi_\nu^{\text{sat}}$ = saturation flux $= \dfrac{P+\Gamma_{21}}{2\sigma(\nu)}$

$$\text{(14.34)}$$

The corresponding expressions for the saturation intensity and photon number are

$$I_\nu^{\text{sat}} = h\nu\phi_\nu^{\text{sat}} = \frac{h\nu(P+\Gamma_{21})}{2\sigma(\nu)} \tag{14.35}$$

and $\quad q_\nu^{\text{sat}} = \dfrac{V}{c}\phi_\nu^{\text{sat}} = \dfrac{P+\Gamma_{21}}{2c\sigma(\nu)}V$

$$\text{(14.36)}$$

For $I_\nu \ll I_\nu^{\text{sat}}$, eq. (14.32) becomes

$$g(\nu) = g_0(\nu) \tag{14.37}$$

Therefore $g_0(\nu)$ is called the small signal gain coefficients. The maximum gain is $g_0(\nu_0)$ i.e., the gain when $I_\nu \ll I_\nu^{\text{sat}}$ and the field frequency matches the line center frequency ν_0, where $\sigma(\nu_0)$ has its maximum value. When the line-shape is Lorentzian with HWHM width $\delta\nu_0$, we have

$$g(\nu) = g_0(\nu_0)\frac{1}{(\nu_0-\nu)^2/\delta\nu_0^2 + 1 + \left(\phi_\nu/\phi_{\nu_0}^{\text{sat}}\right)} \tag{14.38}$$

The cavity frequencies at which there is small-signal gain sufficient to overcome loss in a laser are generally those within about $\delta\nu_0$ of line center $(\nu-\nu_0)$, $\delta\nu_0$ can be called the small signal gain band width.

In the case of a four level laser, we can get a similar expression for gain with the flux dependence (14.32), but with a saturation flux twice that given by (14.34). The gain-saturation formulas (14.32) and (14.38) are applicable to a wide variety of actual lasers.

14.8 Spatial Hole Burning

Spatial hole burning occurs as result of the standing wave nature of the optical modes. Since in most lasers, we have standing waves rather than traveling waves.

The gain saturation formulas derived in last section are written with ϕ_ν assumed to be sum of the fluxes of the two traveling waves.

$$\phi_v \rightarrow \phi_v = \phi_v^{+} + \phi_v^{-} \tag{14.39}$$

This is not quite correct, because it ignores the interference of the two traveling waves, the electromagnetic energy density u is proportional to the square at the electric field, i.e.,

$$u = \varepsilon_0 E^2 (\vec{r}.t) = \varepsilon_0 E_0^2 \cos^2 \omega t \sin^2 kz \tag{14.40}$$

We replace $\cos^2 \omega t$ by $\dfrac{1}{2}$, its average value over times long compared to an optical period $\left(\dfrac{2\pi}{\omega} \approx 10^{-14} \, sec \right)$ i.e., equation (14.40) becomes

$$u = \frac{1}{2} \varepsilon_0 E_0^2 \sin^2 kz \tag{14.41}$$

A cavity standing wave field is the sum of two oppositely propagating traveling wave fields, i.e.

$$E(z,t) = E_0 \cos \omega t \sin kz = \frac{1}{2} E_0 \left[\sin (kz - \omega t) + \sin (kz + \omega t) \right]$$

$$= E_{+} (z,t) + E_{-} (z,t) \tag{14.42}$$

$$= \vec{E} \text{ in (+)ve direction} + \vec{E} \text{ (-)ve direction}$$

where $E_{\pm} (z,t) = \dfrac{1}{2} E_0 \sin (kz \mp \omega t)$ \hfill (14.43)

The time averaged square of the electric field (14.43) gives a field energy density $u = u^{+} + u^{-}$ where

$$u^{\pm} = \frac{\varepsilon_0}{8} E_0^2 \tag{14.44}$$

$$\Rightarrow \quad 2 \left(u^{+} + u^{-} \right) = \frac{1}{2} \varepsilon_0 E_0^2 \tag{14.45}$$

From equations (1441) and (14.45), we get

$$u = 2 \left(u^{+} + u^{-} \right) \sin^2 kz \tag{14.46}$$

Therefore $\quad \phi_v = \dfrac{C}{\lambda v} u = 2 \left(\dfrac{c}{hv} u^{+} + \dfrac{c}{hv} u^{-} \right) \sin^2 kz$

$$\Rightarrow \quad \phi_v = 2\left[\phi_v^+ + \phi_v^-\right]\sin^2 kz \tag{14.47}$$

In terms of the intensity $I_v = hv\phi_v$

$$\therefore \quad I_v = 2\left[I_v^+ + I_v^-\right]\sin^2 kz \tag{14.48}$$

Thus, it is not correct to use equation (14.39) as the flux in the gain saturation formulas given by equations (14.32) and (14.38). We should use eq. (14.47), which accounts properly for the interference of the two traveling wave fields. Then the gain saturation formula for a homogeneously broadened transition is

$$g(v) = \frac{g_0(v)}{1 + 2\left[\left(\phi_v^+ + \phi_v^-\right)/\phi_v^{sat}\right]\sin^2 kz} \tag{14.49}$$

This is saturation formula when the standing wave nature of cavity field is properly accounted. The $\sin^2 kz$ term in equation (14.49) is called spatial hole burning in the gain coefficient $g(v)$. At point z for which $\sin^2 kz = 0$, $g(v)$ takes its maximum value called the small signal value $g_0(v)$. When $\sin^2 kz = 1$, $g(v)$ has its minimum value, *i.e.*, it is most strongly saturated (a hole is burned in the curve of $g(v)$ vs. Z as shown in figure 14.4. These holes are separated by

$$\Delta z = \frac{\pi}{k} = \frac{\lambda}{2}.$$

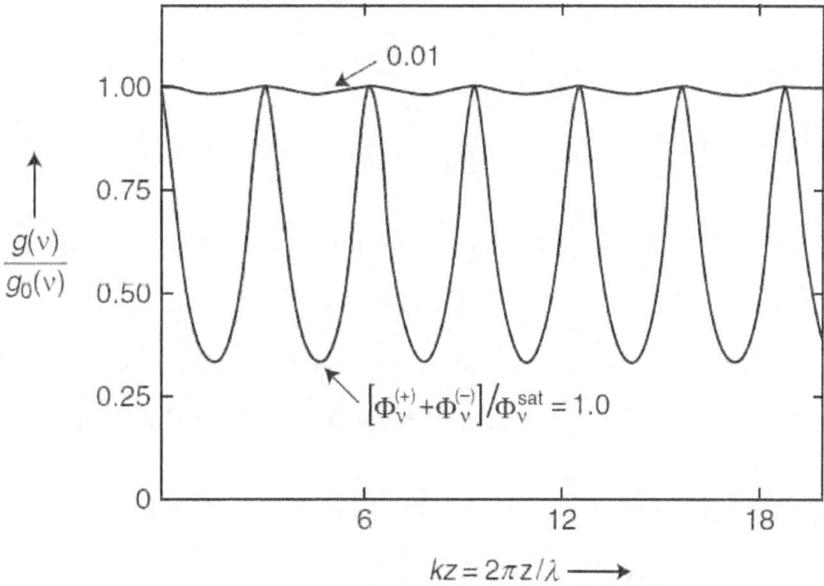

Fig. 8.4 *Spatial hole burning in the gain curve*

Thus, spatial hole burning is a distortion of the gain shape in a laser medium caused by saturation effects of a standing wave.

14.9 Illustrative Examples

Example 1 Calculate the required pump power for a 5-cm long ruby rod of radius 2 mm with following parameter.

$$\Gamma_{21} = \frac{1}{2} \times 10^{-3} \sec, N_T = 1.6 \times 10^{19} cm^{-3} \text{ and } h\nu = 2.25 \text{ eV}$$

Sol. The pumping power density necessary for laser oscillation

$$\left(\frac{\text{Power}}{V}\right)_{min} \approx \frac{1}{2}\Gamma_{21}N_T h\nu$$

$$= \frac{1}{2}\left(\frac{1}{2} \times 10^3\right)\left(1.6 \times 10^{-19}\right)(2.25)\left(1.6 \times 10^{-19}\right)$$

$$= 1.44 \times 10^3 \frac{\text{Joule}}{S \times cm^3}$$

$$\left(\frac{\text{Power}}{V}\right)_{min} = 1.44 \times 10^3 \frac{W}{cm^3}$$

Therefore the required pump power is

$$\text{Power} = \left(1.44 \times 10^3\right)V$$

$$= 1.44 \times 10^3 \times \pi(0.2)^2 (5)$$

$$\text{Power} = 904.32 \text{ W} \qquad\qquad\qquad \textbf{Ans.}$$

Example 2 Estimate the population inversion necessary for a 5 cm long ruby laser of transition 694.3 nm. It has a Lorentzian line shape of width (HWHM) $\delta\nu_0 = 170$ GHz and coefficient for spontaneous emission (A) is 230 s^{-1}. We also assume a resonator with mirror reflectivities $r_1 = 1.0$ and $r_2 = 0.96$, and a scattering loss of 3% per round trip pass through the gain cell.

Sol. Given $\lambda = 694.3$ nm, $\delta\nu_0 = 170$ GHz, A= 230 S^{-1}, $r_1 = 1$, $r_2 = 0.96$, a= 3%, n=1.76 (for ruby) and l = 5 cm.

The line center cross section for stimulated emission is

$$\sigma = \frac{\lambda^2 A}{8\pi n^2} \left(\frac{1}{\pi \delta \nu_0} \right)$$

$$= \frac{\left(634.3 \times 10^{-7} \right)^2 \times 230}{8 \times 3.14 \times \left(1.76 \right)^2 \times 3.14 \times 170 \times 10^9}$$

$$= 2669.3 \times 10^{-23}$$

$$\sigma = 2.66 \times 10^{-20} \, cm^2$$

The threshold gain for laser oscillation is $g_t = -\dfrac{1}{2 \times 5} \ln \left(0.96 \right) + \dfrac{0.03}{2 \times 5}$

$$g_t = 7.1 \times 10^{-3} \, cm^{-1}$$

Therefore threshold population inversion is

$$\Delta N_t = \frac{g_t}{\sigma} = \frac{7.1 \times 10^{-3}}{2.66 \times 10^{-20}}$$

$$\Delta N_t = 2.66 \times 10^{17} \, cm^{-3} \qquad \qquad \textbf{Ans.}$$

Example 3 Calculate the threshold population inversion necessary to achieve lasing in a typical 632.8 nm He-Ne laser with following parameters:

$$g_t = 2.2 \times 10^{-4} \, cm^{-1}, \; A = 1.4 \times 10^{-6} s^{-1} \; \textbf{and} \; s\left(\nu \right) = 6.3 \times 10^{-10} s$$

Solution: The threshold population inversion is given by

$$\Delta N_t = \left(N_2 - \frac{g_2}{g_1} N_1 \right)_t = \frac{8\pi g_t}{\lambda^2 A \, s\left(\nu \right)}$$

$$\Delta N_t = \frac{8 \times 3.14 \times 2.2 \times 10^{-4}}{\left(632.8 \times 10^{-7} \right)^2 \left(1.4 \times 10^6 \right) \left(6.3 \times 10^{-10} \right)}$$

$$= \frac{55.26 \times 10^{-4}}{35.31 \times 10^{-13}}$$

$$\Delta N_t = 1.5 \times 10^9 \, atoms \, / \, cm^3 \qquad \qquad \textbf{Ans.}$$

14.10 Self Learning Exercise

Q.1 Why feedback is required in laser device.

Q.2 Define the term Intensity.

Q.3 What is condition to have a positive steady state population inversion in three-level laser scheme.

Q.4 Write the formula for minimum power per unit volume in case of three-level laser.

Q.5 To achieve Laser oscillation a gain should be greater than the
…………………..

Q.6 Write the condition to have a steady-state population in four-level Laser.

Q.7 The larger the decay rates, the ……………… the saturation flux.

Q.8 When a medium is said to be bleached.

Q.9 The saturation flux is ………………….. to the Einstein coefficient for stimulated emission B.

14.11 Summary

In this unit, we have introduced the concepts of feedback and threshold. The rate equations for photons and populations have been derived. We have used such equations to discuss three and four level lasers. We have also discussed the concept of saturation which is a major consideration in determining how much output power can be obtained with a given laser. In the end of unit, the concept of spatial hole burning is given. It occurs as result of the standing wave nature of the optical modes.

14.12 Glossary

Feedback : To make an oscillator from an amplifier, it is necessary to introduce suitable positive feedback. In the case of a laser, feedback is often obtained by placing the active material between two highly reflecting mirrors.

Threshold : In order to generate radiation or photon, the amplifying medium is placed in an optical resonator, which consists of a pair of mirrors facing each other. Radiation, which bounced back and forth between the mirrors is amplified by the amplifying medium and also suffers losses due to the finite reflectivity of the mirrors, output coupling and other scattering and diffraction losses. If the oscillations have to be sustained in the laser cavity, then the losses must be exactly compensated by the gain. This condition is known as the threshold condition for laser oscillation.

Small Signal Gain Bandwidth : The cavity frequencies at which there is small-signal gain sufficient to overcome loss in a laser are generally those within about of line center, can be called the small signal gain band width.

Spatial Hole Burning : Spatial hole burning occurs as result of the standing wave nature of the optical modes or spatial hole burning is a distortion of the gain shape in a laser medium caused by saturation effects of a standing wave.

Intensity : Intensity refers to the electromagnetic energy flow per unit area per unit time.

14.13 Answer to Self-Learning Exercise

Ans.1: To increase the photon yield.

Ans.2: Intensity refers to the electromagnetic energy flow per unit area per unit time.

Ans.3: The pumping rate into the upper laser level must exceed the decay rate.

Ans.4: $\left(\dfrac{\text{Power}}{V} \right)_{min} = \dfrac{1}{2} \Gamma_{21} N_T h \nu_{31}$

Ans.5: Threshold value

Ans.6: The decay rate for $1 \rightarrow 0$ should be greater than the decay rate for $2 \rightarrow 1$.

Ans.7: Larger

Ans.8: When the absorption coefficient is very small.

Ans.9: Inversely proportional

14.14 Exercise

Section A: Very Short Answer Type Questions

Q.1 How can achieve feedback in a laser?

Q.2 Write the condition for a material to behave as an amplifier.

Q.3 What is gain at threshold value?

Q.4 Write the expressions for the saturation intensity and photon number.

Q.5   What is the order of optical time period?

Section B : Short Answer Type Questions

Q.6   What do you mean by threshold.

Q.7   Is it possible to achieve steady state population inversion in two level laser scheme. Explain.

Q.8   What is advantage of four-level laser over three-level laser.

Q.9   What do you mean by spatial hole burning.

Q.10 Why we use three and four level laser schemes.

Section C: Long Answer Type Questions

Q.11 Derive the expression of rate equation for photons and populations.

Q.12 Determine the steady state values of the three level population.

Q.13 Derive the formula of steady state gain coefficient for a three-level laser and write the expression for the saturation intensity.

Q.14 Derive the formula of population inversion for four-level laser scheme.

Q.15 Derive the expression for spatial hole burning.

References and Suggested Readings

1. A.E Siegman, Lasers, University Science Books, 1986.

2. P.W. Milonni and J.H. Eberly, Laser Physics, John Wiley & Sons, New York, 2010.

3. K. Thyagarajan & A.K. Ghatak, Lasers-Theory & Application, MacMillan, 1981.

4. Hecht & Jeff, The Laser Guide Book-IInd Edition, Tata McGrew-Hill, 1992.

5. O. Svelto and D.C. Hanna, Principles of Lasers, Plenum Press, New York, 2010

UNIT-15
Q-Switching ,Mode Locking

Structure of the Unit

15.0   Objectives

15.1   Introduction

15.2   Q-switching

    15.2.1 Principle of Q-switching

    15.2.2 Evolution of a Q-Switched Laser Pulse

15.3   Methods of Producing Q-switching

    15.2.1 Rotating Mirrors

    15.2.2 Electro-optic Shutter

    15.2.3 Acousto-optic Shutter

    15.2.4Saturable Absorber

15.4   Phased-locked Oscillators

15.5   Mode locking

    15.5.1 Longitudinal modes of the laser cavity

    15.5.2 Mode locking Theory

15.6   Illustrative Examples

15.7   Self-learning Exercise

15.8   Summary

15.9   Glossary

15.10   Answer to Self-learning Exercise

15.11   Exercise

15.12   Answers to Exercise

References and Suggested Readings

15.0 Objectives

A laser is a device that emits highly coherent light through a process of optical amplification based on the stimulated emission of radiation. Under continuous wave (cw) operation, a laser is continuously pumped and emits light continuously. In this case, the population inversion is fixed at its threshold value when oscillation starts. Even under pulsed operating conditions, the population inversion is seen to exceed the threshold value by only a relatively small amount due to the onset of simulated emission and get the output of $1\,msec$ long burst of spikes. Hence, there is a need to develop the special laser cavities for the production of short laser pulses and high-peak power with the same laser gain medium. In this chapter, we will discuss Q-switching, methods of producing Q-switching, pulsed-locked oscillators and mode locking.

15.1 Introduction

A type of laser resonator that can provide useful laser output with a reasonable beam quality without meeting the criteria for stability is known as an unstable resonator. These resonators have been developed to obtain High power output in a nearly Gaussian-shaped beam. This beam is a concentrated spatial and spectral beam of light. It is also possible to qualify the beam as a temporal concentration by considering that the emitted photons are condensed into short energetic pulses. As we know that, the probability of obtaining stimulated emission increases with the number of atoms in excited state and incident photons. Therefore, there are two ways to favour stimulated emission either by raising the number of atoms in excited state or by raising the number of incident photons. The first method to achieve a temporal concentration is to trigger the stimulated emission only when there is a large number of an atom in upper level. This method is referred to as Q-switching. In the second method, the photons in the optical cavity are condensed into a packet or pulse that will bounce back and forth between the mirrors. This method is referred to as mode-locking. Phased-locked oscillator is a simple model of a mode-locked laser. The laser flux associated with a stable resonator and an unstable resonator is shown in fig. 15.1.

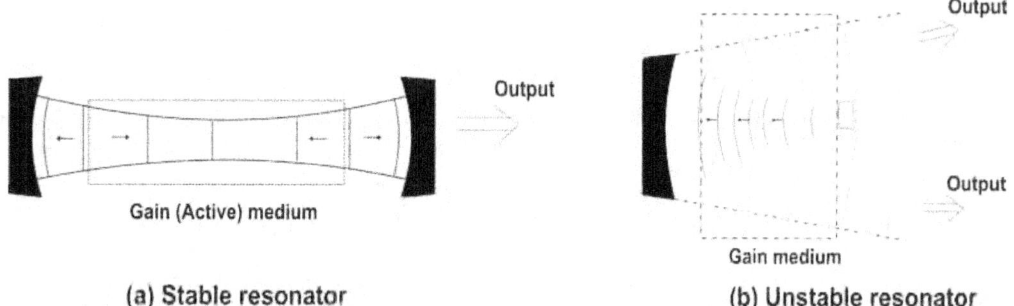

Figure 15.1 Laser flux associated with (a) a stable resonator and (b) an unstable resonator

15.2 Q-Switching

It was first discovered and demonstrated in 1962 by R.W. Hellwarth and F.J. Meclung using electrically switched Kerr cell shutter in a ruby laser.

Q-switching is technique to obtain energetic short pulses (10^{-9} seconds) from a laser by modulating the intra-cavity losses. Since laser is an oscillator, its resonator cavity is characterized by the quality factor (Q), which is defined as 2π times the ratio of the energy stored in the system to the energy losses per cycle, *i.e.*

$$Q = 2\pi \frac{\text{Energy stored in the resonator}}{\text{Energy loss per cycle}}$$

Therefore, a high (Q) factor corresponds to low resonate losses and vice versa.

15.2.1 Principle of Q-Switching

Consider a laser cavity in which a shutter is introduced in front of one of the mirrors. If the active medium is continuously pumped keeping the shutter closed (Q is low) the laser action is prevented and the population inversion will go on increasing and far exceeds the threshold population holding when the shutter is absent. The principle of Q-switching is shown in fig. 15.2.

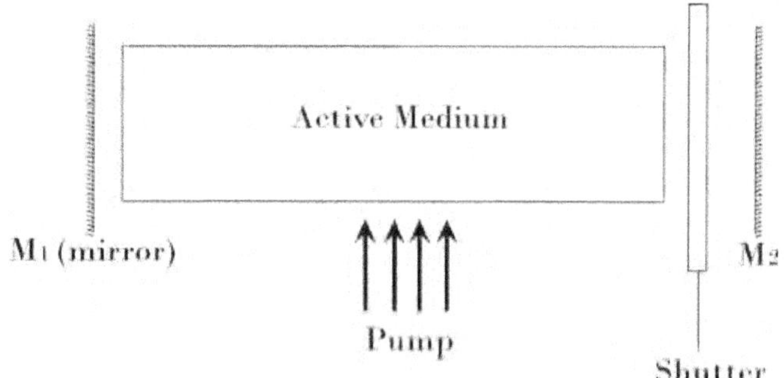

Fig. 15.2 Principle of Q-Switching

If the shutter is now suddenly opened (Q is changed from low to high), then the existing population inversion will correspond to a value much above the threshold value for oscillations and the laser will exhibit a gain that greatly exceeds the losses.

Therefore, the radiation in the cavity mode will build up very rapidly. This rapid increase in the intensity will deplete the population inversion which will go below threshold. The stored energy may then be released in the form of a short and intense light pulse. Since this operation involves switching the cavity Q factor, value changes from low to high. Therefore, this technique is known as Q-switching. If the shutter is opened in a time much shorter than the time required for the building of laser oscillation, the output would be a series of pulses having smaller peak power.

15.2.2 Evolution of a Q-Switched Laser Pulse

The following four requirements must be satisfied to produce the necessary inversion density required for Q-switching.

(1) The life time of the upper energy level must be longer than the cavity build up time (when gain equals losses), so that the upper level can store the extra energy pumped into it over the extended pumping time.
(2) The pumping flux duration must be longer than cavity build up time and preferably at least as long as the upper level life time.
(3) The initial cavity losses must be large during the pumping duration to prevent beam growth and oscillations.
(4) After Q-switching, the cavity losses must be reduced almost instantaneously, so that the beam could evolve and extract the extra energy from the upper level of the gain (active) medium.

Satisfying these requirements would produce a giant pulse laser output. Fig. 15.3 shows a high initial cavity loss during the intense pumping and consequent gain build up over the life time of upper level. The cavity loss is rapidly reduced producing a pulse output within a time of the order of cavity build up time after the cavity Q is changed.

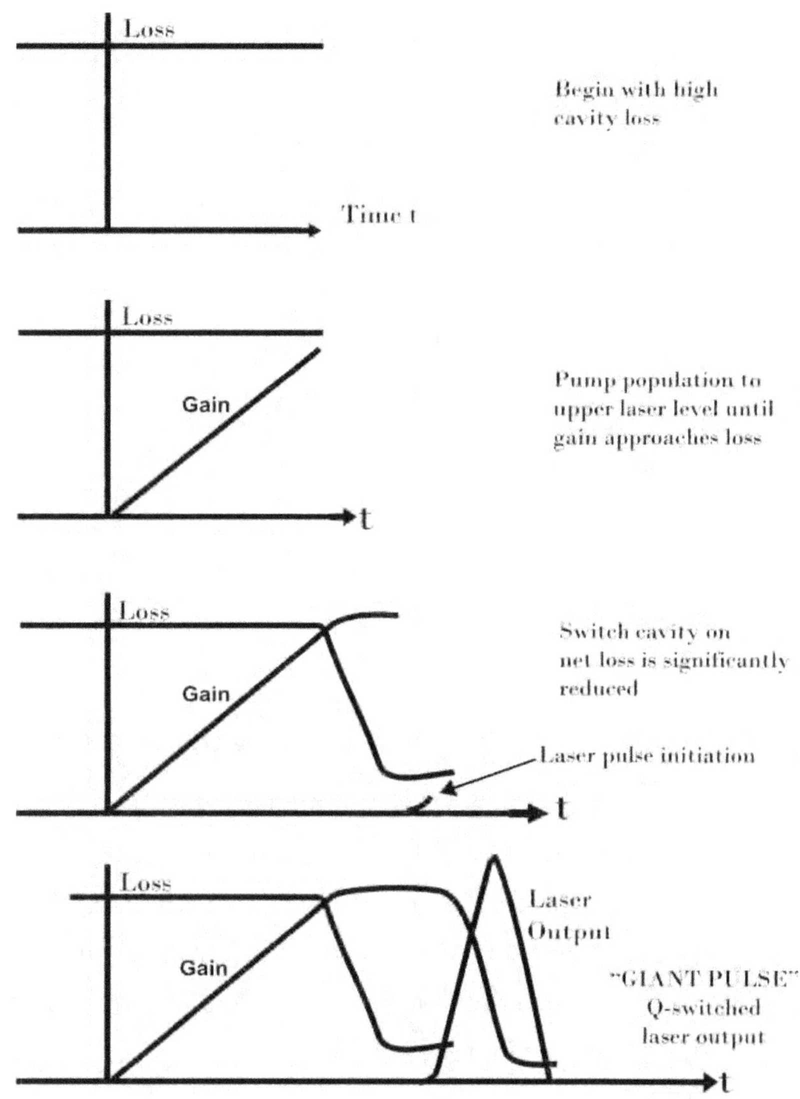

Fig. 15.3 Evolution of a Q-switched laser pulse

15.3 Methods of producing Q-Switching

There are two types of methods to produce Q-switching.

(a) Active Q-switching and (b) Passive Q- switching

(a)Active Q-switching

In this switching, the external devices are used to control Q-factor of cavity. The rotating mirror, electro-optic shutter (Kerr and Pockels effects) and acousto-optic shutter are the examples of active Q-switching. To understand the concept of Q-switching, the working of these methods is given below in detail.

15.3.1 Rotating Mirrors

This was the first method used for Q-switching laser. It consists of mounting hexagonal- shaped mirror assembly on a rotating shaft and aligning the facets of the mirror with the laser cavity such that, for every sixth of a rotation of the shaft, a mirror would be aligned with the laser cavity as shown in figure 15.4.

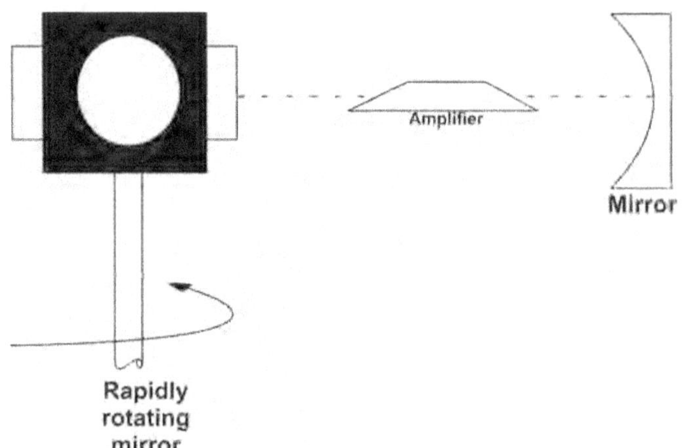

Fig. 15.4 Q-switching with rotating mirror as a shutter

This rotating mirror would serve as the rear mirror of the laser cavity, and an output mirror would be located at the other end of cavity. The mirror rotates at frequency related to upper laser level lifetime. There are two problems with this technique (i) size of the hexagonal mirror should be reasonably small to get high rotation speed without developing vibration problems and (ii) obtaining the ideal mirror alignment for each facet.

15.3.2 Electro-optic shutter

In this method, a shutter typically an electro-optic crystal that becomes birefringent when an electrical field is applied across the crystal. Since these shutters operate by rotating the polarization of beam, the laser must have a polarizing element within the cavity like Brewster angle windows. In this case, Brewster angle windows allow a low-loss beam to transit the cavity polarized in the plane of the paper. When the voltage is on, the shutter rotates the plane of polarization of the beam by 45° as it passes through the cell. The beam reflects from the mirrors and returns through the cavity, travelling in the opposite direction where after its plane is rotated by an additional 45° by the shutter. It thus arrives back at the laser amplifier with a polarization of 90° with respect to that when it left the amplifier. This 90° polarized beam is rejected by the Brewster window and directed out of the cavity. In such a position, the losses in the cavity would be large. When the voltage is turned off, the cell losses its birefringence and the cell does not rotate the plane of polarization. In this position, the losses are small and correspond to an open shutter. Two types of electro-optic shutters that can be used for Q-switching are the Kerr cell and the Pockels cell. The Pokels cell is generally preferred over the Kerr cell because of the lower voltage needed to produce the desired effect. The Q-switching with an electro-optic shutter is shown in fig.15.5.

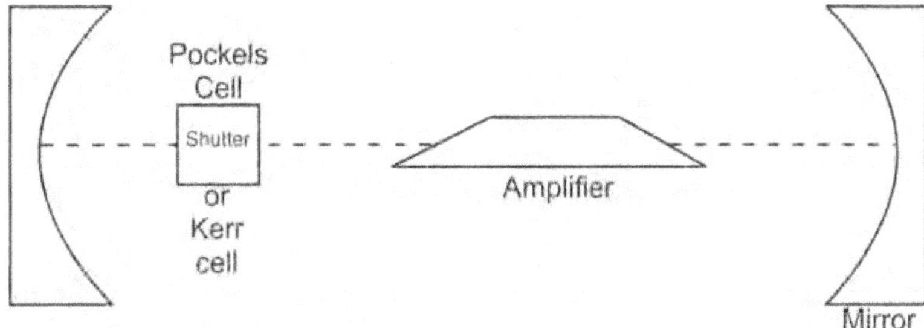

Fig. 15.5 Q-switching with an electro-optic shutter

POCKELS CELL: - The Pockels effect is produced when an electric field is applied to certain kinds of birefringent crystals KDP (potassium dihydrogen phosphate), CDA (cesium dihydrogen arsenate) to after the indices of refraction of the crystals. This effectively provides a rotation of the beam that is proportional to the thickness of the cell and the magnitude of the field. The field is applied in the direction of the optic axis of the crystal and in the direction of the optical beam. The pockels effect is directly proportional to the magnitude of the applied electric field.

KERR CELL: - A Kerr cell uses the Kerr electro-optic effect to rotate the laser beam. This effect is produced when an electric field is applied across a normally isotropic liquid comprising asymmetric module such as liquid nitrobenzenl to make it doubly refracting (birefringence) by aligning the molecules of the liquid. The field is applied in a direction transverse to that of optical beam. This electric field induced birefringence in isotropic liquids is called the Kerr effect. The birefringence is proportional to the square of the applied voltage, *i.e.* in Kerr effect, the change in refractive index is proportional to the square of the electric intensity of the electric field.

15.3.3 Acousto-Optic Shutter

An acousto-optic shutter typically uses a quartz crystal installed within the cavity, either at Brewster's angle or with low-loss antireflection coatings on the optical surfaces of the crystal, as shown in figure 15.6. The quartz crystal has a piezoelectric transducer attached to the crystal that propagates strong acoustic waves within the crystal when the RF (radio-frequency) signal is applied to the transducer. When this field is applied during the time the amplifier is being pumped, the laser beam is deflected out of the laser cavity by an effective diffraction grating that is established by the acoustic waves propagating because of the RF single. When the signal is turned off, the beam passes through the cavity un-deflected and the Q-switched pulse develops in the cavity. The typical RF range is from 25 to 50 MHz.

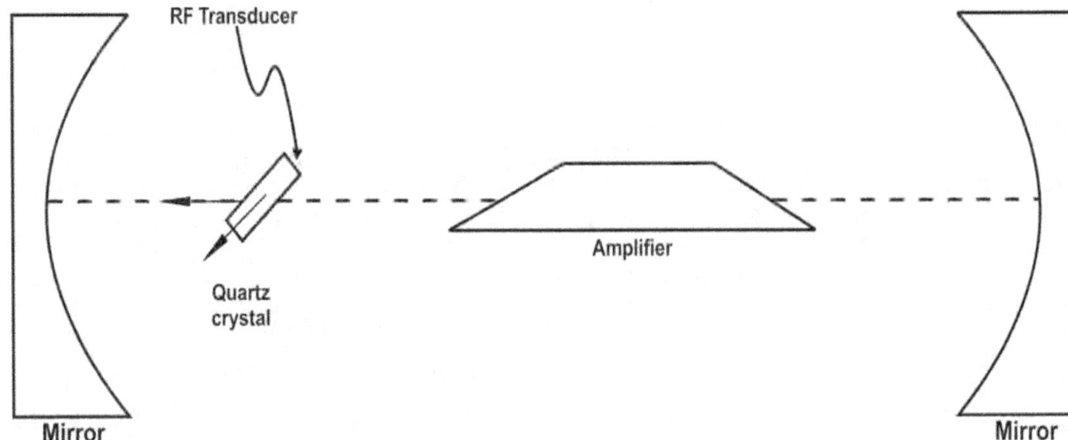

RF on – Beam deflected out of cavity yielding high loss.

RF off – Beam transits the cavity with low loss

Fig. 15.6 Q-switching with an acousto-optic shutter

(b) Passive Q-switching

In this switching, the losses inside the cavity are automatically modulated with a saturable absorber. Therefore, it is also known as self Q-switching. Saturable absorber is a material whose transmission increases when the intensity of light exceeds some threshold. The material may be an ion-doped crystal, a bleachable dye or a passive semiconductor device.

15.3.4 Saturable Absorber

It can be used as a passive Q-switching element by placing it within the laser cavity. The operation of such a device may be understood as follows Because of the small transmittance of the saturable absorber, laser oscillation can not start until the attainment of a large population inversion. As the power level inside the cavity goes on increasing, the dye begins to be bleached. This bleaching result in a larger transmittance which in turn increases, the power level inside the cavity. The increased power becomes almost transparent. Since in this condition, the inversion is much more than the threshold inversion, the gain is much more than the losses and thus a giant pulse is produced. The Q-switching with saturable absorber is shown in fig.15.7.

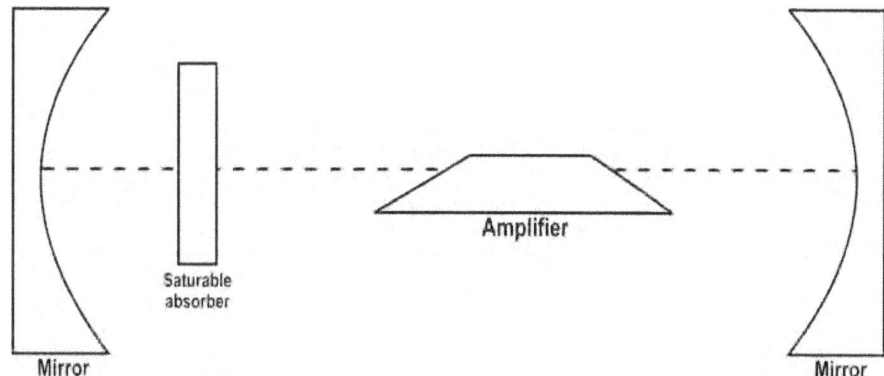

Fig. 15.7 Q-switching with saturable absorber

A good saturable absorber can be determined by ensuring that (the laser intensity at the time arrives at the absorber) × (the absorption cross section of the absorber) *exceeds* (the intensity of the laser as it arrives at the gain medium)×(the stimulated emission cross section of the gain medium). A liquid dye solution such as DODCI is typically used for such a Q-switching element. A thin film coating on one of the mirrors, in the form of a quantum well, can serve as a solid-state saturable absorber. The exact choice of saturable absorber is determined by the laser wavelength and also the gain of the laser medium.

15.4 Phased-Locked Oscillators

The phased locked oscillator (PLO) is a simple model of Mode locked laser. In a Q-switched laser, the light pulse must make several passes through the active medium after the cavity Q is switched. Therefore, feed-back is necessary in order to build up a large field amplitude by stimulated emission. There are some applications where it is desirable to have pulses of light even shorter than the pulses achieved by Q-switching. Such ultrashort and powerful pulses of light can be obtained by the technique called mode locking.

Q-switching may involve either a single mode or many modes but mode locking is a fundamentally multimode phenomenon. In mode locking, there is a phase locking among many longitudinal modes of cavity. The purpose of this topic is to consider a simple analog of a mode locked laser.

Let us consider the addition of the displacement of N harmonic oscillators with equally spaced frequencies, which is given by

$$x_n(t) = x_0 \sin(\omega_n t + \emptyset_0),$$ (15.4.1)

where

$$\omega_n = \omega_0 + n\Delta, \qquad n = \frac{N-1}{2}, \; -\frac{N-1}{2}+1, -\frac{N-1}{2}+2, \dots, \frac{n-1}{2}$$ (15.4.2)

i.e. the amplitudes x_0 and phase $\emptyset_0$ of the oscillators are identical, and their frequencies ω_n are equally spaced by Δ and centered at ω_0, as shown in Fig. 15.8. The sum of the displacement is

$$X(t) = \sum_n x_n(t) = -\sum_{N-1/2}^{N-1/2} x_0 \sin(\omega_n t + \emptyset_0)$$ (15.4.3)

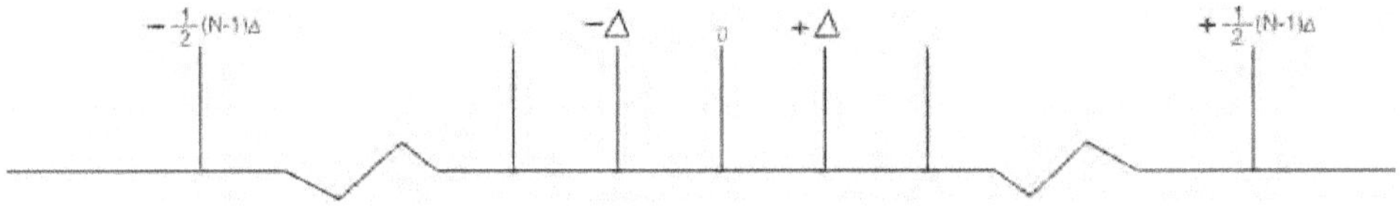

Figure 15.8 A collection of N frequencies running from $\omega_0 - \frac{1}{2}(N-1)\Delta$ to $\omega_0 + \frac{1}{2}(N-1)\Delta$ as in eq. (15.4.2)

Since $\sin x$ is the imaginary part of e^{ix}, we may write this as

$$X(t) = x_0 \mathrm{Im}\left(\sum_n e^{i(\omega_0 t + \emptyset_0 + n\Delta t)}\right) = x_0 \mathrm{Im}\left((e^{i(\omega_0 t + \emptyset_0)} \sum_n e^{in\Delta t}\right)$$ (15.4.4)

The general identity

$$\sum_{-(N-1)/2}^{(N-1)/2} e^{iny} = \frac{\sin(Ny/2)}{\sin(y/2)}$$ (15.4.5)

We prove (15.4.5) as follows. Let the sum be denoted by S_N. For convenience, we will first evaluate

$$S_{n+1} = \sum_{n=-N/2}^{+N/2} e^{iny}$$ (15.4.6)

The first step is to shift the summation label by introducing

$$m = n + \frac{N}{2}$$ (15.4.7)

So that

$$S_{n+1} = \sum_{m=0}^{N} e^{i\left(m-\frac{N}{2}\right)y} = e^{-iNy/2} \sum_{m=0}^{N} e^{imy} = e^{-iNy/2} \sum_{m=0}^{N} (e^{iy})^m$$ (15.4.8)

The second step is to make use of the identity

$$\sum_{m=0}^{N} x^m = \frac{1-x^{N+1}}{1-x} \qquad (15.4.9)$$

Then we can write

$$S_{N+1} = e^{-iNy/2}\frac{1-e^{i(N+1)y}}{1-e^{iy}}$$

$$= e^{-iNy/2}\frac{e^{i(N+1)y/2}}{e^{iy/2}}\frac{e^{-i(N+1)y/2}-e^{i(N+1)y/2}}{e^{-iy/2}-e^{iy/2}} = \frac{\sin(N+1)y/2}{\sin y/2} \qquad (15.4.10)$$

and so we have proved that

$$S_{N=}\frac{\sin Ny/2}{\sin y/2} \qquad (15.4.11)$$

Thus, equation (15.4.4) can be written as

$$X(t) = x_0 \mathrm{Im}\left[e^{i(\omega_0 t+\emptyset_0)}\frac{\sin\left(\frac{N\Delta t}{2}\right)}{\sin(\Delta t/2)}\right] = x_0\sin(\omega_0 t + \emptyset_0)\left[\frac{\sin\left(\frac{N\Delta t}{2}\right)}{\sin\left(\frac{\Delta t}{2}\right)}\right]$$

$$= A_N(t)x_0 \sin(\omega_0 t + \emptyset_0) \qquad (15.4.12)$$

The function $A_N(t)$ is plotted in Fig. 15.9 for $N = 3$ and $N = 7$. In general $A_N(t)$ has equal maxima $A_N(t)_{max} = N$ $\qquad (15.4.13)$

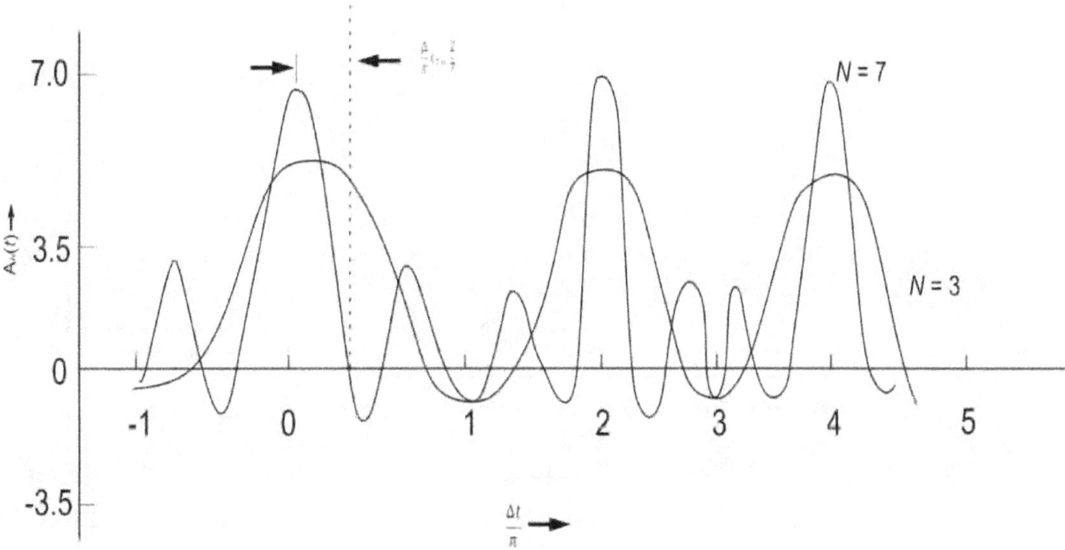

Figure 15.9 The function $A_N(t) = \dfrac{\sin\left(\frac{1}{2}N\Delta t\right)}{\sin\left(\frac{1}{2}\Delta t\right)}$ vs. $\Delta t/\pi$

At values of t given by

$$t_m = m\left(\frac{2\pi}{\Delta}\right) = mT, \qquad\qquad m = 0, \pm 1, \pm 2,\ldots\ldots \qquad (15.4.14)$$

As N increases, the maxima of $A_N(t)$ becomes larger. They also become more sharply peaked. A measure of their width is the time interval τ_N indicated in Fig. 15.9 for $N = 7$.

$$\tau_N = \frac{2\pi}{N\Delta} = \frac{T}{N} \tag{15.4.15}$$

we have thus shown that the addition of N oscillators of equal amplitudes and phases, and equally spaced frequencies (15.4.2), gives maximum total oscillation amplitudes equal to N times the amplitude of a single oscillator. These maximum amplitudes occur at intervals of time T[Eq. 15.4.14]. For large N we have, loosely speaking, a series of large amplitude "spikes." The smaller the frequency spacing Δ between the individual oscillators, the larger the time interval $T = 2\pi/\Delta$ between spikes, and conversely. The temporal duration of each spike is $\tau_N = \frac{T}{N}$, so the spikes get sharper as N is increased.

We have assumed for simplicity that each oscillator has the same phase $\emptyset_0$[Eq. (15.4.1)]. A more general kind of phase locking occurs when the phase differences of the oscillators are constant but not necessarily zero:

$$\emptyset_n = \emptyset_0 + n\alpha,$$

or
$$\emptyset_{n+1} - \emptyset_n = \alpha, \tag{15.4.16}$$

In this case the sum of the oscillator displacements (15.4.3) is replaced by

$$X(t) = \sum_n x_0 \, \mathrm{Sin}(\omega_0 t + \emptyset) = x_0 \, \mathrm{Im}\left(e^{i(\omega_0 t + \emptyset_0)} \sum_{-(N-1)/2}^{(N-1)/2} e^{in(\Delta t + \alpha)} \right) \tag{15.4.17}$$

and this may be evaluated to give the total displacement

$$X(t) = x_0 \sin(\omega_0 t + \emptyset_0) \left[\frac{SinN(\Delta t + a)/2}{\sin\left(\Delta t + \frac{\alpha}{2}\right)} \right] \tag{15.4.18}$$

having basically the same properties as (15.4.12) obtained with $\alpha = 0$.

It is a simple model of a mode- locked laser. The individual oscillators in the model play the role of individual longitudinal mode fields, while their frequency spacing Δ represents the mode (angular) frequency separation $2\pi \, \Delta v = 2\pi \left(\frac{c}{2c}\right) = \frac{\pi c}{L}$. The assumption of equal oscillator phase difference (Phase locking) in the model corresponds to the locking together of the phases of the different cavity modes.

This oscillator model suggests that if there are N longitudinal modes of a laser phase locked together with a frequency separation Δ, then the output light of the laser will consist of a train of pulses separated in time by $T = \frac{2\pi}{\Delta} = \frac{2L}{C}$The temporal duration of each pulse in the train will be $= \frac{T}{N} = \frac{2L}{CN}$. The larger the number N of phase locked modes, greater the amplitude and shorter the duration of each individual pulse in the train.

Mode locking

Mode locking is a technique for producing periodic, high power and extremely short duration (10^{-12} second to 10^{-15} second) laser pulses. Since a laser usually oscillates in a large number of longitudinal modes and under ordinary circumstances, the modes do not oscillate at the same time and their phases have random values. Therefore, it would be interesting to see what would happen if the modes of the resonant cavity are forced to oscillate together with their phases locked Laser. This operation is known as mode locking and then laser is said to be phase-locked or mode locked laser. Interference between these modes causes the laser light to be produced as a train of pulses of very short duration.

15.5.1 Longitudinal Modes of the Laser Cavity

Laser light is not a single pure frequency light but produces light over some natural bandwidth or range of frequencies as shown in fig. 15.10. Therefore the range of frequencies that a laser may operate over is known as the gain bandwidth.

We know that the resonant cavity of the laser consists of two plane mirror facing each other and active medium. Since light is a wave, when bouncing between the mirrors of the cavity the light will constructively and destructively interfere with itself leading to the formation of standing waves between the mirrors. These standing waves form a discrete set of frequencies, known as the longitudinal modes of the cavity as shown in fig. 15.10. These modes are due to constructive interference.

For simple plane mirror cavity, the allowed modes are those for which the separation distance of the mirrors L is an exact multiple of half the wavelength of light λ *i.e.* If η be the refractive index of laser medium then the length of optical path is given by

$$L = \frac{n\lambda}{2} \tag{15.5.1}$$

where n is an integer known as the mode order. From Eq. (4.21)

$$\lambda = \frac{2L}{n} = \frac{c}{v}$$

$$\Rightarrow \quad v = \frac{cn}{2L} \tag{15.5.2}$$

Therefore, frequency separation between any two adjacent modes n and $n+1$ is given by

$$\Rightarrow \quad \delta v = \frac{c}{2L} \tag{15.5.3}$$

and the pulse separation time or pulse repetition rate is

$$\tau = \frac{1}{\delta v} = \frac{2L}{c} \tag{15.5.4}$$

$$\eta L = \frac{n\lambda}{2} \tag{15.5.5}$$

Thus, frequency separation between any two adjacent mode $\delta v = \frac{c}{2L\eta}$ and pulse separation time$= \frac{2L\eta}{c}$

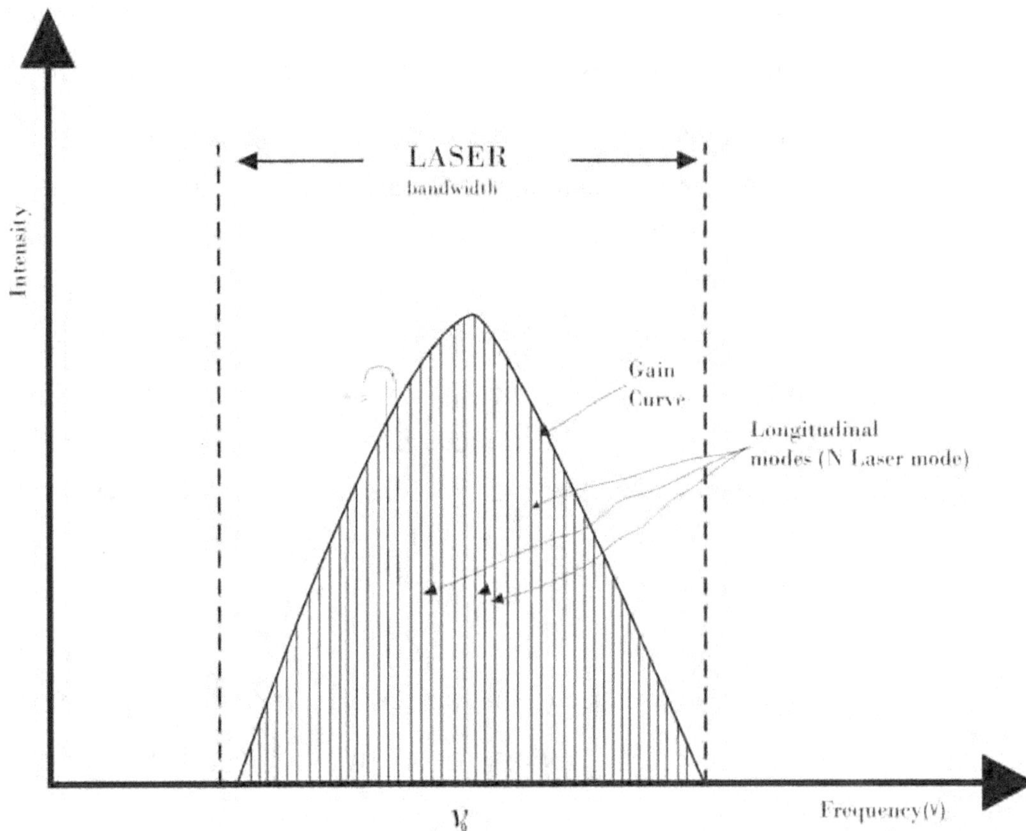

Fig. 15.10 Laser output spectrum

15.5.2 Mode Locking Theory

To achieve the mode locking, a number of distinct longitudinal modes of a laser having slightly different frequencies are combined in phase. The comparison between non-mode locked and mode locked laser output for three different frequencies ω, 2ω and 3ω is shown in Fig. 15.11.

In a simple laser, amplitudes and phases of various longitudinal modes (frequencies) are randomly distributed and these produce randomly varying total field amplitude and intensity (Fig.15.11 (a)). When these three frequencies (modes) are added in phase, then they combine to produce a total field amplitude and intensity output in the form of repetitive pulse of high intensity (Fig. 15.11 (b)). These pulses are called mode locked pulses.

In mode locked laser, the intensity is periodic, *i.e.*, in the time interval $\tau = \dfrac{1}{\delta v} = \dfrac{2L}{c}$, which is time taken for the light to make a complete round trip of the cavity, *i.e.*, the pulse separation time or pulse repetition rate is

$$\tau = \frac{1}{\delta v} = \frac{2L}{c} \tag{15.5.6}$$

The duration of width of each pulse of light is determined by the number of modes which are oscillating in phase. If there are N modes locked with a frequency separation δv, the overall mode locked bandwidth is $N\delta v$. Therefore pulse duration or width is

$$t_p = \frac{1}{N\delta v} = \frac{1}{\Delta v} \tag{15.5.7}$$

$$t_p = \frac{2L}{cN} \tag{15.5.8}$$

From equations (15.5.6) and (15.5.8), it is clear that the ratio of the pulse spacing to the pulse width is approximately equal to the number of modes (N). Thus, to obtain high power short duration pulse there should be a large number of modes. This requires a board laser transition and a long laser cavity.

The methods of mode locking can be divided into two categories:- (i) Active mode locking, in which the mode locking element is driven by an external source and (ii) Passive mode locking, in which the element that induces mode locking is not driven externally but instead exploits some nonlinear optical effect such as saturation of a saturable absorber.

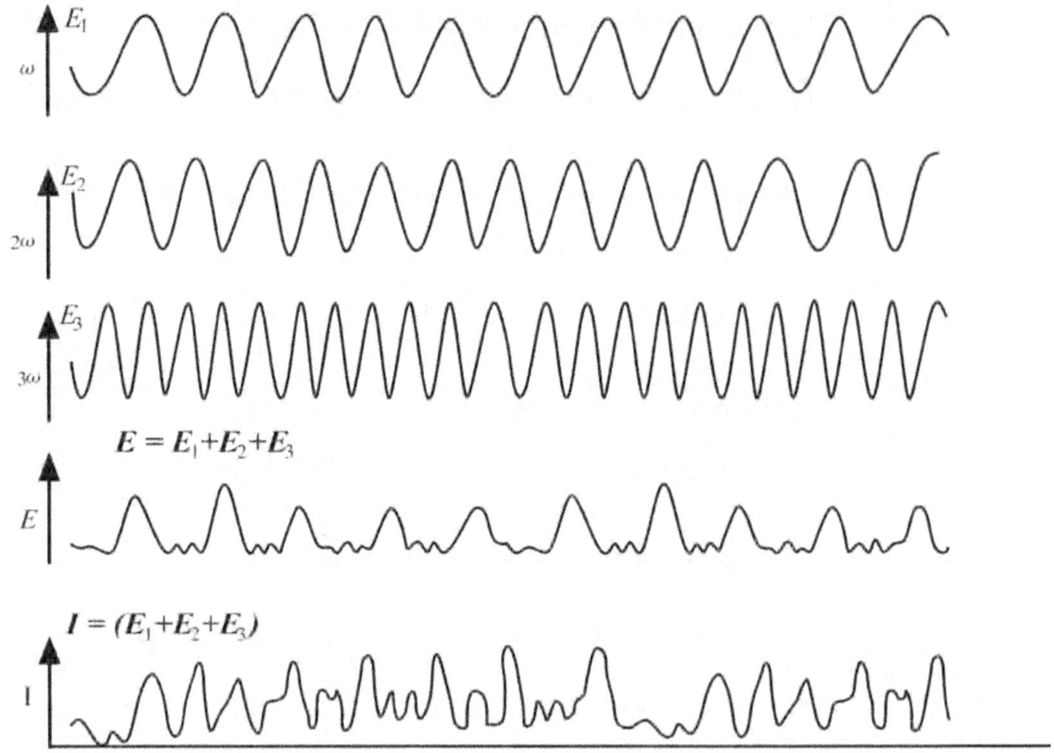

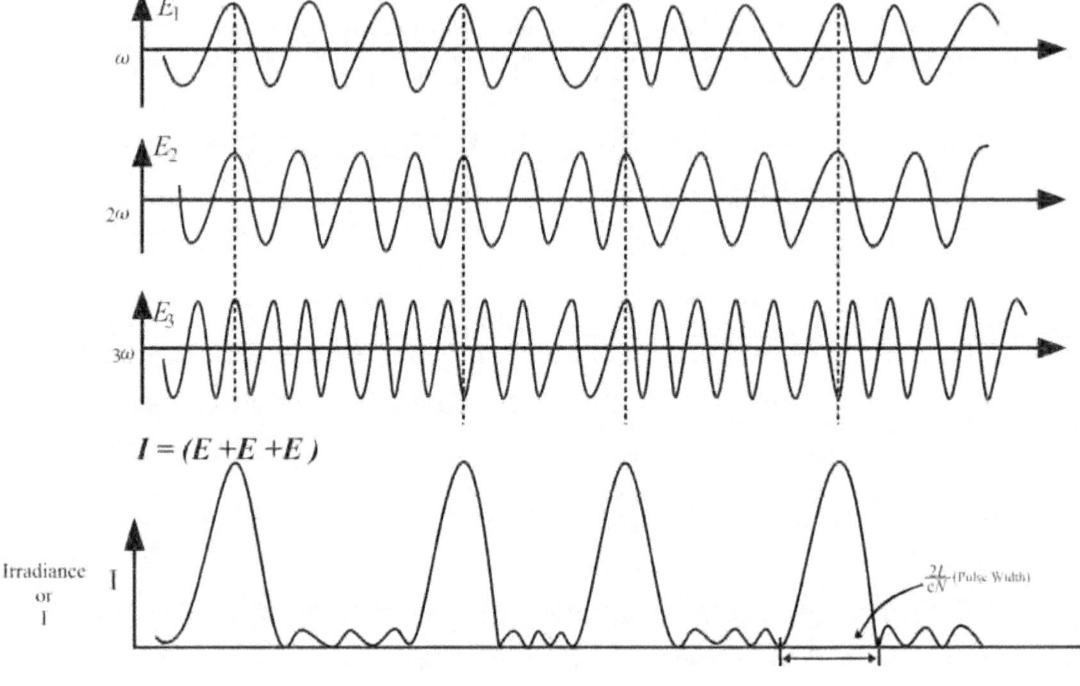

(a)Non-mode Locked (Simple laser) output

(b)Mode Locked output (Mode locked laser)

Fig.15.11 Comparison of non-mode locked and mode locked laser output

15.6 Illustrative Examples

Example 15.1 The transition responsible for ruby laser emission is spread over an energy resulting in wavelength spread of 0.65 *nm* around 694.3 *nm*. If the length of the ruby rod is 3 *cm* and refractive index is 1.75, how many longitudinal cavity modes would be ruby laser emission contain?

Solution: - Given $\Delta\lambda = 0.65 = 0.65 \times 10^{-9}$m, $\lambda = 694 \times 10^{-9}$m

$L = 3cm = 3 \times 10^{-2}$m and $\eta = 1.75$

Mode separation $= \dfrac{c}{\eta L} = \dfrac{3\times10^8}{1.75 \times 3 \times10^{-2}} = 5.71 \times 10^9$Hz

Frequency spread of laser emission

$$\Delta v = \frac{c\Delta\lambda}{\lambda^2} = \frac{3 \times 10^8 \times 0.65 \times 10^{-9}}{(694.3 \times 10^{-9})^2}$$

$$= 404.52 \times 10^9 Hz$$

$\therefore$ Number of cavity mode (N) $= \dfrac{\Delta v}{mode\ separation}$

$$N = \frac{404.52\times10^{+9}}{5.71\times10^9} = 70.84 \cong 71 \qquad\qquad \textbf{Ans}$$

Example 15.2 Calculate the number of modes, pulse separation and pulse duration in a mode locked Nd : YAG laser where the fluorescent line width is 1.1×10^{11}Hz and the laser rod is 0.1 m long.

Solution:- Given that the line width $\left(\dfrac{1}{t_p}\right) = \left(\dfrac{cN}{2L}\right) = 1.1 \times 10^{11}$ Hz and $L = 0.1\ m$

As we know that the pulse duration is

$$t_p = \frac{2L}{cN} = \frac{1}{1.1\times10^{11}}\ \text{sec.}$$

The pulse separation time is

$$\tau = \frac{2L}{c} = \frac{2\times0.1}{3\times10^8} = 0.66 \times 10^{-9} = 0.66\ ns \qquad\qquad \textbf{Ans.}$$

No. of modes $(N) = \dfrac{Pulse\ spacing}{pulse\ duration}$

$$= \frac{0.66\times10^{-9}}{9.09\times10^{-12}} = 0.7333 \times 10^2 = 73.33$$

Example 15.3 A continuous laser of wavelength $\lambda = 5000\text{Å}$ is Q-switched into 0.6 ns pulses. Compute its coherent length, bandwidth and line width.

Solution: Given that $\Delta\tau = 0.6ns = 6 \times 10^{-10}$ S and $\lambda = 5000\text{Å} = 5 \times 10^{-7}$ m

The coherence length $L = c\Delta$ Band width $\Delta v = \dfrac{1}{\Delta\tau} = \dfrac{1}{6\times10^{-10}} = 3 \times 10^8 \times 6 \times 10^{-10} = 0.18$ m

$$\text{Band width } \Delta v = \frac{1}{\Delta\tau} = \frac{1}{6\times10^{-10}} = 1.66 \times 10^9 \text{Hz}$$

$$\text{Line width } \Delta\lambda = \frac{\lambda^2}{c}\Delta v$$

$$= \frac{(5\times10^{-7})^2 \times 1.66\times10^9}{3\times10^8} = 0.013\times10^{-10}\text{m}$$

$$\Delta\lambda = 0.013\text{Å} \qquad\qquad \textbf{Ans.}$$

Example 15.4 Calculate the pulse width produced by a mode locked Nd : glass laser assuming that the laser cavity is 10 *cm* long and there are 2500 participating longitudinal modes. What is the time separation of the pulse?

Solution: Given L= 10 cm = 0.1 m and N = 2500 modes

$$\text{Pulse width} = \frac{2L}{CN} = \frac{2\times0.1}{3\times10^8 \times 2500} = 0.26 \times 10^{-12}\text{ s} = 0.26 \text{ ps}$$

$$\text{Pulse separation time} = \frac{2L}{c} = \frac{2\times0.1}{3\times10^8} = 0.66\times10^{-9} = 0.66 \text{ ns} \qquad \textbf{Ans.}$$

15.7 Self Learning Exercise

Q.1 What is continuous wave operation of laser?

Q.2 What are the pulse duration in Q-switching and mode locking, respectively?

Q.3 Define unstable resonator.

Q.4 What do you mean by Q-switching?

Q.5 Define the Kerr effect.

Q.6 What do you mean by passive Q-switching?

Q.7 What is difference between active and passive Q-switching.

Q.8 What do you mean by mode locking?

Q.9 What is phased locked oscillator?

Q.10 Why we get a train of pulses of very short duration in mode locking.

15.8 Summary

This unit starts with the introduction of some special laser cavities for the production of short laser pulse and high peak power with the same laser gain medium. The chapter introduces the principal of Q-switching and mode-locking. It also describes the methods of production of Q-switching and phased-locked oscillator. In the end, some examples on above concepts are given.

15.9 Glossary

(CW) Operation: continuous wave (cw) operation

RF :(radio-frequency)

15.10 Answer of Self-Learning Exercise

Ans.1: In this operation, light is continuously pumped and emits light continuously.

Ans.2:10^{-9} seconds and 10^{-12} to 10^{-15} second, respectively.

Ans.3: It is a laser resonator that can provide useful laser output with a reasonable beam quality without meeting the criteria for stability is known as an unstable resonator.

Ans.4: The production of a short and intense light pulse involves switching the cavity Q factor value from low to high. Therefore this technique is known as Q-switching.

Ans.5: The electric field induced birefringence in isotropic liquids is called the Kerr effect.

Ans.6: In the passive Q-switching, the losses inside the cavity are automatically modulated with a saturable absorber.

Ans.7: In an active Q-switching, some external active medium operation must be applied to the device to produce Q-switching. In a passive Q-switching, the switching operation is automatically produced by the optical nonlinearity of the element used.

Ans.8: Since laser has a large number of longitudinal modes. The process by which these modes are made to adopt a definite phase relation is called mode locking.

Ans.9: It is a simple model of a mode-locked laser.

Ans.10: The interference between longitudinal modes of laser causes light to be produced as a train of pulses of very short duration.

15.11 Exercise

Section-A (Very Short Answer Type Questions)

Q.1   Write the fundamental mode of laser cavity

Q.2   Write down an example of unstable resonator.

Q.3   Define the Q-factor of resonator cavity.

Q.4   What is the pulse duration in Q-switching?

Q.5   What is the pulse duration in mode-locking?

Section-B (Short Answer Type Questions)

Q.6   What are the active optical devices and when these devices are used.

Q.7   Why a laser with an unstable resonator cavity supports the fundamental mode only.

Q.8   What do you mean by the longitudinal and the transverse modes.

Q.9   What are the important devices used for Q-switching.

Q.10 What are the pulse separation time and pulse duration in made locking.

Q.11 Explain the concept of mode locking in detail.

Section C- Long Answer Type Questions

Q.12 What do you mean by Q-switching? Describe different type of Q-switches.

Q.13 Describe various methods of producing Q-switching. How it is helpful in generating laser pulse.

Q.14 Discuss the construction and working of phase locked oscillator.

Q.15 Discuss a technique for obtaining powerful pulses of short duration for laser source of low power.

15.12 Answers to Exercise

Ans.1: TEM_{00}

Ans.2: Confocal telescopic resonator.

Ans.3: $Q = 2\pi \dfrac{\textit{Energy stored in the resonator}}{\textit{Energy lost in a cycle}}$

Ans.4: 10^{-9} second

Ans.5: 10^{-12} to 10^{-15} second

Ans.6: Rotating mirrors and optical modulators are called active optical devices which do different things to light at different times. These devices are used when it is required to concentrate light energy into short time intervals.

Ans.7: The unstable resonators exhibit high diffraction losses leading to suppression of high-order transverse mode. Thus, a laser with an unstable cavity supports only the fundamental mode.

Ans.8: The laser modes governed by the axial dimension of the resonant cavity are called longitudinal or axial modes. The laser modes determined by the cross-sectional dimension of the optical cavity are called the transverse modes.

Ans.9: Rotating mirrors, electro-optic shutter, acousto-optic shutter and saturable absorbers are important devices used for Q-switching.

Ans.10: Pulse separation time $(\tau) = \dfrac{2\eta L}{c}$ and   Pulse duration $t_p = \dfrac{2\eta L}{cN}$

Where η = Refractive index of laser medium

L = Distance between the mirrors of cavity, c = velocity of light

and   N = Number of modes which are oscillating in phase

References and Suggested Readings

1. B.A. Lengyel, Introduction to Laser Physics, John Wiley and Sons, New York, 1967.

2. W.T. Silfvast, Laser Fundamentals, Cambridge University Press, 1998.

3. O. Svelto and D.C. Hanna, Principles of Lasers IV edition Plenum Press, New York, 1976.

4. P.W. Milonni and J.H. Eberly, Laser Physics, John Wiley and Sons, New York

5. A.E. Siegman, laser, Cambridge University, Oxford, 1986.

6. A.K. Ghatak and K. Thyagrajan, Laser Theory and Applications, Macmillan India Ltd. New Delhi, 1981.

7. M.J. Beesley, lasers and their Applications, Tylor and Francis Ltd. London, 1976.

UNIT-16
Specific Lasers and Pumping Mechanisms

Structure of the Unit

16.0 Objectives

Different types of lasers can be classified into solid, liquid, gas and semiconductor lasers. This classification is based on the laser medium that is chosen. In this unit, we will briefly survey the different types of lasers available today. We will attempt discussing each of them, in detail, in this unit. The main objective of this chapter is to study all the lasers, which are used today in our life.

16.1 Introduction

The solid laser, a representative of which is the ruby laser, was the first one to be designed. This laser was immediately followed by the gas laser, a representative of which is He-Ne laser. Then come Nd-YAG and glass, argon ion, carbon dioxide and the dye lasers. Certain other varieties of lasers such as chemical lasers, excimer laser and gas dynamic lasers also assumed importance owing to their characteristics in terms of wavelengths, method of excitation, tunability and power levels etc.

There are some special types of laser which consists of Solid, Liquid, Gas and Semiconductor lasers. All these lasers are having different properties.

16.2 Pumping Mechanism

It is a process by which population inversion can be produce known as pumping. The population inversion can be achieved by exciting the medium with suitable form of energy.

There are many methods of pumping a Laser and producing population inversion necessary for occurrence of stimulated emission. Some of the commonly used methods are given below:

(1) Optical pumping,

(2) Electric discharge,

(3) Inelastic atom-atom collision,

(4) Direct conversion,

(5) Chemical Reactions.

Production of population inversion by above given method is follow as given description:

16.2.1 Optical Pumping

If luminous energy is supplied to medium for causing population inversion, then the pumping is called the optical pumping. In optical pumping, the luminous energy usually comes from a light source in the form of short flashes of light. This method was first used in Ruby Laser by *Mainman* and is now a days used in solid state lasers. The laser material is simply placed inside a helical xenon flash lamp of the same type as used in photography.

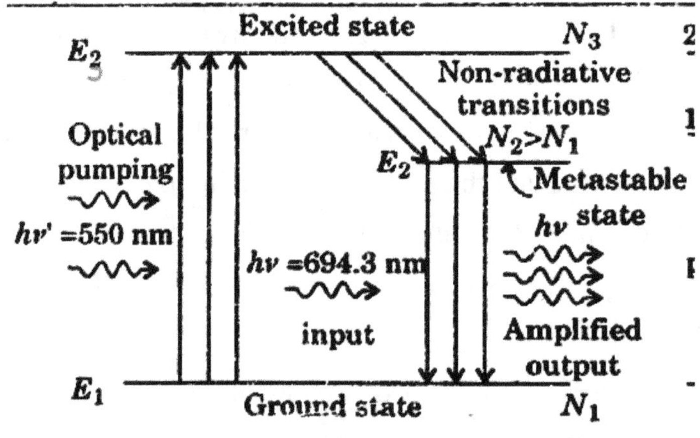

Figure 16.1

Let us consider a material whose atoms can reside in three different states as shown in fig 16.1. Atoms in ground state are pumped to state are state E_3 by photons of energy equal to $E_3 - E_1$. The excited atoms then undergo $non-radiative$ transitions with a transfer of energy to the lattice thermal motion, to the level E_2. They remain in this *metastable* energy state for a comparatively longer time.

Hence, there will be more atoms in the higher *metastable* energy state E_2 than in the ground state E_1, i.e., we have a *"population inversion"*. Atoms in the metastable state E_2 are now bombarded by photons of energy $hv = E_2 - E_1$, resulting in a stimulated emission giving an intense, coherent beam in the direction of the incident photons. This is the method used in the ruby laser. This is show in figure 16.2.

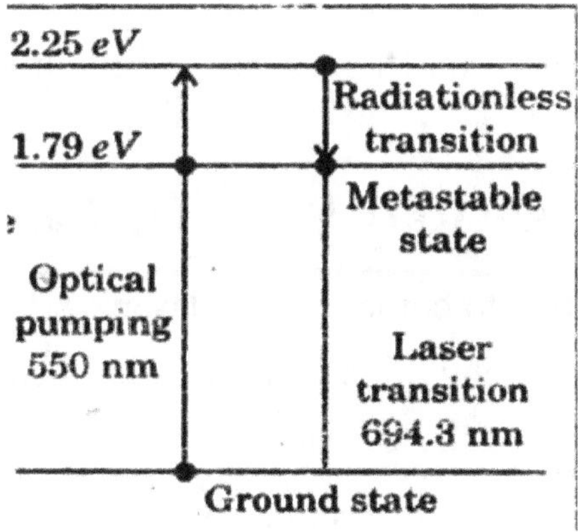

Figure 16.2

16.2.2 Electric Discharge

The pumping by the method of electric discharge is preferred in gaseous-ion lasers for example Argon-ion laser. In discharge tube, when a potential difference is applied between cathode and anode; the electrons emitted from cathode are accelerated towards anode. Some of these electrons collide with atoms of the active medium, ionize the medium and raise it to the higher level. This produces the required population inversion. This mean is also called direct-electron excitation.

16.2.3 Inelastic Atom-Atom Collisions

In electric discharge, one type of atoms is raised to their excited states. These atoms collide *inelastically* with another type of atoms. It is these latter atoms which provide the population inversion needed for Laser emission. The example is helium neon laser.

16.2.4 Direct Conversion

In this type of method, a direct conversion of electrical energy into radiant energy occurs in Light Emitting Diodes ($LEDs$). The example of population inversion by direction by the method of direct collision occurs in semiconductor lasers.

16.2.5 Chemical Conversion

This method follows in only Chemical Laser. In chemical Laser, energy comes from a chemical reaction without any need for other energy sources. For example, hydrogen can combine with fluorine to form hydrogen-fluoride

$$H_2 + F_2 \longrightarrow 2HF$$

This above given reaction is used to pump a CO_2 laser to achieve population inversion.

16.3 Solid-State Rare Earth Ion Laser

It is the first one to come into the laser world. Solid state lasers are part of the laser systems involving high-density gain media. For the action of Laser the material should have following features:

(a) It should have strong absorption bands.

(b)The material must have a high degree of quantum efficiency for fluorescent transits.

Crystals or glasses, which have these characteristics, are doped with small amounts of $dopants$. Such materials have optical transition between inner, in complete electron orbits.

$Dopant$ Materials having these characteristics are mainly:

(a) Metals such as transition elements. Example Cr^{3+}.

(b) Rare earths ion. Example Nd^{3+}.

(c) Actinide Series. Example U^{3+}.

Host materials of solid state nature can be categorized into crystalline solids and glasses. The host material should have the following properties:

(a) It should possess excellent optical, mechanical and thermal properties.

(b) The material should have sufficient hardness.

(c) The material should be chemically inert.

(d) There should be no internal strain, voids or impurities.

(e) There should be no variation in the refractive index, throughout the material.

(f) The material should be resistant to damage due to radiation.

(g) The material should be in such a state that it is easy to fabricate the system with it.

When compared to hosts of glasses, crystalline materials have clearer and sharper emission of fluorescence. This is due to the fact that the corresponding transitions are akin to the transitions of free ion. An example of this Nd doping in the crystal of YAG or glass. The threshold of lacing action gets lowered by sharper fluorescent lines. Crystalline hosts have better thermal conductivity.

16.3.1 Neodymium Lasers

It is a type of solid-state rare-earth ion laser, which falls into the group of lasers having narrow line-width. In this laser, the rare earth Nd^{3+} ion is doped in different host materials. Neodymium is considered to be the most useful laser material so far. It is most commonly doped in YAG or glass hosts.

16.3.1.1 Nd-YAG Lasers

The energy level diagram of Nd^{3+} doped in the host material YAG is shown in figure 16.3. Here, the laser levels are slightly less for Nd^{3+} doped in glass than that in YAG.

For the further discussion of ND-YAG laser now we will see another simplified energy diagram which is shown in figure 16.4.

Fluorescence effect can be seen between $4F_{3/2}$ to the four $multiplets$ of levels to the ground state. It is important to note here that the $F \longrightarrow I$ transition is not permitted in the dipole approximation due to the fact that the orbital quantum numbers get changed by 3. As a result, F levels are $metastable$.

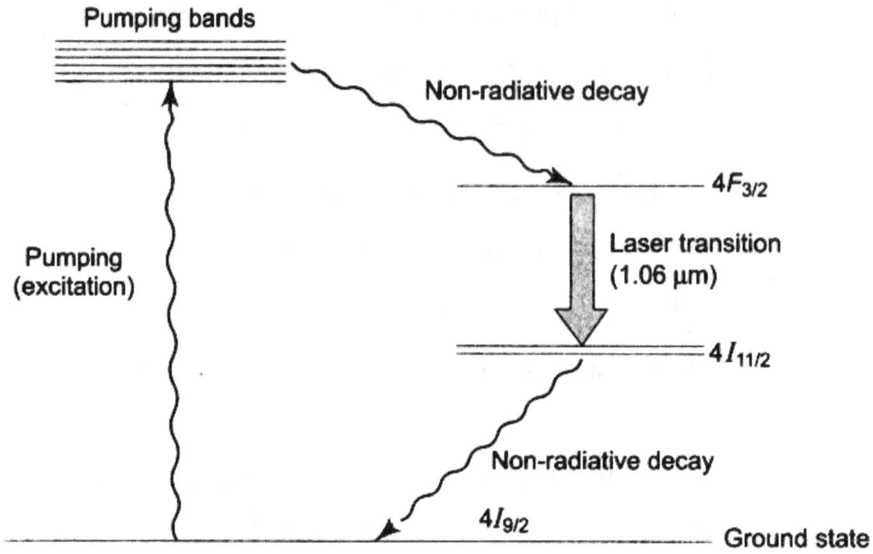

Figure 16.3: Energy level diagram of neodymium YAG laser

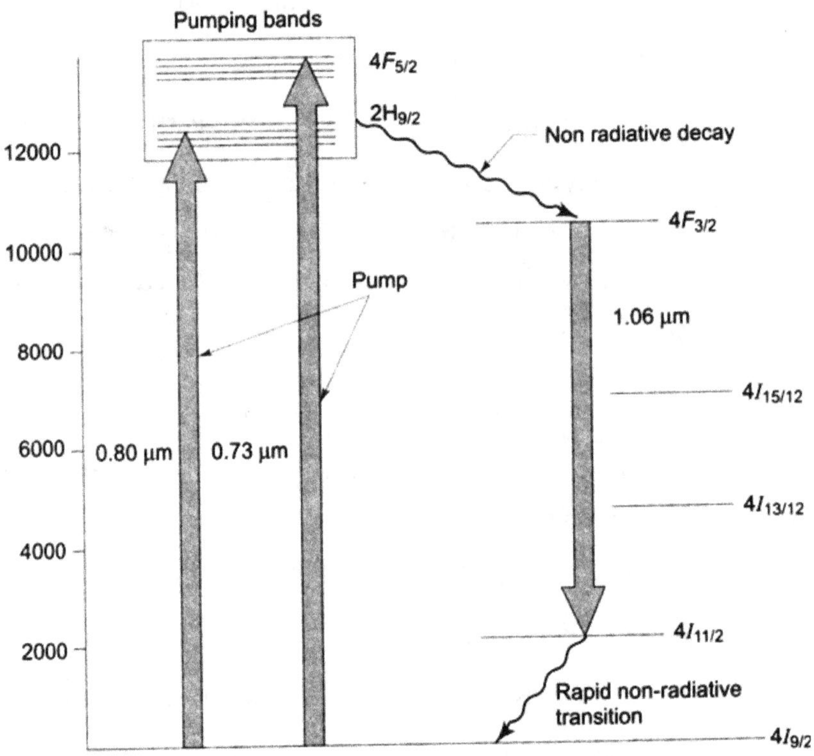

Figure 16.4: Simplified energy level diagram of Nd^{3+} in YAG

The transition probability to the level $4I_{11/2}$ has an order of magnitude, which is higher than the probability of transition to other *multiplets*. The level $4I_{11/2}$

has an order of magnitude, which is higher than the probability of transition to the other *multiplets*. The level $4I_{11/2}$, as can be seen in figure 16.4 is about $2000\ cm^{-1}$ $(f = 6 \times 10^{13})$ above the ground state. Hence, at ambient temperature, $\hbar f$ is much greater than kT. Therefore, in a state of thermal equilibrium, the $4I_{11/2}$ level is practically empty. Therefore, in a state of thermal equilibrium, the $4I_{11/2}$ level is practically empty. A population in level $4F_{3/2}$ will initiate a process of inversion.

Transition from $4I_{11/2}$ to the lowest level occurs through *non − radiative* decay in a very rapid pace. The corresponding wavelength λ of the transition $4F_{3/2} \longrightarrow 4I_{11/2}$ is $1.06\mu\ m$. At ambient temperature, this can be seen with a *Lorentzian* line-shape at $195\ GHz$ $(\delta f = 6.5^{-1})$. The upper level lifetime is $0.23 \times 10^{-3}\ seconds$.

The lower lasers level $4I_{11/2}$ decays with a lifetime of approximately 30 ns. However, this radiation is strongly absorbed by the lattice of the host and the resultant energy is converted into heat. Since YAG has excellent thermal conductivity, the heat energy is removed by conduction. The $Nd − YAG$ system can be worked either in pulsed or in CW mode. It can also be operated in mode-locked $Q − switched$ condition.

$Nd − YAG$ laser is a four-level system. Because of the narrow line-width, the cross section of stimulated emission is large and the threshold of pumping is low. But the absorption bands are also narrow. Hence, the excellent radiation emitted by the flash lamp is not fully utilized. As a result, attempts have been made to use gases like krypton in the pumping lamp, which matches the emission bands.

A recent development in $Nd − YAG$ system is by using semiconductor laser output to pump the $4F_{3/2}$ level directly. The ability of semiconductor lasers to convert electrical energy to photons is quite high. As such, YAG laser can be pumped with simple power supplies-even torch batteries.

The wavelength of the output can be converted to green $(\lambda_0 = 532\ nm)$ with a TEM_{00} mode, which is limited by the diffraction of the output, by adopting frequency-doubling techniques. The system is so compact and portable that it can be even contained in a very small box one forth the size of a brief case.

In several commercial operations, the emission in the infrared region of the $Nd-YAG$ laser is frequency-doubled, to bring it to the visible region, using a non-linear interaction in the YAG Crystal.

16.3.1.2 *Nd*-Glass Laser:

Glass has been under study from ancient times and has found application in many fields. Glass is smooth, sufficiently strong and can be made in any shape. It can be polished to very fine surface finish.

The materials of glass consist of a mixture similar to molasses, frozen to solid form. This leaves various small areas of the material oriented in slightly varying directions. As such, different levels of energy, which are dependent on the strains that are reposed upon them in the matrix that is frozen. The different areas having different energy levels in the excited state result in different radiating frequencies in the different areas. The emission line consists of the sum total of all the individual lines. Therefore, a much broader emission spectrum is obtained which is more than that could be achieved, from one single crystalline structure.

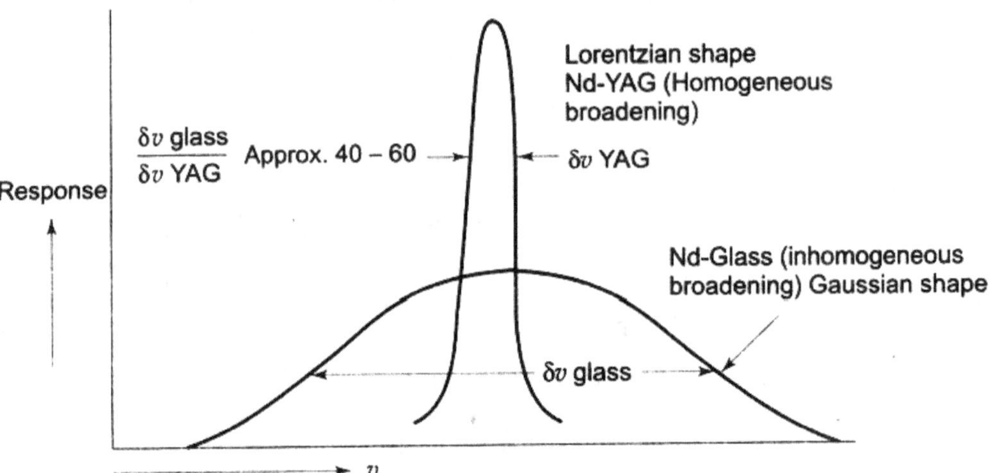

Figure 16.5: Comparison of emission line widths of radiating $Nd-YAG$ and $Nd-glass$ materials.

The emission schemes such as $Nd-YAG$ is broader *inhomogeneously* having a Gaussian shape and is generally much broader than the emission line in a crystalline material such $Nd - YAG$ where the broadening takes place homogeneously giving a $Lorentzian$ form. As a result, Nd doped in glass gives a $\delta f = 7.5 \times 10^{12}\ Hz$, as compared to Nd doped in crystalline matrix such as $Nd - YAG$ which gives a broadening $\delta v = 1.2 \times 10^{11}\ Hz$. Hence, if we consider the same concentration of doping, the maximum emission that occurs at the centre frequency of $Nd - YAG$ is much larger than that of $Nd - glass$. This is due to the fact that there are the same numbers of radiating species and the $radiative$ rates from the upper laser levels are similar. However, they are not really identical since the upper level lifetimes are different. This comparison between $Nd - YAG$ and $Nd - glass$ is shown in given figure 16.5.

There is large use of Nd doped glass as a laser medium for Pulsed lasers. $Nd - glass$ And $Nd - YAG$ oscillate at almost the same frequency $(f = 2.83 \times 10^{14}\ Hz)$. However, while $Nd - YAG$ laser can be utilised in $CW\ mode$, due to low thermal conductivity of glass, its utilisation in continuous oscillation is limited and hence it is usually worked in single pulses.

There are many types of glasses used in Lasers and all the glasses are different in properties with each other. For example Phosphate Glass, Silica Glass, Borate Glass etc.

The main characteristics of glass laser are as follows:

(a)Glass rods having good quality can be made with almost any diameter starting from fibres to conventional rods. Hence, we can have high level of powers with comparatively low power density. It is rather easy to fabricate glass in various forms with good optical quality. Therefore, glass has become a more common candidate for lasers.

(b)Glass does not absorb the emission of laser.

(c)Glass is better than crystalline systems since the latter develops problems mainly due to the fact that during laser operations the crystals get distorted due to excessive heating, which cannot be avoided. However, neodymium-glass lasers are tough and highly resistant to damage from high power densities. This is due to the fact that glass can be made totally free of internal strain and stress.

(d)Glass is optically homogeneous and hence it van be doped in high level of concentration of impurities.

(e)$Nd - glass$ lasers produce high-energy outputs, per unit volume, of the material. Thousands of joules per millisecond can be achieved with larger sizes of glass rods.

(f) Nd^{3+} doped in glass has the broad $linewidth$ of laser transition, which is of the order of 30 to 40 nm. This broadening happens because of the $non - homogenity$ of the ion environment. Therefore, this laser usually does not operate in a single spectral mode.

The importance of $Nd - glass$ are that, glass has isotropic properties and can be doped in high concentration $(about\ 3\%)$ and can be made in large sizes with good optical quality as mentioned earlier. In fact, $Nd - glass$ laser was made even before $Nd - YAG$ laser.

The laser transition occurs at about 1.06 microns and the transition varies from one type of glass to another type of glass, which also depends upon the constitution and structure of the glass. Different glass hosts are available and the $dopant$ concentration also can be changed. The glass doped in Nd is also a four-level system. The line-width of fluorescence ranges about $300cm^{-1}$ and is approximately 50 times larger than that of $Nd - YAG$. This is because of the amorphous structure of glass. Because of wider line-width, there will be higher threshold of laser action. The upper laser level lifetime depends on the Nd^{3+} concentration and type of the host glass.

A large number of glasses have been in use. Some of them are mentioned below:

$(A)\ LG - 60$ of Schott (Germany).

$(B)\ ED - 2$ of Owen Illinois (USA).

$(C)\ Q - 88$ of $Kigre$ (USA).

(D) $LHG - 5$ And $LCG - 11$ of Hoya (Japan).

$LG - 60, ED - 2$ and $LCG - 11$ are Silicate glasses. $LHG - 5$ and $Q - 88$ are phosphate glasses. The gain coefficient of a phosphate glass is higher than that of a silicate glass due to its line-width, which is narrower.

The CGCRI has developed an indigenous glass, which is found to be as efficient as the $ED - 2$ glass. This glass is reported to have given a fluorescence decay time of 240 ± 20 microseconds at $1/e$ point. In two studies carried out, emission of the order $8 \times 3 \times 10^{-20} cm^2$ and $3.2 \times 10^{-20} cm^2$ have been reported.

Silicate glasses have the less importance of less thermal expansion, lower refractive index and hence smaller non-linear effects and better chemical stability, which means, resistance to etching action by water, as compared to phosphate glasses. But phosphate glasses is have the importance of narrower fluorescent band, higher gain and larger lifetime in excited state, which gives better output and better pumping efficiency as compared to silicate glasses. Phosphate glasses have the problem of getting etched by water. In order to prevent this, water is mixed with about 50% ethylene glycol.

Fluorophosphates glasses are now being made which produce a high gain with lower non-linear refractive index. It may be recalled that the non-linearity is connected with refractive index μ and it is larger for larger refractive indices. If we take, for example, a high power cascaded chain of oscillator/amplifier, we can do a matching of wavelengths of $Nd - glass$ oscillator with the wavelength of $Nd - YAG$ amplifier. Higher level of concentration of doping is the main concern in systems with high power.

Flash lamps are beset with the problem of intense radiation of ultraviolet lights emitted by them. These radiations enter the rod, used in the laser. This is known in the laser parlance as solarisation. It is essential to get this effect reduced. There are different methods and techniques to get the solarisation minimized.

We are used a method in laser, or flash lamp which method is to shroud the rod, with a tube having a material, which can absorb the ultraviolet rays. An example is a glass doped with samarium. We can also use a filter of glass, which is coloured. This filter cuts off the unwanted ultraviolet rays. However, it permits the radiation in the normal band of pumping. Another method is by exploiting the spectral properties of the coolant in the pumping cavity and then getting the absorption of the ultraviolet rays improved using potassium dichromate in the coolant. In yet another method, use of fluorescent in the required band for useful pumping action 6G can be used. The dependence pattern for thermal energy in case of ruby, YAG and glass lasers indicate that a ruby laser is highly dependent on temperature due to high level of changes in line-width with temperature. $Nd - YAG$ also is temperature dependent but as compare to the ruby laser not much, it is much less; whereas $Nd - glass$ is not sensitive to thermal energy.

Apart from having smaller line-width, crystalline materials have the advantage over glass that they have better thermal conductivity. However, crystals have the limitation of being and cannot be grown into large dimensions. Owing to comparatively poor conductivity in glass, the pulse repetition rate is limited as compared to YAG. Moreover, the line-width of fluorescence of Nd in glass is quite sensitive to variation in temperature. Glasses find more suitable application in high energy pulsed operations because of larger line-width. Formulations of glass could be improved by dense suspension of tiny crystallites, which can be embedded in the matrix of glass.

We saw that one of the major problems in solid state lasers is the distortions in the laser rods, which are thermally induced. This result is variation in refractive index within the laser medium. The thermally induced distortion also leads to alteration of effective cavity dimensions. These distortions in turn create distortion of the mode structure of the laser and the wave front. Therefore, it is necessary to cool the system in high power solid state lasers. For cooling, either air is forced into the system or a liquid coolant is circulated in the scheme.

One important requirement of an optical amplifier is the storage of energy in the inversion of population. As mentioned earlier, glass lasers have a very high figure of merit. Amplification of very short pulses is possible due to wide bandwidth over which gain is prevalent.

Glass, cut into discs, is used in high power chain amplifiers to study the features of fusion. These discs are placed at Brewster's angle in the beam, which reduces reflection loss and increases pumping efficiency, by a better pump illumination. Laser radiation with high intensity at power densities of the order of many *gigawatts* causes cracks in the glass. This is due to heating and absorption, inclusions and self-focusing of the laser beam by non-linear effects. By using discs, self-focusing of laser radiation can be avoided.

16.4 Dye Lasers

16.4.1 General Characteristics

These lasers are produced in an organic liquid media. These lasers have unique characteristics that they do not have *inhomogeneities* as solid lasers. They also do sustain permanent damage. The emission spectra and gain of dyes are very broad, which offers tunability and mode locked short pulses at output. The laser gain medium consists of strongly emitting and absorbing organic dyes which are dissolved in a solvent.

The molecules of the organic dyes are very complex belonging to one of several classes which are given below:

Xanthene dyes: This radiates in the visible region from 500 to 700 nm,

Polymethine dyes: This emits in the red or near infrared region at 700 to 1500 nm,

Coumarin dyes: This radiates in the green and blue region at 400 to 500 nm

Scintillator dyes: This radiates in the ultraviolet region at 320 to 400 nm.

For making a laser medium, organic dyes are dissolved in solvents like ethylene glycol, methyl and ethyl alcohol and water. In these solvents, the concentration of the molecules of the dye is of order of one in ten thousand. Therefore, the dye molecules are quite apart from one another and because of this, each molecule is surrounded only by molecules of the solvent.

A specific dye laser operates over a range of wavelengths between 30 to 40 nm. The region of gain is narrower than the bandwidth of emission due to absorption at ground state. There are more than 200 dye lasers. When these are used in a sequential manner, they can produce *tunable* laser beams over a range of wavelength between 320 and 1200nm.

16.4.2 Types of Dye Lasers

There are three types of dye Lasers, which are given below:

(a)Pulsed dye lasers,

(b)Continuous Wave (CW) lasers,

(c)Mode-locked dye lasers.

When pumped with other lasers such as frequency multiplied $Nd-YAG$, nitrogen or *excimer*, pulsed dye lasers can produce output pulses up to tens of *millijoules* in 10 nm pulses. The diameter of the beam will be a few mm and the pulse repetition rate, up to 1 KHz. Dye lasers pumped by flash lamp can produce up to 400 Joules of output in $10\mu m$ pulses. Continuous wave dye lasers produce power outputs up to 2 watts. These lasers can be made to have a very thin *linewidth* of emission of less than 1 KHz. Pulses of $200\ fs$ have been produced by mode-locked dye lasers, without having to introduce gratings or dispersive prisms in the cavity. The pulses can be as short as $6\ fs$. Dye lasers have been made in sizes ranging from small (of the order of 10 cm in length) to massive mode-locked systems, which would extend too many optical tables.

16.4.3 Geometry of Dye Lasers

There are four types of lasers which are given below:

(a) Pulsed dye lasers pumped by flash lamps.

(b) Tunable pulsed dye lasers pumped by other lasers.

(c) Tunable continuous wave lasers.

(d) Mode-locked dye lasers.

16.4.3.1 Pulsed Dye Lasers Pumped By Flash Lamps

A pulsed dye lasers, pumped by flash lamp, is shown in figure 16.6, given below:

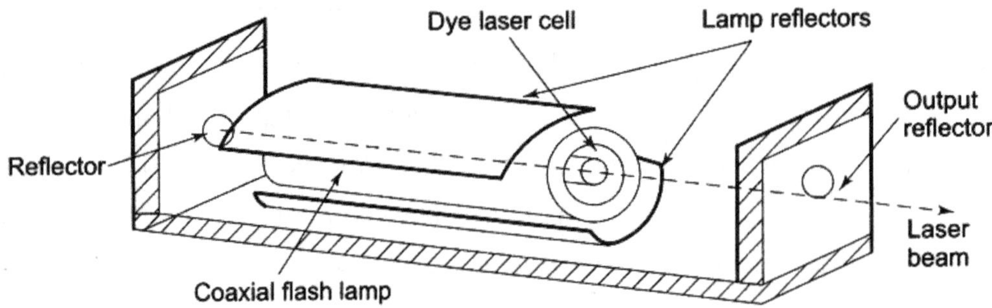

Figure 16.6: Pulsed dye laser pumped by a flash lamp

The dye laser cell is surrounded by a coaxial cylindrical flash lamp. Between the flash lamp and the dye cell, there is a provision for water-cooling. The reflector around the flash lamp redirects its output to dye cell. This laser can produce pulses of 5 joules in pulses duration of 1.5 μs and a peak power of more than 3 megawatts per pulse. The input power of the laser is 1000 joules and the pulse repetition rate is 0.5 Hz. The dye used is $rhodamine$ 6 G. The laser can be tuned and operated at a wavelength of 585 nm.

16.4.3.2 Tunable Pulsed Lasers Pumped By Other Lasers

This category of dye lasers produces *tunable* pulsed laser outputs of extremely narrow frequency of the order from 190 nm to 4.5 μm. The system of a basic dye laser is given below in figure 16.7.

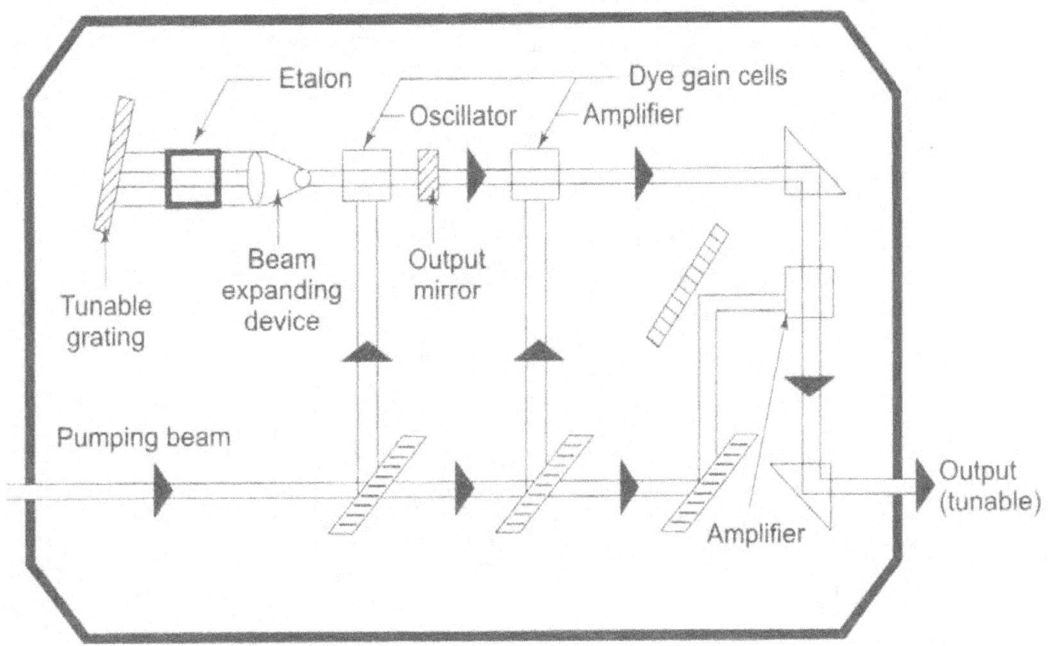

Figure 16.7: Arrangement of a tunable dye laser pumped with another laser.

It covers a frequency range of the order of 380 to 900 nm, in which pumping is done either by a *excimer* laser or $Nd - YAG$ laser, whose frequency is doubled or tripled. in this type of laser scheme, the extended frequencies at both the longer and shorter wavelengths are produced by the sum and difference frequency mixing of the basic range of wavelengths. This is done by another optical component which is not indicated in the above given figure.

The energies of the output pulse range from tens to hundreds of millions joules, over pulse duration of the order of 10ns. And this depends upon the wavelength. The pump beam, either *excimer* or frequency doubled or tripled $Nd - YAG$, gets divided into several separate beams as shown in figure 16.7. One of the beams pumps a *tunable* oscillator, with narrow frequency output. The *tunable* oscillator contains a beam grating, an etalon and beam-expanding unit. The other beams are used for end pumping or side pumping of several dye amplifiers. The output laser of the dye can be very narrow-as narrow as 10 GHz. If we use an air-spaced etalon, it cam further gat reduced to 1.5 GHz.

16.4.3.3 Tunable Continuous Wave Dye Lasers

These types of lasers are usually pumped by other CW lasers like the Argon ion laser. The arrangement of this type of lasers is given below in figure 16.8:

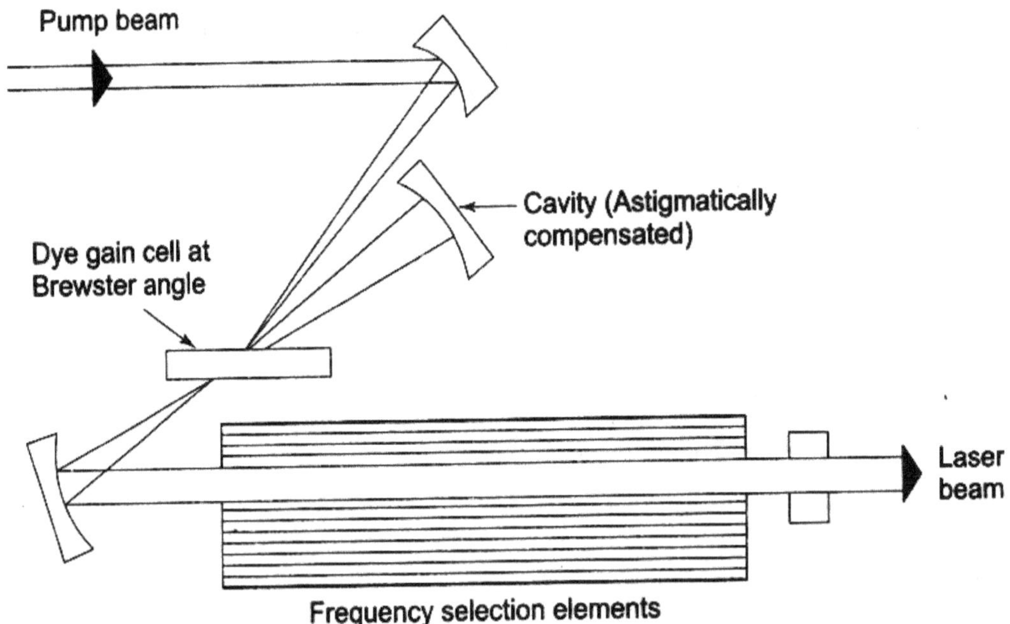

Figure 16.8: CW tunable dye laser-laser pumped.

Here, the cavity consists of 3 reflectors. Two of these reflectors focus the beam into the jet stream, which is the thin following dye region. The jet stream is oriented at Brewster angle. The flow of the dye is in a direction normal to the plane of this page. The end-pumping method is used to pump the dye. The pumping is usually at a small angle to the jet stream with an additional reflector. The operation can also be done by pumping directly through the third reflector. Tuning can be done by inserting a prism and rotating the laser reflector located beyond the prism as shown in figure 16.9.

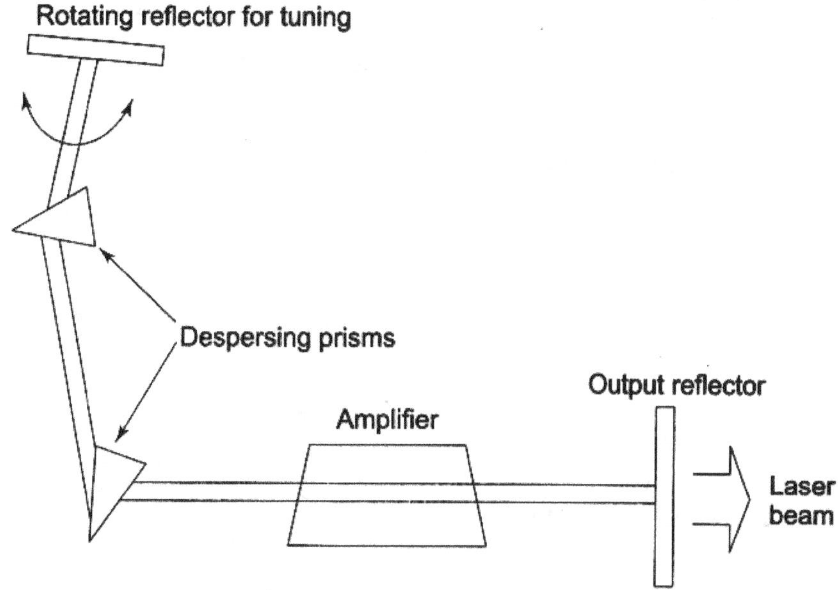

Figure 16.9: Tuning of dye laser using prisms

The pump beam enters the cavity through the prism since it is at a shorter wavelength as compared to output of the dye. Hence, it enters the prism at a different angle then the laser beam.

16.4.3.4 Mode Locked Ring Dye Laser

The schematic arrangement in a passively mode-locked ring dye laser is shown in below given figure 16.10.

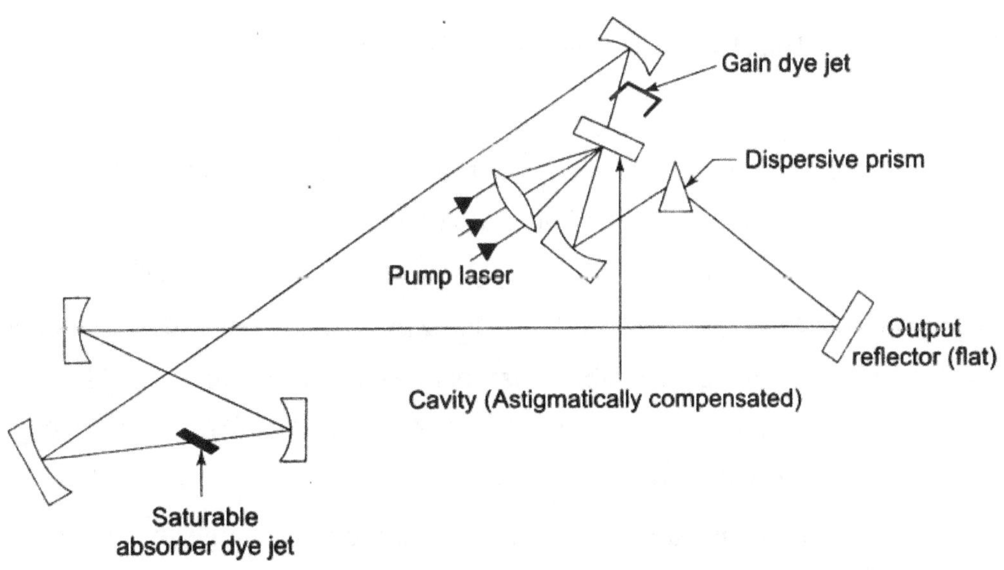

Figure 16.10: Schematic arrangement of a mode-locked ring dye laser

Here, we see the production of a laser pulse of *femtosecond* duration. The design of the cavity is such that, within the cavity, there are odd numbers of beam waists are formed by the five concave mirrors, which ensures uniformity in the spectral distribution of the laser bandwidth within the resonator. The optical path-length between the two jet streams is one fourth of the optical paths surrounding the ring.

There are two counter-propagating pulses inside the cavity. As a result, both pulses have identical characteristics. This happens by making the time required for recovery of gain to be the same for each pulse. Maintaining the critical dimensions or geometry of such a cavity is extremely important. Then only we will get the shortest pulse.

16.5 He-Ne Lasers

It is gaseous laser beam. Solid state laser (Ruby laser) does not generate a continuous laser beam. For this difficulty in solid state laser, *Javan, Bennett and Harriot* gave a gas laser, which emits continuous laser beam rather than in pulses. It uses a mixture of helium (He) and neon (Ne) gases. Its operation involves four energy levels, three in neon and one in helium. The excitation of helium and neon atoms to higher energy states is performed by means of radio (high) frequency electromagnetic field. The construction of the He-Ne laser is given in figure 16.11.

16.5.1 Construction of He-Ne Laser

It consists of:

(i) A working substance in the form of a mixture of helium and neon gases in the ration 7:1 at a total pressure of 1 mm of Hg.

(ii) A resonant cavity of quartz tube is of about 0.5 m length and 5 mm diameter. There are two windows W_1 and W_2 made optically flat and cemented at Brewster' angle to the tube axis for specific wavelength λ. The ends of the cavity are enclosed by two concave mirrors, M_1 and M_2, one perfectly reflecting and other partially reflecting.

(iii) An exciting source for creating a discharge in the tube. It is generally a radio frequency high voltage source such as a *tesla* coli and is applied by means of metal bands around the outside of the tube.

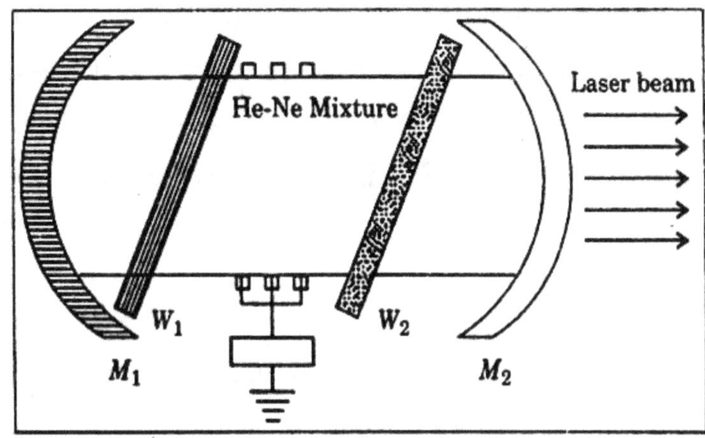

Figure 16.11: He-Ne Laser.

16.5.2 Operation of He-Ne Laser

The working of the He-Ne laser is based on the fact that the neon has energy levels very close to *metastable* energy of helium. He-Ne gas lasers can operate into three distinct spectral regions in the red 6328 Å, in the near infrared around 1.15 μm and in the infrared at 3.39 μm. The partial energy level diagram of he-Ne is given in the figure 16.12, which explains the origin of these lines.

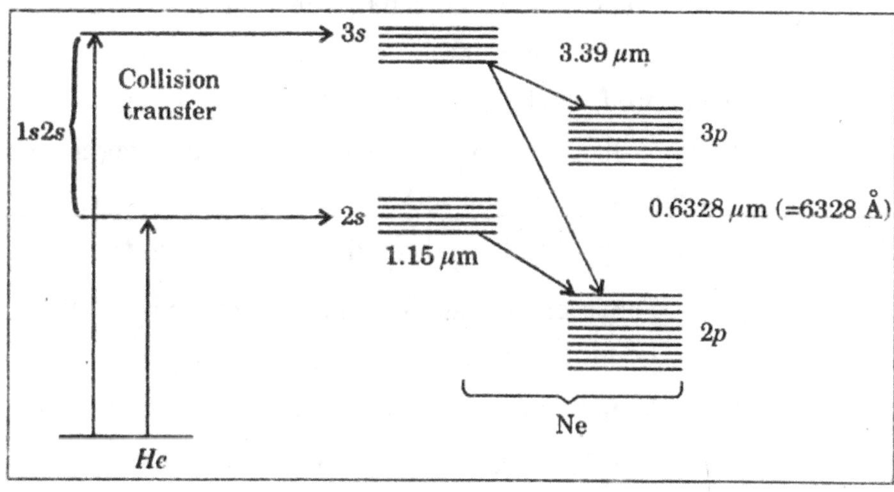

Figure 16.12: Partial energy level of He-Ne gas LASER along with transitions.

When electromagnetic energy is injected into the tube through metal bands by means of a radio frequency high voltage source, helium atoms get excited to *metastable* state. The excited helium atoms collide with unexcited neon atoms and resonant energy transfer takes place so that neon atoms get excited to a specific energy level. Helium atoms after transferring energy return to the ground state. The laser action takes place only in neon-atoms while helium in the mixture serves the only purpose to enhance the excitation process.

When population inversion has occurred in Ne-atoms, they return to lower energy states emitting the photons. The photons emitted parallel to the axis of tube bounce back and forth between polished mirrors and stimulate emission of the same wavelength from other excited Ne-atoms. Hence, the photons get multiplied and a powerful, coherent, parallel laser beam emerges from the partially reflecting end of the tube.

16.6 CO_2 Electric Discharge Lasers

In all gaseous lasers, carbon dioxide laser is considered to be most efficient and powerful laser. This laser uses a mixture of carbon dioxide, nitrogen and helium. Oscillations occur between two vibrational levels, in carbon dioxide, while the efficiency is greatly improved by nitrogen and helium. The laser operates in the middle infrared in rotational-vibrational transitions, in the 10.6 μm and 9.5 μm regions of wavelengths. Both CW and pulsed output is available in different configurations of gas discharge in a mixture of CO_2, N_2 and He gases. A typical mixture has a ratio of $CO_2: N_2$ of approximately 0.8:1. The ratio of helium is more than nitrogen. This laser gives pulsed energy of 10 KJ and CW power of the order of more than 100KW. The gain takes place, over a range of rotational vibrational transitions, which are dominated by either Pressure-broadening or Doppler-broadening. This depends on the pressure of the gas.

The lasers operate over a range from small CW modes, of the types in wave-guides, with lengths up to 35 cm to larger pulsed modes. A man could even walk through such large tubes designed for these lasers. There is a CW version with a cavity having a length of 1.2 m, which gives one or more KW of output power.

16.6.1 Geometry of Carbon Dioxide Lasers

There are different structures of carbon dioxide lasers, which are given below:

(a) Wave-guide carbon dioxide laser.

(b) Transverse Excitation Atmospheric Lasers (TEA lasers).

(c) Longitudinally excited laser.

(d) Gas dynamic laser.

16.6.1.1 Wave-Guide Carbon Dioxide Laser

For the construction of this laser, the first attempt was with a hollow structure having a small bore. A practical scheme of this configuration was made about a decade after that. This geometry is known as wave-guide laser. These lasers are made with gases like helium-neon, helium-xenon, nitrogen, carbon dioxide, xenon-fluorine and a large number of other molecules, which have emission in the far infrared region of the spectrum.

These are considered to be the most efficient of schemes, which can produce a compact, continuous wave, CO_2 laser. The scheme consists of two transverse radio frequency (RF) electrodes separated by insulating sections, which form the bore region as shown in figure 16.13.

A radio frequency power supply is connected to the electrodes, which provides a high-frequency alternating field, across the electrodes inside the bore region. The bore is usually square in shape. The bore should be able to propagate the beam in wave-guide modes, which gets reflected from insulating materials (or insulators). The modes reflect, off the walls of the discharge, in zigzag manner and hence, the modes can access the entire gain medium.

The small bore regions make the scheme work in high pressure. It also helps in quick removal of heat. As a result, a high output power and high gain is available.

The main property of this category of lasers to produce CW powers, up to about 100 W.

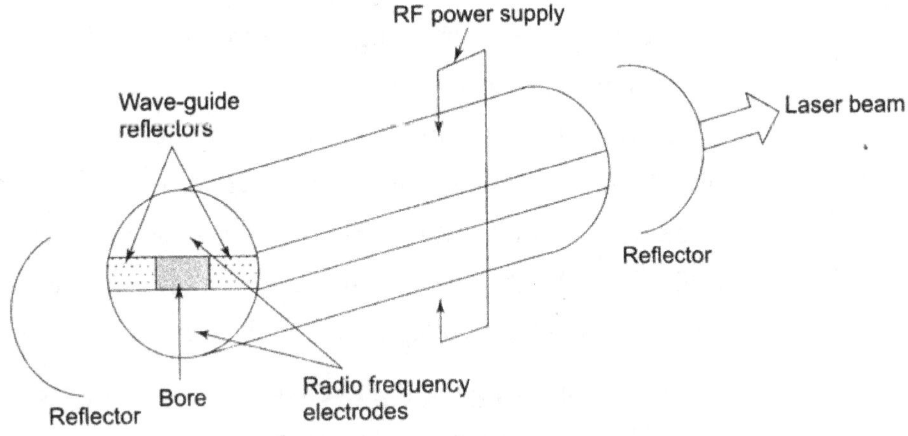

Figure 16.13 Scheme of Wave-guide CO_2 laser

16.6.1.2 Transversely Excited Laser At Atmospheric Pressure (Tea Laser)

In a TEA laser, atmospheric pressure is maintained inside the discharge tube. However, the gas discharge is not maintained by applications of an electric field longitudinally. The discharge is maintained by applying the field in the transverse direction as showing in figure 16.14.

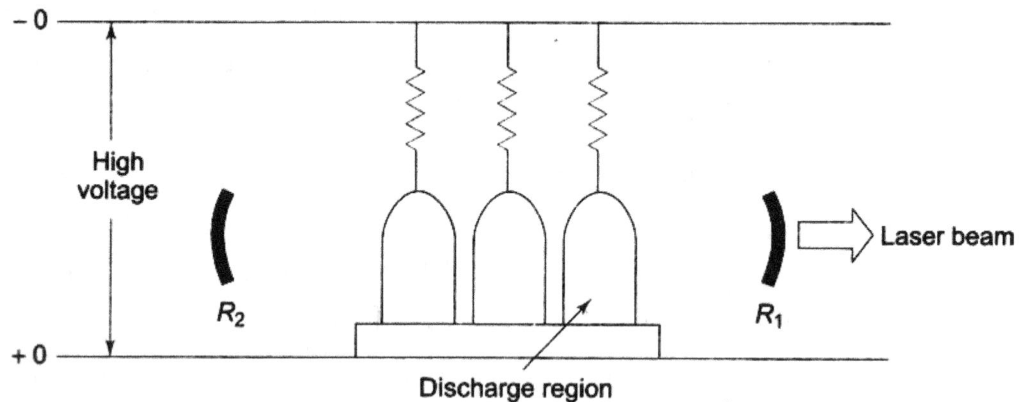

Figure 16:14: Arrangement of TEA laser showing discharge taking place normal to laser cavity.

The voltage, required for transverse excitation, is quite less since the gas discharge occurs at a critical electric field. The transverse directions are of the order of 10 mm, which requires a voltage of about 0.12 KV. The discharge has to be maintained uniformly through the total length of the discharge tube.

The TEA lasers operate at a gas pressure of 1 atmosphere or more. The energy output is function of the volume of gas. There are more laser species at higher pressure.

Operating the laser at high pressure, with a discharge that is longitudinal, is very difficult because, this operation demands very high voltages for initial ionization of the gas; moreover arcing can occur within the discharge, which will make the current to flow in a random and irregular direction, as in the case of a thunder and lightning. Whereas, in a transverse discharge, the electrodes are placed parallel to each other, over entire length of the discharge tube, separated by a few centimeters and high voltage is applied as shown in figure 16.15.

Before applying the high voltage, a process of pre-ionization is done to ionize the region between the anode and cathode, which results in filling up of the area with electrons. The pre-ionization enables the discharge to be uniform through the entire electrode assembly and arcs due to generation of high current are avoided.

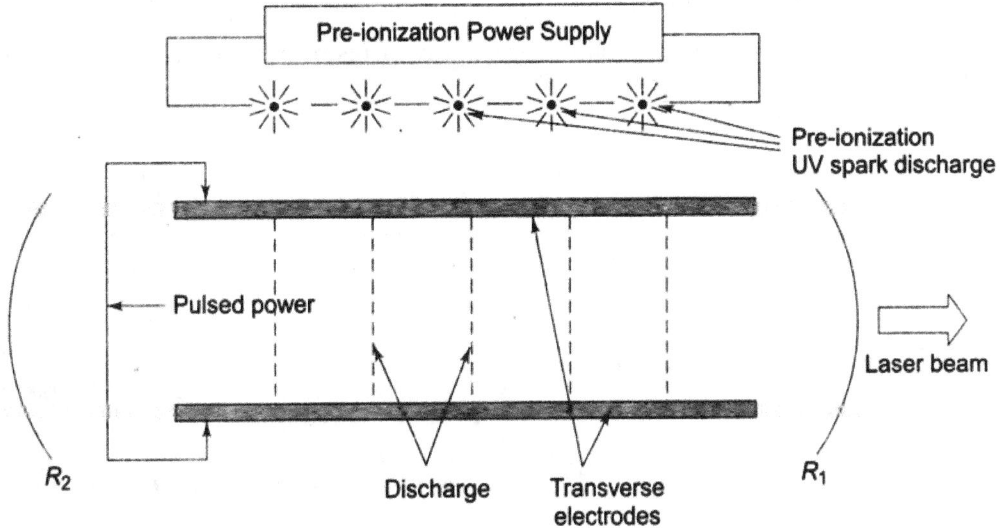

Figure 16.15: Arrangement of TEA CO_2 laser.

The process of pre-ionization involves flashing of ultraviolet rays from an array of pre-ionizing ultraviolet spark discharges. This ionizes a portion of the gas uniformly between the electrodes. These lasers can produce a large amount of energy (many joules). With this configuration the best form of high-energy pulses can be obtained.

16.6.1.3 Longitudinally Excited CO_2 Laser

The arrangement of the system of longitudinally excited CO_2 laser is given below in figure 16.16.

These schemes work as conventional gas discharge lasers having long cylindrical, narrow glass enclosures with electrodes at both ends, which provides the current for excitation of the discharge. These lasers can operate in CW or pulse modes. These tubes can be very long-several meters. In some configurations, the enclosures of the discharge are sealed. This necessitates refilling of the gases, because in the sealed tube, the CO_2 molecule can get disintegrated. The CO_2 molecules can break up and produce oxygen, which can result in corrosion of the electrodes.

In some versions, the gas is introduced into the tube longitudinally. This is done in order to conserve the gas. Water vapour is considered to improve the power output and operating life of a CO_2 laser. Water produces OH radical in the discharge which when combined with CO, produces CO_2 and H. The enhanced output power of CO_2-N_2-He-H_2O (or H_2) laser is believed to be due to the effective relaxation of the lower laser level by impact with H_2.

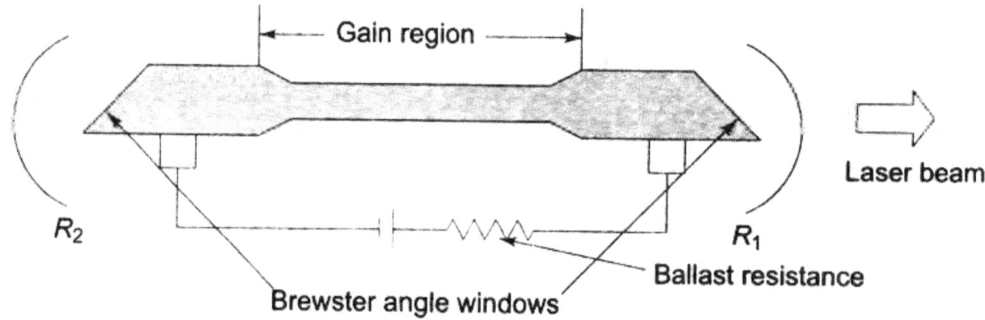

Figure 16.17: Longitudinally Excited CO_2 laser.

Sealed off lasers have the advantage of being compact and portable. The output spectrum is stable. Conservation of costly gases like helium is also possible. However, they suffer the disadvantage of having smaller operating lifetime. The tube is required to be cooled with water to keep the temperature at about $20°C$ because the output of the laser decreases as the walls of the tube becomes hotter.

16.7 Gas Dynamic Lasers

This laser is different with compare to other laser. In this laser, population inversion is achieved by application of thermodynamic principles and not by standard electrical discharge. A mixture of carbon dioxide and nitrogen is first heated, compressed and then passed through a specially designed nozzle, into a low-pressure region. While heating and compressing, the population of the energy states touches the Boltzmann distribution adequate for higher temperature. A large portion of the energy gets stored in the vibrational mode of the nitrogen molecule.

The resonant impact of the excited nitrogen molecules, with the carbon dioxide molecules, populate the 001 state of the CO_2, which results in population inversion which can be increased by adding a small quantity of water vapour as we have seen in the previous section. The water vapour has to be added at the points, where the gas expansion occurs. This accelerates the relaxation of the 100 state, to the ground state. An optical cavity is introduced inside the region where the maximum amount of population inversion occurs. By this, CW powers, of the order of tens of kilowatts, can be obtained. However, the system is quite voluminous and operation is noisy.

In these lasers, the gas flows in the transverse direction, to the axis of the laser, as is done in the TEA laser. Excitation takes place due to the heating, which is done thermally. In some versions, the heating is also done electrically. The upper laser level thus gets populated. The arrangement of gas dynamic laser is shown in figure 16.18.

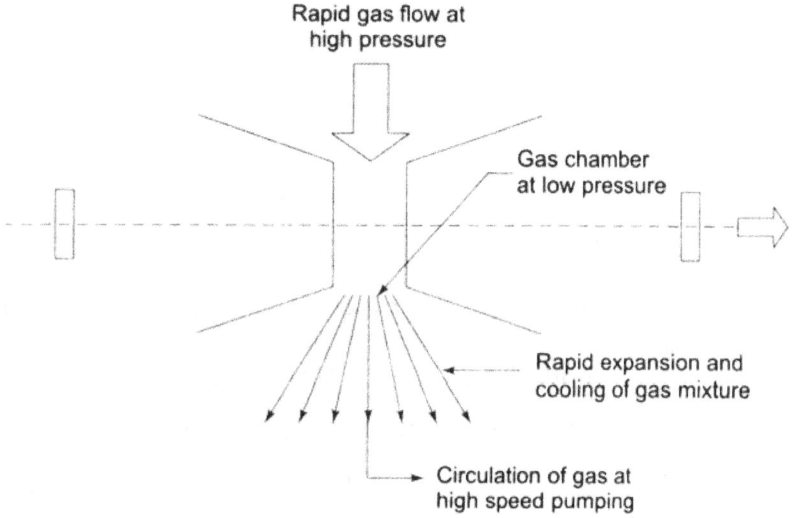

Figure 16.18: Gas Dynamic Laser

As mentioned earlier, a small amount of water vapour is added to the gas mixture. The rapidly flowing gas is then expanded at a very high speed through an expansion nozzle into the low-pressure zone, using high speed pumps. This sudden expansion cools the gas. This provides a rapid relaxation of the lower laser level from the highest rotational state. This leaves a population inversion of the unoccupied higher rotational state, as compared to the upper laser level.

16.8 Excimer Lasers

For the principle of *excimer* lasers, let us consider a situation, which is given below:

Consider a diatomic molecule M_2, having curves of potential energy as shown in figure 16.19. These curves show the ground state and excited states. The ground state is repelling and therefore the molecule does not exist in this state; which means that the species M, exists only in the 'monomer' form M in the ground state. But the curve of potential energy for excited state touches a minimum. Hence, the molecule M_2 exists in the excited state; which means, species M exists in the excited state. Such a molecule M_2^* is called an 'Excimer' which is evolved from the words *Excited Dimer*.

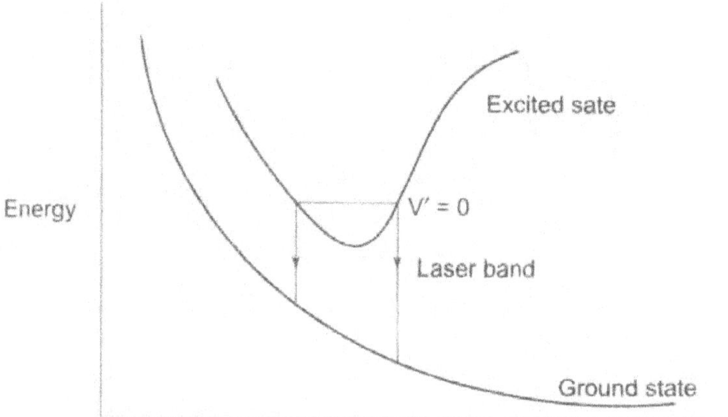

Figure 16.19: Energy levels in an Excimer language.

A similar situation can happen with molecules of more complex nature. Some complex molecules exist only in excited state and they dissociate when they are in the ground state. This type of a molecule is called *Exciplex* derived from the words 'Excited Complex'.

Let us consider that a large number of *excimers* have been created within a particular volume. Lasing can take place on the transition between upper bound state and the lower unbound level. Such a laser is called an *excimer* laser. The *excimer* laser has unique characteristics indicating the condition, that the ground state is repelling in nature. They are:

(a) After undergoing laser transition, once the molecule reaches the ground state, it quickly dissociates. This indicates that the lower laser level will always be unoccupied.

(b) There is no *clear − cut* or well defined *rotational − vibrational* transition in existence. The transition is also broadband. As a result, a *tunable* laser radiation over the broad band can be obtained.

Excimer lasers are pulsed lasers of short duration. These lasers are made by combining a rare gas atom of argon, krypton, xenon, etc and a halogen atom of fluorine, chlorine, bromine or iodine. The excimer molecule, as mentioned earlier, exists only in the excited state because of the fact that the ground state is of extremely short duration, due to the repulsive force between the two atoms of the molecule in the ground state.

A typical excimer lasers emit in the ultraviolet region of the spectrum. However, some of them do operate in the visible region also.

Examples of excimer lasers are:

(a) F_2 with transition at 153nm.

(b) $Ar - F$ at 193 nm.

(c) $Kr - F$ at 248 nm.

(d) $Xe - Cl$ at 308 nm.

(e) $Xe - F$ at 353 nm.

A typical excimer laser has gain medium with length of 0.5 to 1.0 m. It would also have a transverse discharge configuration for its electrodes similar to CO_2 lasers. The halogen species have tendency to disintegrate and form other unwanted species during the operation of the laser. Therefore, the schemes require a *recirculating* system of gas, to provide a constant and regular supply of purified constituents of the gas, to the gain region. Because of very high order of gain, these lasers produce a multi-mode output, of a very high order. Since they can give out high pulse energy, and ultraviolet output, they can be effectively used in material processing.

16.8.1 Structure of Excimer Lasers

The arrangement of a typical excimer laser is shown in figure 16.14. The halogens are corrosive in nature. So, the entire structure and components are made of stainless steel with corrosion resistant materials, such as Teflon or polyvinyl, like in TEA laser, the discharge takes place in the transverse direction. The electrodes are metal pieces that are flat and long in size. The metal pieces are rounded in the corners to make the electric field uniform between the electrodes, when the voltage is applied. So, arcing is avoided, and there will be a uniform excitation. There is an initial electron seeding which is achieved by a pre-ionization pulse; in the area between the electrodes, which also ensure uniform excitation. Just as in the case of a TEA CO_2 laser, the pre-ionization pulse is produced by an array of tiny ultraviolet spark charges; which is called a 'flash board'. These sparks emit sufficient ultraviolet radiation to produce ionization in the gain region, which substantially increases and electrical conductivity of the gas medium

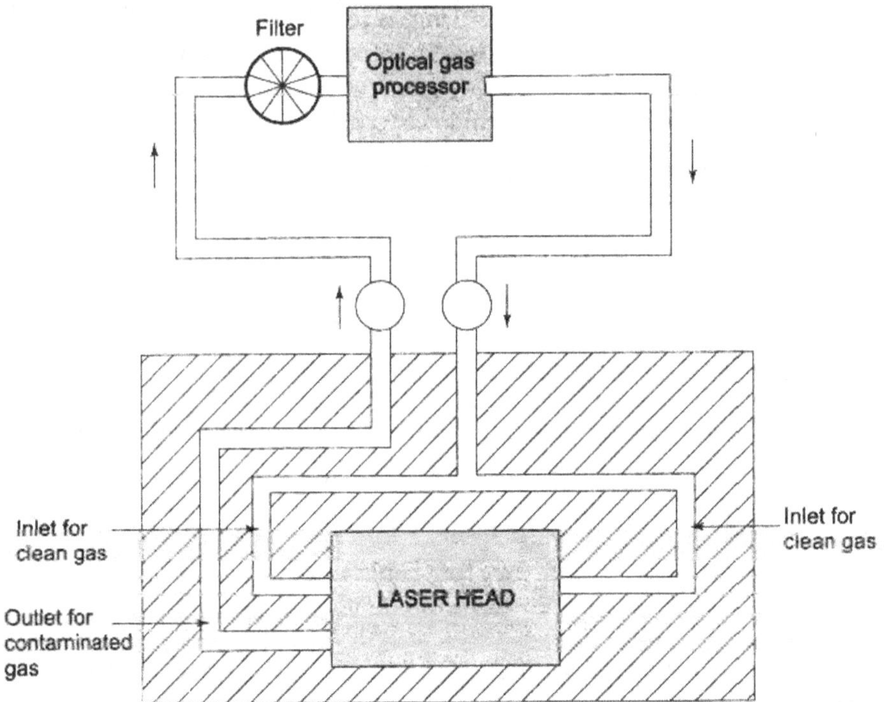

Figure 16.20: Structure of Excimer Laser

A high-voltage capacitor, coupled with a thyratron-switching device, provides the discharge current. A mirror with high reflectivity is placed at the rear of the laser system. This arrangement is not shown in given diagram. A flat quartz is placed at output-coupling mirror. The low reflectivity of the quartz is adequate to give the required optical feedback, within the medium, for efficient working, because the gain is typically of the order of $0.8 m^{-1}$.

In these lasers also, like in CO_2 lasers, a wave-guide can be used.

In waveguide excimer lasers, electric discharge tubes, with bore regions, having a lateral dimension of less than 1 mm are used. The metallic transverse electrodes are fixed externally to the bore. The electrodes provide pre-ionizing pulse, of high voltage and a radio frequency main pulse to the tube.

16.9 Self Learning Exercise

Q.1 Give the application of rare-earth ion laser.

Q.2 Give the application of He-Ne laser.

Q.3 Give the application of Dye lasers.

Q.4 What is metastable state?

Q.5 What is the principle of Laser?

Q.6 Give the application of Excimer Laser.

Q.7 Give the characteristics of Laser rays.

16.10 Summary

A solid state laser was the first one to come into the laser world. Solid state lasers are part of the laser systems involving high-density gain media. Neodymium is a typical example of an optically pumped rare-earth laser system.

After solid state laser, liquid lasers are designed. These are the lasers in which the active medium is formed by solutions of certain organic dyes dissolved in liquids such as alcohols or water.

Then gas laser fall in the category of lasers involving low-density gain media. The gas media are used in almost half of the commercial lasers that are currently available.

In chemical laser, the energy pumping is obtained from a chemical react ion. Typical chemical lasers operate on molecular transitions.

16.11 Glossary

Pumping: It is a process by which population inversion can be produce known as pumping. The population inversion can be achieved by exciting the medium with suitable form of energy.

Direct Conversion: In this type of method, a direct conversion of electrical energy into radiant energy occurs in Light Emitting Diodes (*LEDs*). The example of population inversion by direction by the method of direct collision occurs in semiconductor lasers.

Dye Lasers: These lasers are produced in an organic liquid media. These lasers have unique characteristics that they do not have *inhomogeneities* as solid lasers.

16.12 Answer of Self Learning Exercise

Ans.1: *Nd* laser are commonly used than other similar lasers. They find various applications as compared to other lasers. The major application is in the different forms of material processing such as spot welding, drilling and making other lasers. Since they can be focused to tiny spots, these lasers are used for resister trimming, memory repair and in circuit mask. They are also used in cutting out specialized circuits.

It is also used in the field of military exploitation. They are used for target designation, range finding etc. Laser gyros are used in many aircraft and in weapon systems. Attempts are being made to use them in detection of submerged inertial confinement fusion.

Nd lasers also find wide application in general laboratory work and in scientific fields. Since frequency multiplication into ultraviolet and green can be done with these lasers, they are good sources for pumping tunable dye lasers and other varieties of diagnostic and laser probes.

Ans.2: There are some fields, where helium-neon laser is applied, which are given below:

(1) Interferometry   (2) Laser printing   (3) Bar-code reading.

(4) As pointing and directional reference beams.

In the first field, the helium-neon laser gives a very stable and single transverse mode reference beam, which is essential in identifying optical characteristics of material like smoothness (Δ) and the figure of the surface.

In the field of laser printing, the laser beam is used as a source for writing on the photosensitive material, which give print patterns in detail.

It is common now days to see departmental stores using He-Ne lasers to check out or for scanning the merchandise having digital codes, which are imprinted on the product.

In optical fiber communication the laser having 1.523 micrometer is used for measuring the lines, which have minimum loss in those wavelengths.

Ans.3: There are several important uses of Dye lasers, which are given below:

(1) Dye lasers are ideal for research work where either tunable laser of ultrafast pulses or tunable narrow bands are required.

(2) They find vide application in spectroscopy, especially in absorption spectroscopy in solid materials and in photochemistry. The most important application in the field of spectroscopy is in the separation of isotopes. Here, a tunable laser, pumped by a copper vapour laser is used for selective excitation of specific isotopes of uranium, which gives an enriched version of uranium.

(3) Dye lasers are also used to retard atoms to very slow speeds, which is a new breakthrough in atomic and laser science. This is achieved by tuning the lasers to specific frequencies of absorption in atoms.

(4) Ultrafast pulses are also used in the study of dynamic excited states of semiconductors and some other classes of solids.

(5) In the field of medical science, dye lasers are used to remove tattoos and birthmarks. They are also used in treatment of cancer by applying selective absorption in a similar way as is done with gold vapour laser.

Ans.4: When all the constituent photons of a light beam have the same energy, the same direction of momentum, and identical polarization, the light beam is said to be perfectly coherent.

Ans.5: Normally, the atoms of a substance are in their ground quantum state. When they are given energy by some external source, they are excited and reach some higher-energy state. An atom can persist in an excited state only for 10^{-8} second after which it returns to its normal state. In this process, the atom emits light-photons of frequency υ, where $h\upsilon = E_2 - E_1$

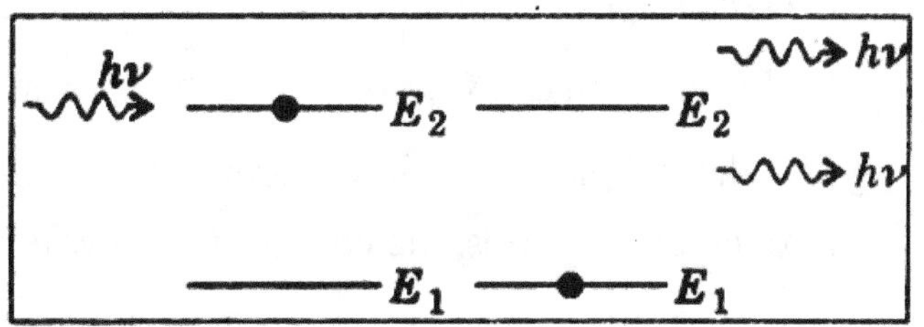

Where E_1 and E_2 are the energies in the lower energy and higher energy states respectively. The process is called 'Spontaneous emission'. This is an irregular emission and takes place at different times for different atoms. So, the light obtained by spontaneous emission from different atoms is incoherent.

Ans.6: Excimer lasers are used in medical field, material processing, pumping of dye lasers and in photolithography. Excimer lasers can provide pulse intensities of the order of $10^{13} W/m^2$. Moreover, the absorption coefficient for most of the materials, is much greater for wavelengths of ultraviolet, than for most of the materials, is much greater for wavelengths of ultraviolet, than for visible and near infrared. Hence, excimer laser beams are absorbed over very short depths, when materials are irradiated with excimer lasers. They produce sharp edges in cutting process. This property can be gainfully employed in medical applications, such as surgery and cornial sculpting which can provide optical correction to the eye without using glasses.

In lithography, the excimer laser can provide an excellent source of ultraviolet illumination, at 248 nm and 193 nm, which can produce features of microchips as small as 0.18 to 0.25 μm.

Excimer lasers are also used for pumping organic dye lasers, because all such dyes have extended absorption, into the ultraviolet region.

Ans.7: There are some characteristics of laser rays which are given below:

(1) Laser rays are completely coherent.

(2) The laser light is almost perfectly monochromatic.

(3) Laser rays are directional. Hence, a laser beam is very narrow.

(4) These rays can go to long distances without absorption. They are not absorbed in water.

(5) Laser light is very intense.

(6) Laser beam can vaporize even the hardest metal.

(7) The colour of the laser light can be changed. If laser red light be passed through quartz strips, the colour of light will change.

16.13 Exercise

Q.1   What is Laser? Explain the Salient features of a laser radiation. Mention some of its principal uses.

Q.2   Describe He-Ne laser. How is population inversion achieved in this type of laser?

Q.3   What is pumping? Give its mechanism also.

Q.4   Define population inversion and Give the classification of the Pumping.

Q.5   Describe the solid state rare earth ion laser and its application also.

Q.6   Give the brief summary of Dye lasers with its construction.

Q.7   Give the brief summary of excimer lasers with its uses and its geometry also.

Q.8   What type of Gas dynamic laser? Give its application and its construction also.

Q.9   Give the theory of any one Solid state laser and uses of Solid state laser.

Q.10   Give the theory of any one CO_2 electric discharge laser and its construction.

References and Suggested Readings

1. Laser and Non-linear Optics by B.B. Laud.

2. Laser Physics by Peter W. Milonni and Joseph H. Eberly.

3. Laser physics and its application by L.V. Tarasov.

4. Atomic and Molecular Spectra: Laser by Raj Kumar.

UNIT-17
Lasers Resonators

Structure of the Unit

17.0 Objectives

In this unit we shall examine the properties of optical resonators consisting two plane parallel, flat mirrors placed a distance apart. First of all, the properties of standing electromagnetic waves in such a system and the way in which their stored energy is lost if the mirrors are not totally reflecting will be considered. Also we shall see some very useful and quite simple techniques for analysing optical systems and a wave standpoint at how narrow beams of light travel through optical systems.

17.1 Introduction

When a Fabry-Perot resonator is filled with an amplifying medium, laser oscillation will occur at specific frequencies if the gain of the medium is large enough to overcome the loss of energy through the mirrors and by other mechanisms within the laser medium. In Paraxial ray analysis, we will be confronted with the problem of how the components of the system affect the passage of light. Here also we shall see that the special solutions to the electromagnetic wave equation exist that take the form of narrow beams called Gaussian Beams

17.2 Laser Resonators

A device with population inversion, convert oscillator into amplifier, known as Resonator. When it is used by the help of Laser beams, then it is known as Laser Resonators. There are many type of Laser Resonator, which are given below:

(1)$Plane - Parallel$ or $Fabry - Perot$ Resonator,

(2) $concentric$ or$Spherical$ resonator,

(3) $Confocal$ Resonator,

(4) *Hemiconfocal* Resonator system,

(5) *Unstable* and *Stable* Resonator,

17.3 Basic Resonators

In very simple terms, we can say that a medium, with population inversion, can get amplified. However, if the medium is to function as an oscillator, a part of the output energy has to be given as a feedback in to the laser system.

The authors sometimes noted that in the initial stages of development of lasers, some researchers wanted to name the optical phenomenon as LOSER, which is an acronym for Light Oscillation by Stimulated Emission of radiation. But, the term 'loser' indicates a person who 'loses'! Not a very good expression. The move was subsequently dropped.

The feedback is affected by placing two mutually facing reflectors or mirrors at the ends of the active medium.

The resonators used in lasers are different from those used in masers. The main differences are that:

(*a*) In lasers, the dimensions of the resonators are much larger than the wavelengths of the lasers. The wavelength usually is over a range from a fraction of a micron to tens of microns. As a result, the cavity having dimensions akin to the wavelengths actually have very low gain to sustain oscillations.

(*b*) In lasers, no lateral surface is used and the resonators are generally open; for convenience. For a laser, pumped by a flash lamp, the lateral surface can rather be a nuisance for pumping.

These two characteristics have a great influence on the performance of optical resonators. Since the resonator is open for any cavity mode, there will certainly be some losses taking place. This is because of a part of the energy, escaping from the sides of the cavity, due to diffraction of the electromagnetic field. These losses, therefore, are aptly termed as 'Diffraction Losses.'

We had seen earlier that the mode of a resonator is having the configuration of an electromagnetic field, which could satisfy both Maxwell's equation and the boundary conditions. The electric field of such a configuration could be written as:

$$E(r, t) = E_0 u(r) \exp(ift) \tag{17.1}$$

Here, the mode frequency is $f/2\pi$.

When we take into account the losses taking place, the definition of the mode, as indicated by equation (17.1) above, cannot be implemented in case of a resonator, which is open, and as such, true modes having a configuration, which is stationary, are not in existence for such a resonator. However, the configuration of standing electromagnetic waves with aluminium losses can be seen in open resonators.

Therefore, we can have new definitions for mode with some modification and we will call this quasi mode. We could also call the mode an electromagnetic configuration having an electrical field as:

$$E(r,t) = E_0 u(r) \exp\left[(ift) - \left(\frac{t}{2t_d}\right)\right] \tag{17.2}$$

Where t_d is the 'time for decay' of the square of the amplitude of the electric field. This time t_d is termed as the 'Cavity **Photon Decay Time**'(CPDT).

The attribute of the laser beam as mentioned at (a) above (where we compared laser and maser resonators) indicates that the resonant frequencies of the cavity are spaced very closely. This basically is the reason for the tendency of the lasers to oscillate on different modes.

The optical resonators are therefore called 'Multimode Resonators'; a term, which is in fact, is not scientifically correct, according to some authors.

In all the laser resonators we have either plane or spherical reflectors of rectangular or spherical shape, separated by a selected distance. The distance between the mirrors may be over range from a fraction of a centimetre to metre, whereas the dimensions of the mirrors may range from a fraction of a centimetre to a few centimetres.

17.4 Fabry-Perot or Plane-Parallel Resonator

In this set-up, two plane reflectors placed parallel and facing each other are used. The modes of this optical resonator, according to a first approximation, can be considered as the superimposition of two plane electromagnetic waves traversing in opposite directions along the axis of the cavity. This arrangement is shown in figure 17.1.

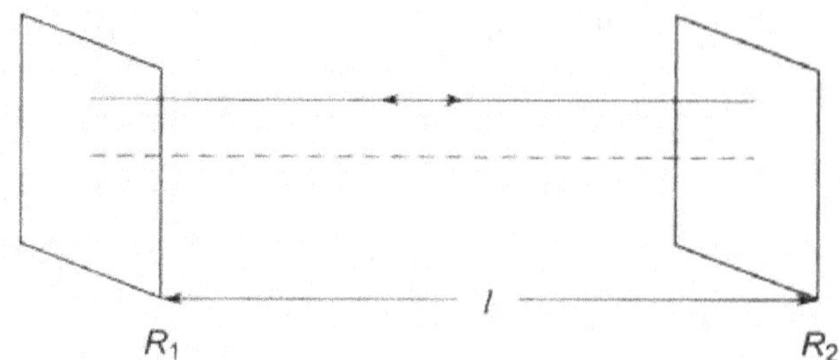

Figure 17.1: Plane Parallel Cavity Resonator

Within this condition, we can get the resonant frequencies by dictating that the length of the cavity l must be an integral number of half of the wavelength λ.

That is

$$l = n\left(\frac{\lambda}{2}\right) \tag{17.3}$$

Here n is a positive integer. This is essential for the electric field of the electromagnetic standing wave to be zero on the two reflectors R_1 and R_2. Naturally, the resonant frequencies can be obtained by the equation

$$f_r = n\left(\frac{c}{2l}\right) \tag{17.4}$$

Here, it is apparent that the equation (17.4) can also be derived by superimposing the conditionality, that the phase shift of a plane wave through one round trip, along the length of the cavity, must be equal to an integral number of times of 2π. That is:

$$2kl = 2\pi n \tag{17.5}$$

The mathematical analysis of this configuration done by *SCHAWLOW* and *TOWNES*.

17.4.1 Mathematics Analysis Of Fabry-Perot Resonator By Townes And Schawlow

Now we will see what kind of mathematical treatment these great scientists gave us. Here, we have to have a bit of practice on matrix analysis and integration.

Townes and Schawlow were the first pair of scientists who proposed an extension to the concept of maser into optical frequencies. What they used, in order to derive the first laser approximation, was an analogy with a closed rectangular resonator with known modes. One may say that the treatise they gave was rather intuitive than a rigorous work up. But some cardinal expressions in a lucid manner could be obtained.

Townes can be considered *as the father of practical lasers* and *no less a position* is occupied by *Schawlow.*

Let us consider a figure: 17.2.

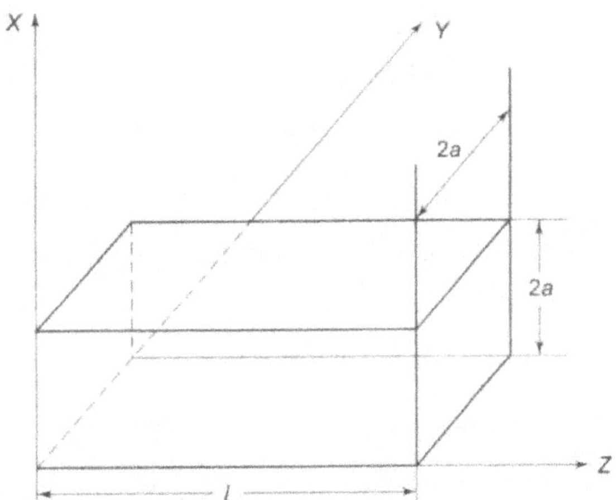

Figure 17.2: A cavity of rectangular shape having perfectly conducting walls at temperature T.

The electric field (E-field) components of the modes of the cavity can be written as:

$$\left.\begin{array}{l} E_x = e_x cosk_x x\ sink_y y\ sink_z z\ sin\omega t \\ E_y = e_y cosk_x x\ cosk_y y\ sink_z z\ sin\omega t \\ E_z = e_x sink_x x\ sink_y y\ cosk_z z\ sin\omega t \end{array}\right\} \tag{17.6}$$

Here,

$$k_x = \frac{p\pi}{2a}, k_y = \frac{q\pi}{2a}, k_z = \frac{r\pi}{l}$$

And here p, q, r are positive integers.(here r is not radius of curvature.). The notation ω is angular velocity. The resonant frequencies are given as

$$f_r = \frac{c}{2}\left[\left(\frac{r}{l}\right)^2 + \left(\frac{q}{2a}\right)^2 + \left(\frac{q}{2a}\right)^2\right]^{1/2} \tag{17.7}$$

Here, c is the velocity of light in the active medium.

The expressions of equation (17.6) assumes a complex form when the sine and cosine functions are expressed in the realm of exponential functions. By doing this, the component of the $E - field$ is put as the sum of eight plane waves travelling in the directions of eight wave vectors with components $\pm k_x$, $\pm k_y$, $\pm k_z$. Hence, the direction cosines of the vectors are, $\pm \left(\frac{p\pi}{4a} \right)$, $\pm \left(\frac{q\pi}{4a} \right)$ $and \pm \left(\frac{r\pi}{4a} \right)$ $and \pm \left(\frac{r\pi}{2l} \right)$. We know that λ is the wavelength of a specific mode. When we superimpose these eight plane waves, we get the standing wave as indicated in the set of equations at (17.6).

The assumption arrived at by Townes and Schawlow was that the modes the open cavity as shown at fig: 17.1 are described by those modes of the rectangular shown at figure 17.2. The fig. 17.1 is actually a modified version of figure 17.2 by simply eliminating the lateral surface.

The assumptions of Townes and Schawlow can be justified in the light of the fact that the modes of such a cavity may be shown as the superimposition of plane waves travelling at a minute angle to the axis Z. Removal of the lateral surface, therefore, will not alter these modes greatly. In fact, the modes, which correspond to values of p and q, which are not small as compared to r, will be drastically affected when we remove the sides of the resonator. When the sides are eliminated, we see that these modes undergo heavy diffraction losses.

When we assume that $(p \text{ and } q)$ are much less than r, the resonant frequencies of the plane-parallel cavity can be derived from equation (17.7). For each set of values of the three quantities $p, q \text{ and } r$, there exists a well-defined cavity mode with a well-specified resonant frequency. The difference in frequency Δf_n between two modes having the same values for p and q and whose values of quantity r differs by unity (1) is given by the expression

$$\Delta f_n = \frac{c}{2l} \tag{17.8}$$

When we assume that (p, q) is much less than r, the resonant frequencies can be derived from equation (17.7) using a power series expansion within the square root:

$$f \simeq \frac{c}{2}\left[\frac{r}{l} + \frac{1}{2}\frac{(p^2+q^2)}{r}\frac{1}{4a^2}\right] \tag{17.9}$$

Since the modes differ only in their field distribution longitudinally along the z-axis, Δf_n is generally termed as a difference in frequency between two successive longitudinal modes.

The difference in frequency between two modes which vary only by having a difference of 1 in their p value is:

$$\Delta f_q = \frac{cl}{4}\frac{\left[(q+1)^2-q^2\right]}{4a^2r} = \frac{cl}{8ra^2}\left(q+\frac{1}{2}\right) \tag{17.10}$$

And for equation (17.9), we can get

$$\Delta f_q = \Delta f_r \frac{l^2}{4ra^2}\left(q+\frac{1}{2}\right) \tag{17.11}$$

In an analogous manner, two modes having a difference of 1 in p values will have a frequency separation of

$$\Delta f_p = \Delta f_r \frac{l^2}{4ra^2}\left(p+\frac{1}{2}\right) \tag{17.12}$$

Those modes, which only differ in their p or q value by unity, will in turn have only a difference in the transverse distribution of field along the plane, which lies orthogonal to the $z-axis$, that is, transversely.

As a result Δf_q and Δf_p are termed as the difference in frequency between two transverse modes. We can write (17.11) and (17.12) as

$$\Delta f_q = \Delta f_r \frac{\left(q+\frac{1}{2}\right)}{8N} \tag{17.13}$$

$$\text{and } \Delta f_p = \Delta f_r \frac{\left(p+\frac{1}{2}\right)}{8N} \tag{17.14}$$

Here, we have used N. N is equal to $\frac{a^2}{l\lambda}$. This component N is the Fresnel number which is used as a dimensionless factor.

It then becomes apparent that the Fresnel number N is equal to $\frac{D_g}{2Dl}$. Therefore, the Fresnel number is half the ratio between geometrical half angle D_g and the angle of Diffraction Dl of the plane electromagnetic eave with the same transverse dimensions as the cavity resonator. Higher orders of Fresnel number therefore indicate a small spread of diffraction as compared to the geometrical angle.

In the treatise by Townes and Schawlow, so far, the losses in the cavity resonator have not been taken into account. The cavity resonances have been taken as infinitely narrow. But we have seen earlier, that there are many types of losses due to diffraction in a resonator. As such, the mode can be expressed in terms of equation (17.2). Hence, its resonance will have a Full Width at Half Maximum (FWHM) given by the expression:

$$\Delta f_e = \frac{1}{td} \tag{17.15}$$

The photon decay time td is given by:

$$td = 1/c\upsilon \tag{17.16}$$

Where υ is the fractional loss per pass of the cavity resonator and contained in it, are losses due to diffraction, the mirror transmission and internal loss of the active material.

The magnitude of υ can range from a small percentage of 1 to 3 in case of a gas laser having low level of gain. For solid state or dye lasers having higher order of gain, the percentage may be 10-30 or more.

17.5 Spherical or Concentric Resonator

In this type of resonator, two spherical mirrors with exactly the same radius r are in operation. The reflectors are separated by a distance l. The centres of curvature of the reflectors $C_1 and\ C_2$ are coincident. Hence, we can say that $l = 2r$.

This system is shown in figure 17.3 below:

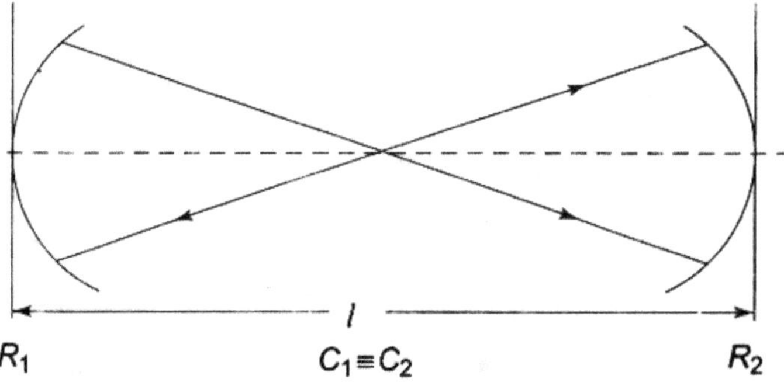

Figure 17.3: Configuration of a concentric resonator.

The mathematical analysis of this configuration as done by **Fox and Li,** which is given below:

17.5.1 Mathematics Analysis of Spherical Or Concentric Resonator By Fox And Li:

This mathematic analysis is based on a scalar approximation. In this analysis, these scientist made an assumption that the field is almost linearly or circularly polarised, uniformly, and is transverse.

The field in this case may be expressed by a scalar quantity U representing the magnitude of the electromagnetic field. Let us take an arbitrary distribution of field on reflector R_1 as U_1.

This distribution due to diffraction will result in a consequent field distribution on reflector R_2. the expression for this distribution can be derived by using the Kirchhoff diffraction integral.

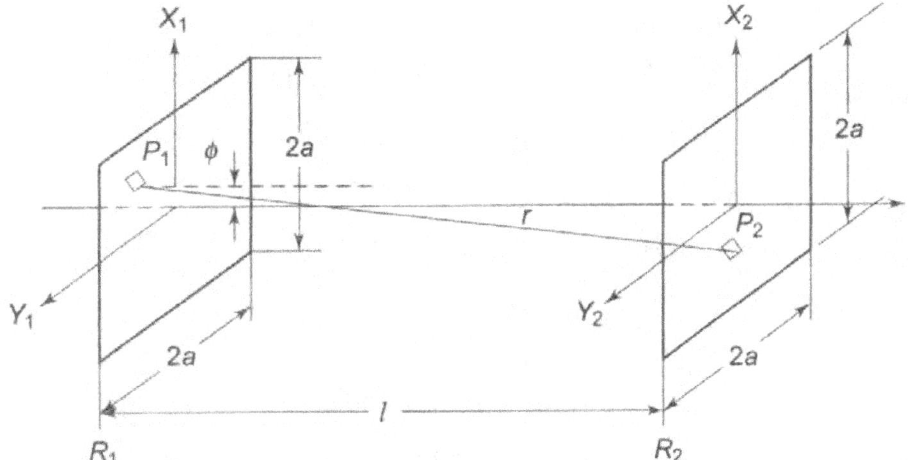

Figure 17.4: Calculation of mode for a plane-parallel optical resonator using Kirchhoff's diffraction integral.

The electromagnetic field $U_2(P_2)$ at an arbitrary point P_2 on the reflector R_2 may then be written as:

$$U_2(P_2) = -\frac{i}{2\lambda} \int_1 \frac{U_1(P_1)\exp(ikr)(1+cos\phi)}{r} lS_1 \qquad (17.17)$$

Where r is distance between P_1 and P_2, ϕ is the angle made by the line P_1P_2 with the perpendicular to the surface P_1, lS_1 is a surface element around P_1 and

$k = 2\pi/\lambda$. The integral shown at (17.17) has to be calculated for the entire surface of reflector R_1.

We may take a distribution U corresponding to the optical resonator mode instead of the general distribution U_1. Then we can show the distribution on reflector R_2 as derived by equation 17.17 to be equal to U in addition to some constant factor.

$$\sigma U(P_2) = -\frac{i}{2\lambda} \int_1 \frac{U(P_1)\exp(ikr)(1+\cos\phi)}{r} lS_1 \tag{17.18}$$

Where, σ is constant.

When the field distribution U on the reflectors is known, we can calculate the field distribution at any point inside or outside the cavity resonator.

When a is much less than l or in other words, when the length of the cavity is much greater than its transverse dimension, equation (17.18) gets simplified considerably. We can, in fact, write $\cos\phi \simeq 1 \ or \ l \simeq r$ in the amplitude component appearing under the integral sign.

A suitable expression to the nearest approximation can be obtained for the phase factor kr by writing the equation as

$$r = [l^2 + (x_1 - x_2)^2 + (y_1 - y_2)^2]^{1/2} \tag{17.19}$$

$$= l + \left(\frac{1}{2l}\right)[(x_1 - x_2)^2 + (y_1 - y_2)^2] + b \tag{17.20}$$

Here, we make a power expansion of the expression coming under the square root. The remainder b of the power series can be ignored only if kb is much less than 2π. The term b consists of a converging series with terms having signs that are alternating. We can have a corollary, that it has a value smaller than the magnitude of the first term.

So, in order to have a condition that kb is much less than 2π, a situation in which Ka^4/l^3 is much less than 2π, is required. Taking Fresnel's number $N = a^2/l\lambda$, the condition required is that N should be much less than l^2/a^2.

Hence, if we consider that l is much greater than a and N is much less than $\frac{l^2}{a^2}$. We can derive an expression as shown below:

$$\exp(ikr) \simeq \exp\left\{(ikl) + l\left(\frac{\pi N}{a^2}\right)[(x_1 - x_2)^2 + (y_1 - y_2)^2]\right\} \tag{17.21}$$

Now, if we introduce two dimensionless quantities E and F, we have

$$F = \left(\frac{\sqrt{N}}{a}\right)x \ and \ F = \left(\frac{\sqrt{N}}{a}\right)y \tag{17.22}$$

Using the equation (17.21), we can put equation (17.18) as follows:

$$\sigma^* U(E_2, F_2) = -i \int_1 (E_1, F_1)\exp\left\{i\pi\left[(E_1 - E_2)^2 + (F_1 - F_2)^2\right]lE_1 lF_1 \tag{17.23}$$

$$\sigma^* = \sigma \exp(-ikl) \tag{17.24}$$

In case of reflectors of rectangular or square shape, the variables appearing in equation(17.23) can be eliminated. A simplified expression is as follows:

$$U(E, F) = U_E(E)U_F(F) \tag{17.25}$$

$$\sigma^* = \sigma^*_E \sigma^*_F$$

With the above input, we can write equation (17.23) in the following forms for $U_E(E)$ and $U_F(F)$:

$$\sigma^*_E U_E(E_2) = \exp\left[-i\left(\frac{\pi}{4}\right)\right]\int_{-\sqrt{N}}^{+\sqrt{N}} U_E(E_1)\exp\left\{i\pi(E_1 - E_2)^2\right\}lF_1 \tag{17.26}$$

$$\sigma^*_F U_F(F_2) = \exp\left[-i\left(\frac{\pi}{4}\right)\right]\int_{-\sqrt{N}}^{+\sqrt{N}} U_F(F_1)\exp\left\{i\pi(F_1 - F_2)^2\right\}lF_1 \tag{17.27}$$

Here, now we can show that the function U_E gives the field distribution of an optical resonator consisting of two plane reflectors having a dimension of 2a in the x direction and infinitely long in the y-direction. Exactly in the same manner U_F also can give the field distribution. We can identify the Eigen functions and Eigen values shown at equations (17.26) and (17.27) by corresponding q and p values respectively. So according to equation (17.25):

$$U_{qp}(E, F) = U_{Ep}(E)U_{Eq}(E) \tag{17.28}$$

$$\sigma^*_{qpo} = \sigma^*_{Eq}\sigma^*_{Fp} \tag{17.29}$$

Although for circular reflectors also the treatment is identical, it is rather easy to put the equation (17.18) as a function of cylindrical co-ordinates rather than of rectangular co-ordinates.

We see that equations, (17.26) and (17.27) are more easily handled but it is not possible to find an analytical solution for them. In 1961 Fox and Li became successful to find an analytical solution for these equations with the help of computer.

For the solution of these equations Fox and Li showed that, if the initial distribution of field on one reflector is considered as uniform and if phase is constant across that reflector, the relative field distribution after about 300 transits within the optical resonator steadies and settles down to a value having maximum amplitude at the centre and the amplitude starts reducing as it moves further and further away from the centre. The relative distribution of amplitude across the reflector after a single transit and 300 transits in case of a cavity resonator with two infinite strip mirrors is shown in figure 17.5. Strip mirrors have the dimension 2a in the x-direction and infinite length in the y-direction.

In this diagram (a) is half the width of the side of the resonator with a value 25λ ,l is the separation between the reflectors having a value 100λ. We get Fresnel Number N as $N = \dfrac{a^2}{l\lambda} = 11.25$

Similarly, we can show the relative distribution of phase across the reflectors during first transit and after 300 transits as in Figure 17.6.

Fox and Li with the aid of computers solved several values of Fresnel number N.

In fact these scientists took up an interactive procedure on the basis of the following argument.

Consider a wave travelling up and down in the resonator. Let us assume that, at a given time, the distribution of the field $U_1(E_1)$ on the reflector R_1 is known to us. The distribution of the field $U_2(E_2)$ on the reflector R_2, which is borne from the distribution of field U_1, can subsequently be calculated using the formula (17.26). In fact, if the function $U_E(E_1)$ in the right hand of the equation (17.26) is replaced by the function U_1 and if we carry out integration, we can obtain the function $U_2 = U_E(E_2)$ which again is an offspring of the first round about trip.

Once we know U_2, we are in a position to calculate the new field distribution on reflector R_1 due to the second transverse and it continues further.

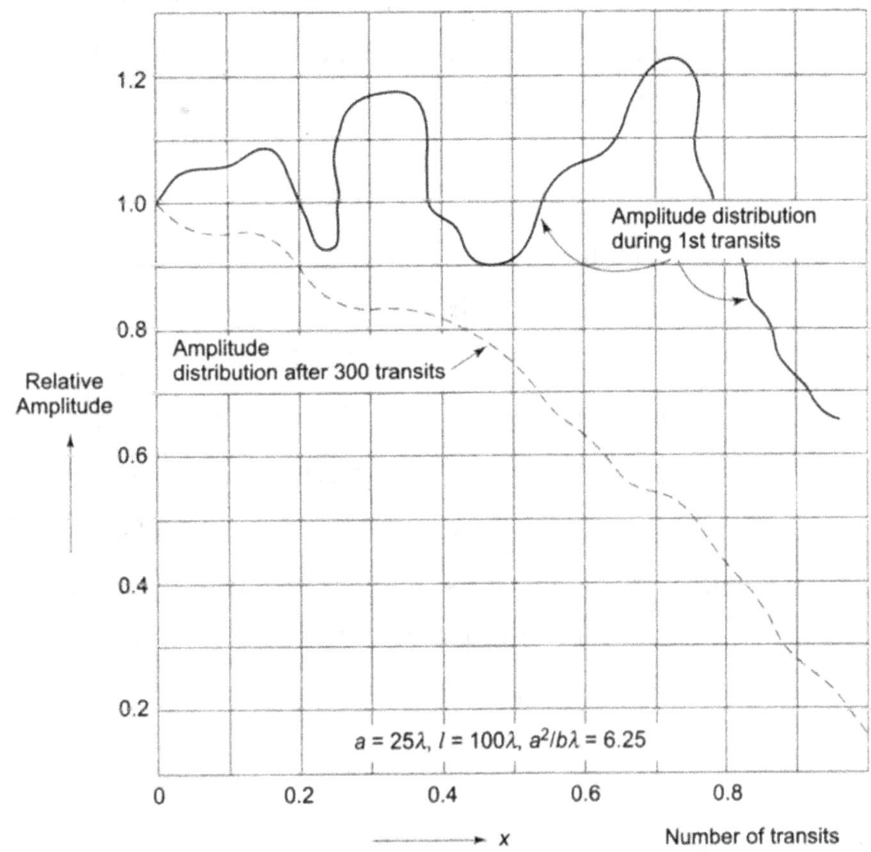

Figure 17.5:   Relative amplitude distribution

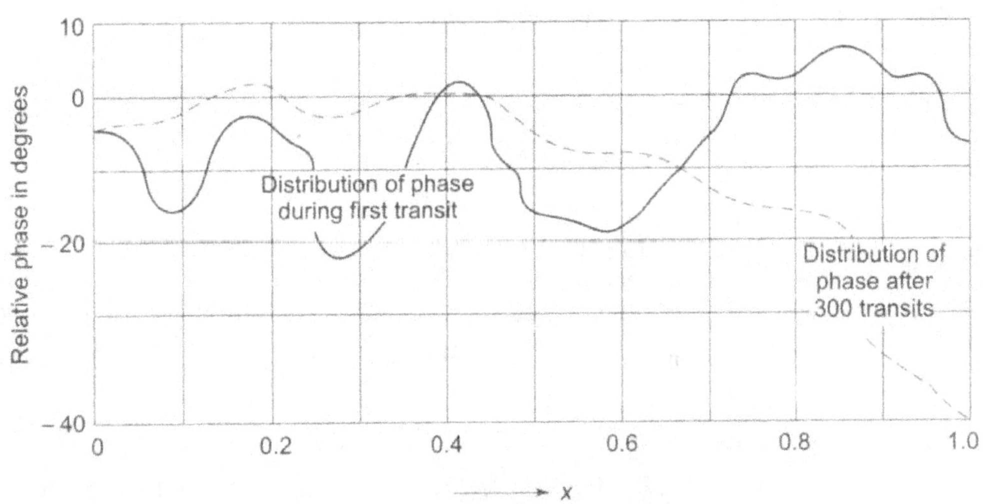

Figure 17.6:   Relative distribution of phase.

Fox and Li have proved that after enough number of passes, disregarding the formative distribution on reflector R_1, we reach a distribution of the field which does not undergo any change from traverse to traverse.

This pattern of distribution which is made by the results of Scientist Fox and Li, will then be an Eigen solution of equations (17.26) and (17.27).By this procedure, we can calculate the Eigen values as well as the loss due to diffraction and the resonant frequency of the chosen mode.

If we choose the initial distribution of the field as an even function of E, we will have a mode having even value. Conversely, if we choose the initial distribution of the field as an odd function of E, we get odd modes.

When we examine the diagrams at Fig 17.5 and 17.6, we get the amplitude and phase of $U = U\left(\frac{x}{a}, N\right)$ provided, we also choose the initial values of U_1 as symmetric and of uniform field distribution. That is, U_1 becomes a constant. If we take N=6.25, about 200 transmits are required to attain the steady stationary form of the curve as shown in Figure 17.7 and 17.8.

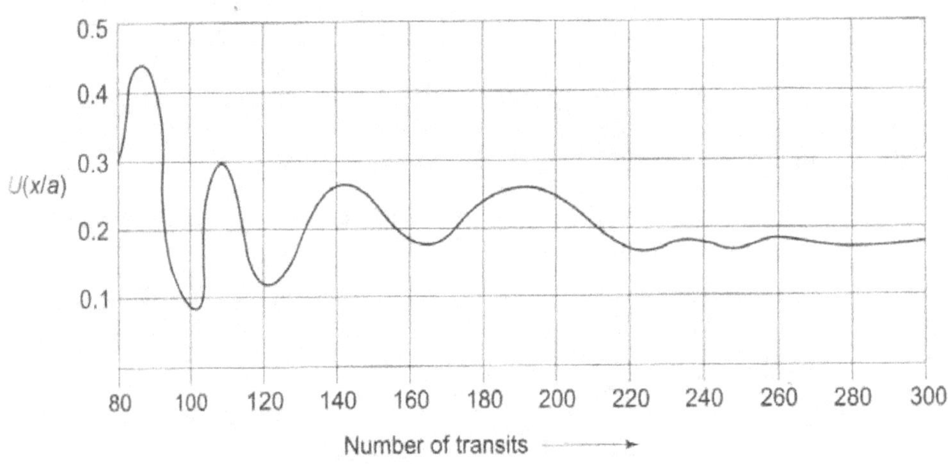

Figure 17.7: Field amplitude U Vs number of transits.

In the same way, we can get an asymmetric mode of the lowest order if we chose a uniform and asymmetric initial distribution. The condition for the same is:

$$a > x > 0 \; and \; U_1 = -1 \; for - a < x < 0$$

Figure 17.7 gives the field distribution $U\left(\frac{x}{a}, N\right)$ obtained in the above described manner for two values of Fresnel number. As regards phase, it is shown in figure 17.8.

As per equations (17.28) and (17.29), the total field distribution $U_{qp}(x, y)$ is given by multiplying $U_q(x)$ with $U_p(y)$. The mode which is related to the situation when both $U(x)$ and $U(y)$ are given by the lowest order solution, that is $q = p = 0$, as shown in figure 17.5, is called TEM_{00} mode. The mode TEM_{01} can be obtained when $U(x)$ is given by the lowest order solution as indicated in figure 17.5 and 17.6 when q=0. If $U(y)$ is given by the next higher order solution, we get TEM_{10} mode when p=1 as shown in figure 17.8 and 17.9.

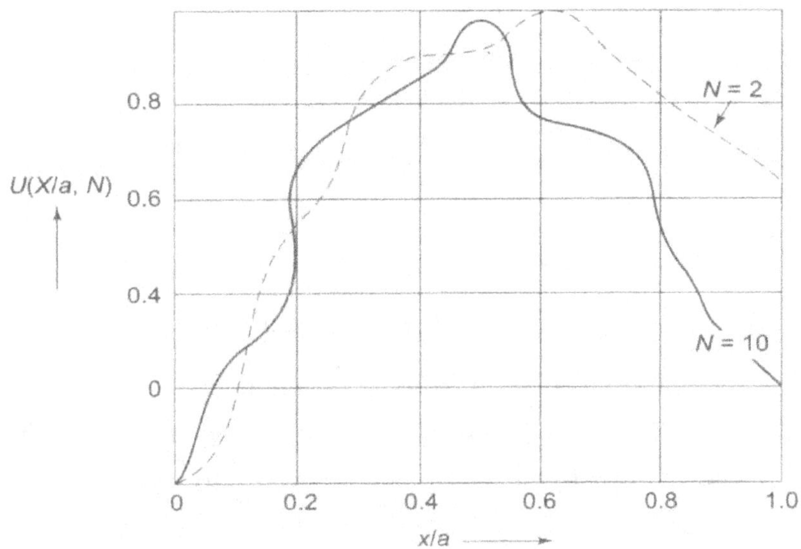

Figure17.8: Amplitude of the asymmetric mode of the lowest order for a plane-parallel optical resonator.

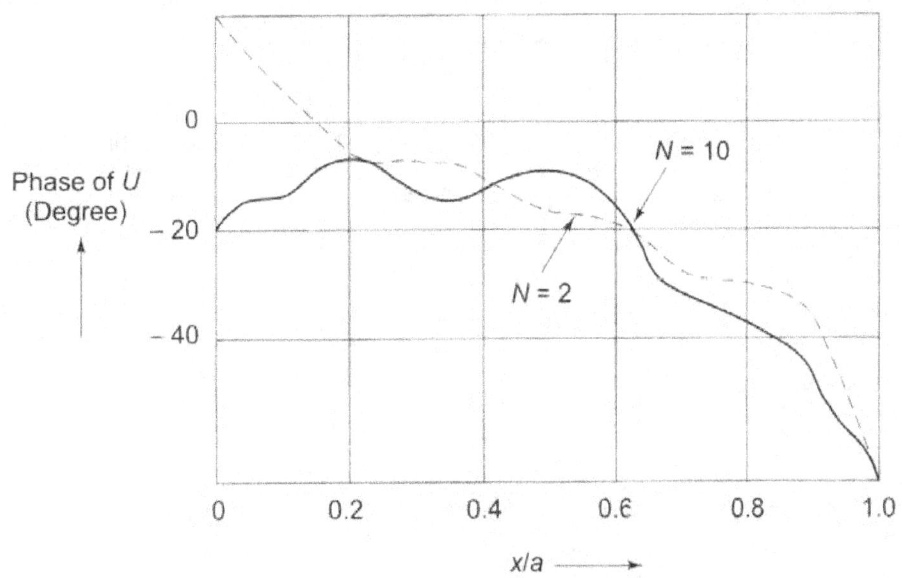

Figure 17.9: Phase of the lowest order asymmetric mode

17.6 Confocal Resonator

This type of arrangement consists of two spherical reflectors having the same radius of curvature r and placed facing each other at a distance l are used. The foci F_1 and F_2 of the reflectors R_1 and R_2 are coincident. Hence, the centre of curvature C of reflector R_1 lies on the surface of the Reflector R_2. By this, we found that $l = r$. From the point of geometrical optics, the paths of the reflecting rays are shown in figure 17.10.

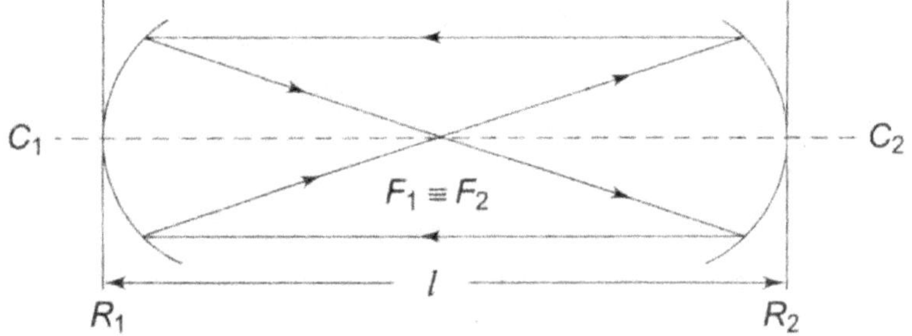

Figure 17.10: Confocal configuration.

From the path of the reflecting rays, we do not have any ides as to what would be the configuration of the modes. We will see that this arrangement will not be described by a plane wave or any spherical wave. As such a treatment of geometrical-optics does not easily give us the form of resonant frequencies.

17.7 Unstable and Stable Resonator

The entire known resonators which are used in Laser can be always grouped in two categories, named as 'Stable' and 'Unstable'.

An unstable resonator can be defined as the resonator where an arbitrary ray of light, while impinging on mirror 1 and bouncing back to mirror 2, would get deflected, or get deflected, or get diverged indefinitely, away from the axis of resonator. The arrangement of an unstable resonator is shown in Fig. 17.11.

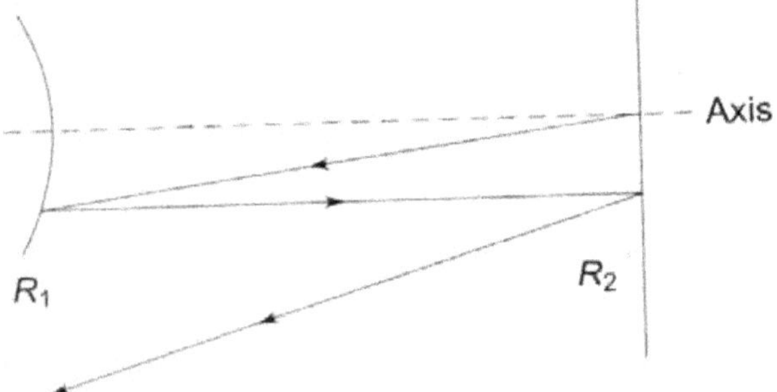

Figure 17.11: Unstable resonator.

Now the stable resonator is defined as the resonator in which the beam of light strictly remains within the cavity. The arrangement of the stable resonator is given in figure 17.12.

A resonator can be termed as stable when $0 < EF < 1$.

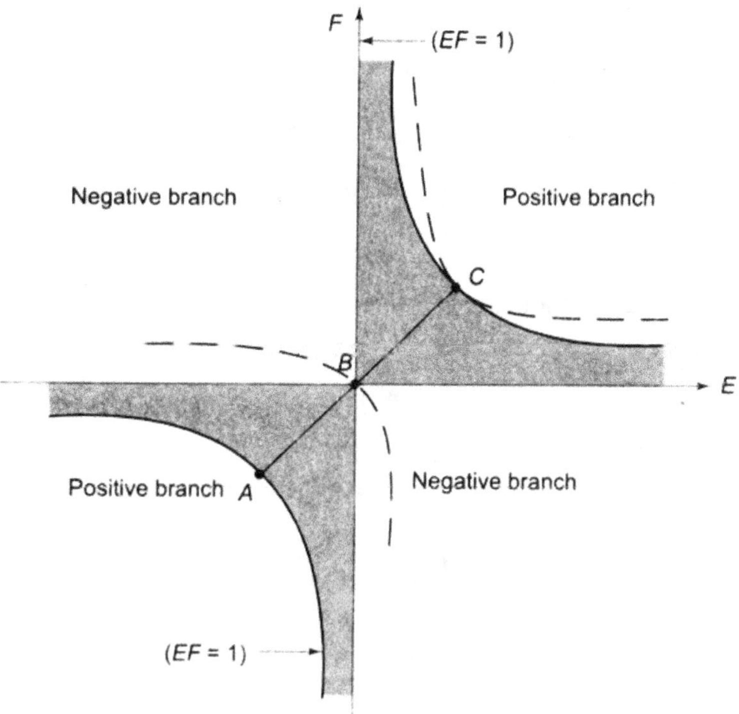

Figure 17.12: Diagram of stability for a general spherical resonator.

17.8 Laser Stability

In the geometrical sense, the optical stability is here, concerned with the question as to whether any light ray has been trapped inside the resonator or not.

For this analysis, we can consider an optical cavity as an infinite series of lenses with alternating focal lengths equal to the optical length of the cavity.

If the rays propagating near the axis of such an infinite series of lenses are refocused periodically at regular intervals, then the system is considered stable.

If all the rays diverge then the condition is violated and the resonator is termed as an unstable resonator.

Now we consider r_1 and r_2 meaning the radii of the reflectors and l the distance between them. Now, let us introduce two dimensionless quantities   E and F such that:$E = 1 - \dfrac{l}{r_1}, F = 1 - \dfrac{l}{r_2}$

A geometrical analysis will indicate that all stable cavities are governed by the condition:

$$0 \leq EF \tag{17.30}$$

$r_1 \ and \ r_2$ are the radii of curvature of the reflectors $R_1 \ and \ R_2$ respectively and l is the optical length of the resonator.

The factors $E \ and \ F$ are introduced for describing the frequencies of the TEM modes.

The diagram of stability of Laser is shown below which has been constructed by plotting $EF = l$.

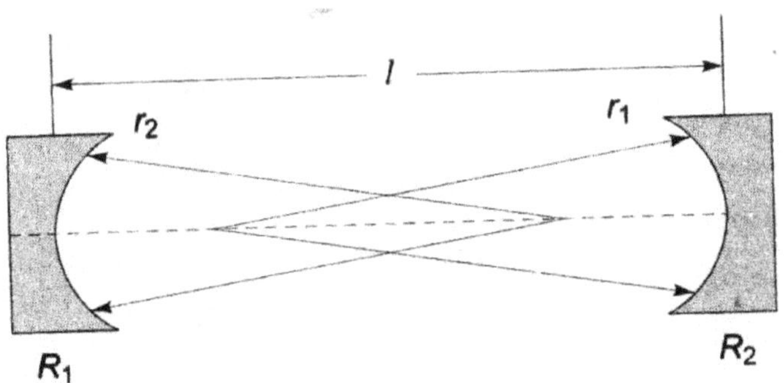

Figure 17.13: Stability Diagram

Some features of the stability diagram which is given below:

(a) The plot of $E = 1 \ and \ F = 1$ represents the Fabry-Perot Interferometer (Plane parallel Configuration).

$(b) E = 0 , F = 0$ is a symmetrical confocal resonator. This configuration gives TEM_{00}.

$(c) E = -1 , F - 1$ Gives a symmetrical concentric configuration.

The configuration $E = 0 , F = 0$ gives the lowest of diffraction losses in the TEM_{00} mode as compared with the higher TEM modes.

17.9 Paraxial Wave Equation

A plane wave is characterized by a unique propagation direction given by the wave vector $\mathbf{k}$. All fields associated with the wave are, at a given time, equal at all points in infinite planes orthogonal to the propagation direction. In real optical systems such plane wave do not exist, as the finite size of the element of the system restricts the lateral extent of the wave.

Nonplanar Optical components with cause further deviations of wave from planarity. Consequently, the wave acquires a ray direction which varies from point to point of the phase front. The behaviour of the optical system must be characterized in the deviations its element cause to the bundle of rays that comprise the propagating, laterally-restricted wave. This is most easily done in terms of paraxial rays. In a cylindrically-symmetric optical system, for example, a coaxial system of spherical lenses of mirrors, *paraxial rays* are those rays whose directions of propagation occur at sufficiently small angles θ to the symmetry axis of the system that it is possible to replace $sin\theta$ or $tan\theta$ by θ - in other words paraxial rays obey the small angle approximation.

17.9.1 Matrix Derivation:

In an optical system whose symmetry axis is in the z direction, a paraxial ray in a given cross-section ($z = constant$) is characterized by its distance r from the z axis and the angle r' it makes with that axis. If the value of these parameters at two planes of the system (*an input and an output plane*) are $r_1 r_1'$ and $r_2 r_2'$ respectively, as shown in Fig. (17.14a), in the paraxial ray approximation there is a linear relation between them of the form

$$r_2 = Ar_1 + Br_1'$$
$$r_2' = Cr_1 + Dr_1' \tag{17.31}$$

Or, in matrix notation

$$\begin{pmatrix} r_2 \\ r_2' \end{pmatrix} = \begin{pmatrix} A & B \\ C & D \end{pmatrix} \begin{pmatrix} r_1 \\ r_1' \end{pmatrix} \tag{17.32}$$

$\begin{pmatrix} A & B \\ C & D \end{pmatrix}$ is called the ray transfer matrix, **M**; its determinant is usually unity, i.e., $AD - BC = 1$.

Optical systems made of isotropic material are generally reversible – a ray which travels from right to left with input parameters $r_2 r_2'$ will leave the system with parameters $r_1 r_1'$ – thus;

$$\begin{pmatrix} r_1 \\ r_1' \end{pmatrix} = \begin{pmatrix} A' & B' \\ C' & D' \end{pmatrix} \begin{pmatrix} r_2 \\ r_2' \end{pmatrix} \tag{17.33}$$

Where the reverse ray transfer matrix satisfies

$$\begin{pmatrix} A' & B' \\ C' & D' \end{pmatrix} = \begin{pmatrix} A & B \\ C & D \end{pmatrix}^{-1}$$

The ray transfer matrix allows the properties of an optical system to be described in general terms by the location of i $focal\ points\ and\ principal\ planes$.ts

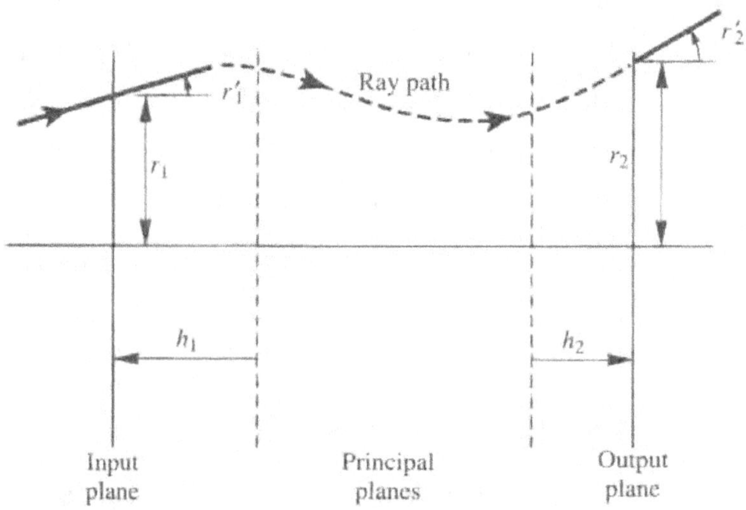

Figure 17.14: (a) Paraxial ray path from input to output plane.

The location of focal points and principle planes is determined from the element of the matrix. The significance of these features of the system can be illustrated with the aid of Fig.(17.14b). An input ray which passes through the $first\ focal\ point, F_1$ (or would pass through this point if it did not first enter the system) emerges travelling parallel to the axis. The intersection point of the extended input and output rays, point H_1 in Fig. (17.14b), defines the location of the $first\ principal\ plane.$ Conversely, an input ray travelling parallel to the axis will emerge at the output plane and pass through the $second\ focal\ point, F_2$ (or appear to have come from the point). The intersection of the extension of these rays, point H_2, defines the location of the $second\ principal\ plane.$Rays 1 and 2 Fig. (17.14b) are called $principal\ rays$ of the system.

Now The location of the principal planes allow the corresponding emergent ray paths to be determine as shown in Fig (17.14b). The dashed lines in the figure, which permit geometric construction of the location of the output rays 1 and 2, are called $virtual\ ray\ paths.$

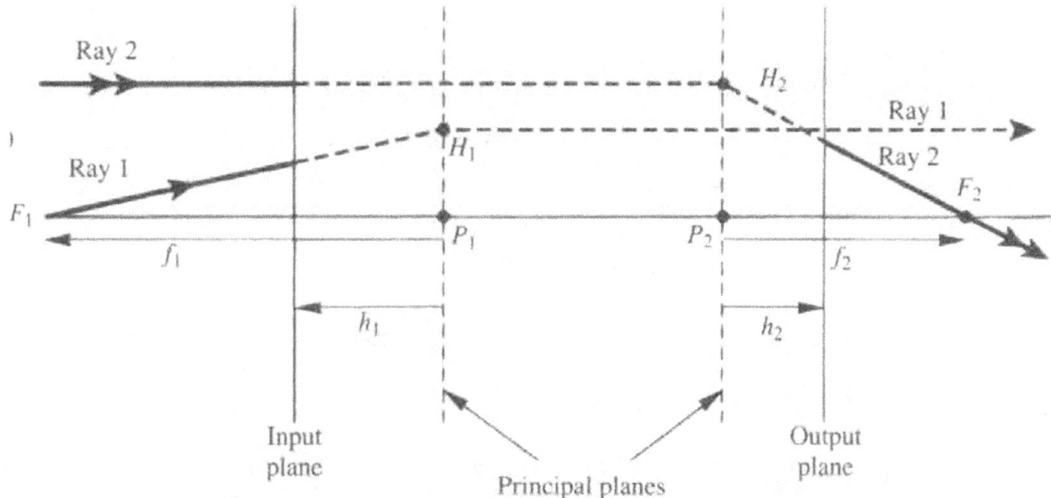

Figure 17.14: (b) Principal planes, points, and rays of a paraxial optical system.

Both F_1 and F_2 lie on the axis of the system. The axis of the system intersects the principal planes at the *principal points*, P_1 and P_2, in Fig. (17.14b).

The distance f_1 from the first principal plane to the focal point is called the *first focal length*; f_2 is the *second focal length*.

In most practical situations, the refractive indices of the media to the left of the input plane (*the object space*) and to the right of the output plane (*the image space*) are equal. In this case we can derive simple relations between the focal lengths f_1 and f_2 and h_1 and h_2; the distance of the input and output planes from the principal planes, measured in the sense shown in Fig. (17.14b).

We can break up the system shown in Fig. (17.14b) into three parts, the region from the input plane to the first principal plan, the region between the two principal planes, and the region from second principal plane to output plane.

If we write the transfer matrix from the left to the right principal planes as $\begin{pmatrix} A'' & B'' \\ C'' & D'' \end{pmatrix}$, then the overall transfer matrix is

$$\begin{pmatrix} A & B \\ C & D \end{pmatrix} = \begin{pmatrix} 1 & h_2 \\ 0 & 1 \end{pmatrix} \begin{pmatrix} A'' & B'' \\ C'' & D'' \end{pmatrix} \begin{pmatrix} 1 & h_1 \\ 0 & 1 \end{pmatrix} \tag{17.34}$$

Which gives

$$\begin{pmatrix} A'' & B'' \\ C'' & D'' \end{pmatrix} = \begin{pmatrix} 1 & -h_2 \\ 0 & 1 \end{pmatrix} \begin{pmatrix} A & B \\ C & D \end{pmatrix} \begin{pmatrix} 1 & -h_1 \\ 0 & 1 \end{pmatrix} \tag{17.35}$$

and therefore

$$\begin{pmatrix} A'' & B'' \\ C'' & D'' \end{pmatrix} = \begin{pmatrix} A - h_2 C & B - h_2 A - h_2(D - h_1 C) \\ C & D - h_1 C \end{pmatrix}$$ (17.36)

Clearly, $C'' = C$ for the principal ray in Fig. (17.14b), the distance of the ray from the axis r, does not change, therefore $A'' = 1$ giving

$$r = (A - h_2 C)\, r,$$ (17.37)

$$\text{So } \quad h_2 = \frac{A-1}{C}$$ (17.38)

Furthermore, for the first principal ray, whose input angle is $\underline{r'}$

$$r = A''\, r + B''\, r'$$ (17.39)

So $B'' = 0$.   Consequently, if the media to the left of the input plane and the right of the output plane are the same, using $det(M) = 1$, gives

$$h_1 = \frac{D-1}{C}$$ (17.40)

Note that both A'' and D'' are equal to unity.

For the second principal ray in Fig.(17.14b)

$$\begin{pmatrix} r_2 \\ r_2' \end{pmatrix} = \begin{pmatrix} A'' & B'' \\ C'' & D'' \end{pmatrix} \begin{pmatrix} r_1 \\ 0 \end{pmatrix}$$ (17.41)

$$\text{So } \quad r'_2 = C''r_1$$ (17.42)

From Fig. (17.14b), it is easy to see that

$$-r'_2 = \frac{r_1}{f_2}$$ (17.43)

Now from combining Equations. (17.42) and (17.43) gives

$$C'' = -1/f_2$$ (17.44)

By a similar procedure using the first principal ray it can be shown that

$$C'' = -1/f_1$$ (17.45)

We have already seen from Eq. (17.36) that $C'' = C$ therefore,

$$f_1 = f_2 = f$$ (17.46)

If the media to the left and right of the input plane are the same, both principal focal length are equal. If the elements of the transfer matrix are known, the locations of the focal points and principal planes are determined. Graphical construction of ray paths through the system using the methods of *ray tracing* is then straightforward.

17.9.2 Ray Tracing:

Practical implementation of paraxial ray analysis in optical system can very conveniently carried out graphically by *ray tracing*. In ray tracing a few simple rules allow geometrical construction of the principal ray paths from an object point, although these constructions do not take into account the non-ideal behaviour, or *aberrations,* of real lenses. The first principal ray from a point on the object passes through (or its projection passes through) the first focal point. From the point where this ray, or its projection, intersects the first principal plane the output ray is drawn parallel to the axis. The actual ray path *XX'* in the thick lens shown in Fig. (17.15b). The second principal ray is directed parallel to the axis; from the intersection if this ray, or its projection, with the second principal plane the output ray passes through (or appears to have come from) the second focal point. The actual ray path between input and output planes can be found in simple cases, for example, the path YY' in the thick lens shown in Fig. (17.15b). The intersection of the two principal rays in the image space produces the image point which corresponds to the same point of the object. If only the back-projections of the output principal rays appear to intersects, the intersection point lies on a *virtual image*.

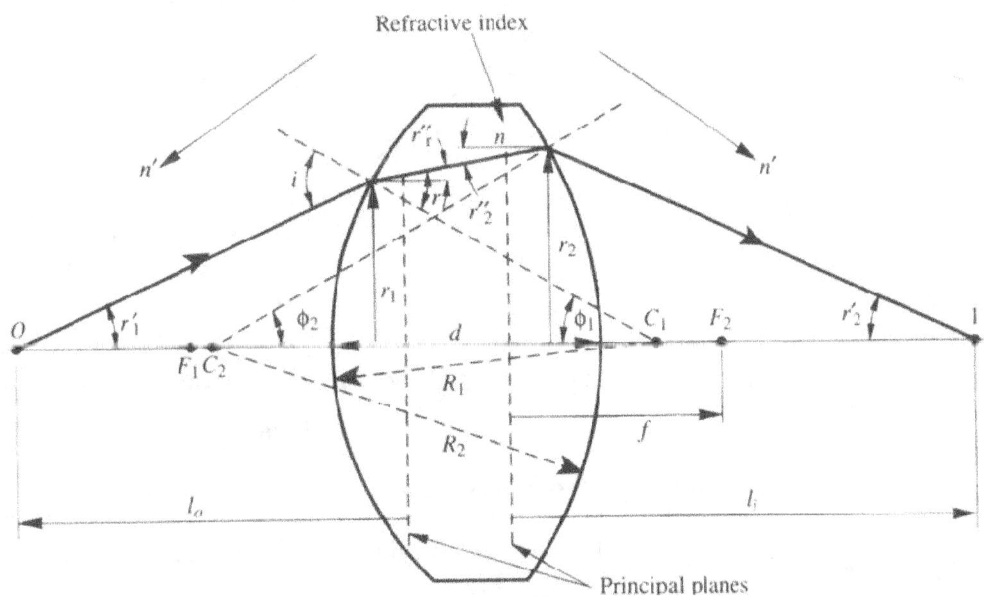

Figure 17.15 (a): Ray path through a thick lens. C_1 and C_2 are the centre of curvature of the two spherical surface of which the faces of a lens are a part.

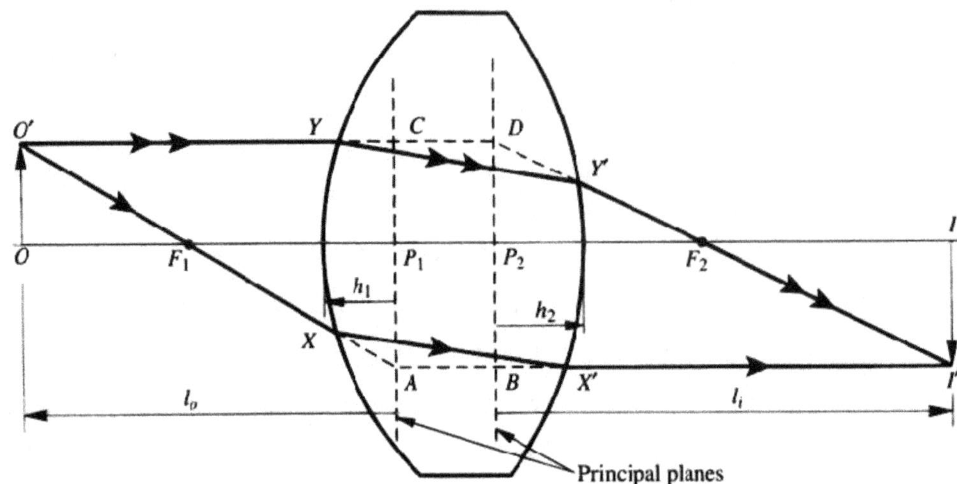

Figure 17.15 (b): Principle rays and planes for a thick lens. The dashed lines show the use of the principle plane for which determining the exit trajectories for the input principle rays.

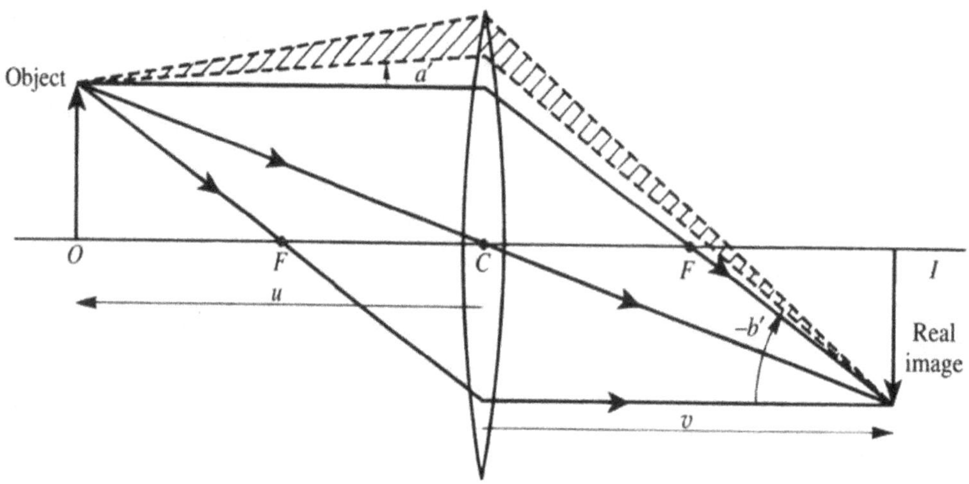

Figure 17.16: Ray tracing Diagrams Real images production by a converging lens.

17.9.3 Imaging And Magnification:

In Fig. (17.16) the ratio of the height of the image to the height of the project is called the *magnification m*. In the case of thin lens

$$m = \frac{b}{a} = \frac{v}{u} \tag{17.47}$$

For a more general system the magnification can be obtained from the ray transfer matrix equation

$$\begin{pmatrix} b \\ b' \end{pmatrix} = \begin{pmatrix} A & B \\ C & D \end{pmatrix}\begin{pmatrix} a \\ a' \end{pmatrix} = M\begin{pmatrix} a \\ a' \end{pmatrix} \qquad (17.48)$$

Where a' is the angle of a ray through a point on the object and b' through the corresponding point on the image. The matrix **M** in the case includes the entire system from object **O** to image **I**. So, in Fig. (17.16a)

$$M = \begin{pmatrix} matrix\ for\ uniform \\ medium\ of\ length\ v \end{pmatrix}\begin{pmatrix} matrix\ for \\ lens \end{pmatrix}\begin{pmatrix} matrix\ for\ uniform \\ medium\ of\ length\ u \end{pmatrix} \qquad (17.49)$$

For imaging, b must be independent of a'.so B = 0.

$$\text{The magnification is } m = \frac{b}{a} = A \qquad (17.50)$$

So the ray transfer matrix of the imaging system can be written

$$M = \begin{pmatrix} m & 0 \\ -1/f & 1/m \end{pmatrix} \qquad (17.51)$$

Where it should be noted the result, $det(M) = 1$, has been used.

The *angular magnification* of the system is defined as

$$m' = \left(\frac{b'}{a'}\right)_{a=0} \qquad (17.52)$$

Which gives $m' = 1/m$.

Note that mm' = 1, a useful general result.

17.10 Gaussian Beams

Here we shall look from a wave standpoint at how narrow beams of light travel through optical systems and also we see that a special solution is exist for the electromagnetic wave equation that take the form of narrow beams called as *Gaussian Beams*. These beams of light have a characteristic radial intensity profile whose width varies along the beam. Because these Gaussian beams behave somewhat like spherical waves, we can match them to the curvature of the mirror of an optical resonator to find exactly what form of beam will result from a particular resonator geometry.

17.10.1 Beam-Like Solution of The Wave Equation

Here we consider that the transverse modes of a laser system will take the form of narrow beams of light that propagate between the mirrors of the laser resonator and maintain a field distribution that remains distributed around and near the axis of the system. We shall therefore need to find solutions of the wave equation that take the form of narrow beams and then see how we can make these solutions compatible with a given laser cavity.

Now, the wave equation is, for any field or potential component U_0 of an electromagnetic wave

$$\nabla^2 U_0 - \mu\epsilon_r\epsilon_0 \frac{\partial^2 U_0}{\partial t^2} = 0 \tag{17.53}$$

Where ϵ_r is the dielectric constant, which may be a function of position. The non-plane-wave solutions that we are looking for are of the form

$$U_0 = U(x, y, z)e^{i[\omega t - k(r).r]} \tag{17.54}$$

Now we allow the wave vector $k(r)$ to be a function of $\mathbf{r}$ to include situations where the medium has a non-uniform refractive index. From eq.(17.53) and (17.54)

$$\nabla^2 U - \mu\epsilon_r\epsilon_0\omega^2 U = 0 \tag{17.55}$$

$\mu\epsilon_r$ may be the function of $\mathbf{r}$. we have seen previously that the propagation constant in the medium is $k = \omega\sqrt{\mu\epsilon_r\epsilon_0}$, so

$$\nabla^2 U - k(r)^2 U = 0 \tag{17.56}$$

This is the time-independent form of the wave equation, frequently referred to as the *Helmholtz* equation.

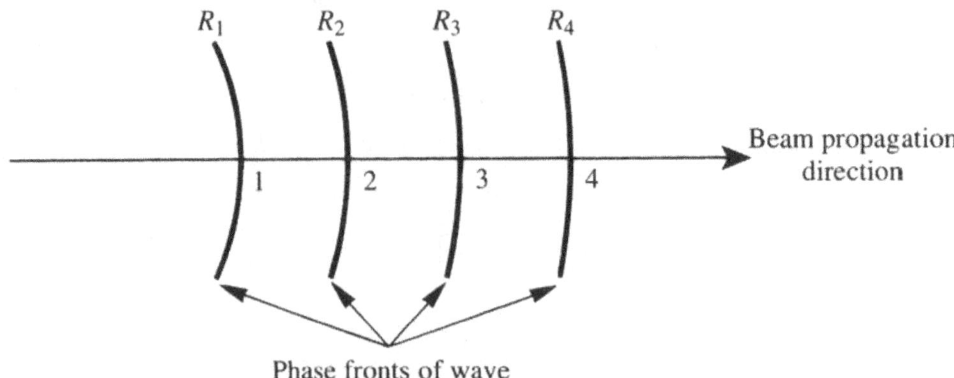

Figure 17.17: The changing curvature of a spherical wave as it propagates in the z-direction.

In general, if the medium is absorbing, or exhibits gain, then its dielectric constant ϵ_r has real and imaginary parts.

$$\epsilon_r = 1 + \chi(\omega) = 1 + \chi'(\omega) - i\chi''(\omega) \tag{17.57}$$

And $k = k_0\sqrt{\epsilon_r}$ (17.58)

Where $k_0 = \omega\sqrt{\epsilon_0\mu}$

If the medium were conducting, with conductivity σ, then its complex propagation vector would obey

$$k^2 = \mu\epsilon_r\epsilon_0\omega^2\left(1 - i\frac{\sigma}{\epsilon_r\epsilon_0\omega}\right) \tag{17.59}$$

So conductivity can be included as a contribution to the imaginary part of the dielectric constant.

We know that simple solutions of the time-independent wave equation above are transverse plane waves. However, those simple solutions are not adequate to describe the field distributions of transverse modes in laser systems. Let us look for solutions will be of the form $U = \varphi(x, y, z)e^{-ikz}$ for waves propagating in the positive z direction. For functions $\varphi(x, y, z)$ which are localized near the z axis the propagating wave takes the form of a narrow beam. Further, because $\varphi(x, y, z)$ is not uniform, the surfaces of constant phase in the wave will no longer necessarily be plane. If we can find solutions of the wave equation where $\varphi(x, y, z)$ gives phase fronts that are spherical then we can make the propagating beam solution $U = \varphi(x, y, z)e^{-ikz}$ satisfy the boundary conditions in a resonator with spherical reflectors, provided the mirrors are placed at the position of phase fronts whose curvature equals the mirror curvature. Thus the propagating beam solution becomes a satisfactory transverse mode of the resonator. For example in figure 17.17 if the propagating beam is to be a satisfactory transverse mode, then a spherical mirror of radius R_1 must be placed at position 1 or one of radius R_2 at 2 etc. Mirrors placed in this way lead of the wave back on itself.

Substituting $U = \varphi(x, y, z)e^{-ikz}$ in equation (17.56), we get

$$\left(\frac{\partial^2\varphi}{\partial x^2}\right)e^{-ikz} + \left(\frac{\partial^2\varphi}{\partial y^2}\right)e^{-ikz} + \left(\frac{\partial^2\varphi}{\partial z^2}\right)e^{-ikz} - 2ik\left(\frac{\partial\varphi}{\partial z}\right)e^{-ikz} -$$
$$k^2\varphi(x,y,z)e^{-ikz} + k^2\varphi(x,y,z)e^{-ikz} = 0 \tag{17.60}$$

which reduces to

$$\left(\frac{\partial^2\varphi}{\partial x^2}\right) + \left(\frac{\partial^2\varphi}{\partial y^2}\right) + \left(\frac{\partial^2\varphi}{\partial z^2}\right) - 2ik\left(\frac{\partial\varphi}{\partial z}\right) = 0 \tag{17.61}$$

If the beam-like solution we are looking for remains paraxial then φ will only vary slowly with z, so we can neglect $\left(\dfrac{\partial^2\varphi}{\partial z^2}\right)$ and get

$$\left(\frac{\partial^2\varphi}{\partial x^2}\right) + \left(\frac{\partial^2\varphi}{\partial y^2}\right) - 2ik\left(\frac{\partial\varphi}{\partial z}\right) = 0 \tag{17.62}$$

We try as solution

$$\varphi(x,y,z) = \exp\left\{-i\left[P(z) + \frac{k}{2q(z)}r^2\right]\right\} \tag{17.63}$$

Where $r^2 = x^2 + y^2$ is the square of the distance of the point x, y from the axis of propagation. P (z) represents a phase shift factor and q(z) is called the beam parameter. We shall see the significance of these parameters shortly.

Substituting in equation (17.62) and using the relations below that follow from equation (17.63)

$$\frac{\partial\varphi}{\partial x} = \frac{-ik}{2q(z)}\exp\left\{-i\left[P(z) + \frac{k}{2q(z)}r^2\right]\right\}.\,2x \tag{17.64}$$

$$\left(\frac{\partial^2\varphi}{\partial x^2}\right) = \frac{-ik}{q(z)}\exp\left\{-i\left[P(z) + \frac{k}{2q(z)}r^2\right]\right\} - \frac{k^2}{4q^2(z)}\exp\left\{-i\left[P(z) + \right.\right.$$
$$\left.\left.\frac{k}{2q(z)}r^2\right]\right\}.\,4x^2 \tag{17.65}$$

$$\frac{\partial\varphi}{\partial z} = \exp\left\{-i\left[P(z) + \frac{\dot{k}}{2q(z)}r^2\right]\right\}\left[-i\left(\frac{dP}{dz} - \frac{k}{2q^2}\frac{dq}{dz}r^2\right)\right] \tag{17.66}$$

We get

$$-2k\left[\frac{dP}{dz} + \frac{i}{q(z)}\right] - \left[\frac{k^2}{q^2(z)} - \frac{k^2}{q^2(z)}\frac{dq}{dz}\right](r^2) = 0 \tag{17.67}$$

Since this equation must be true for all values of r the coefficients of different powers of r must be independently equal to zero so

$$\frac{dq}{dz} = 1 \tag{17.68}$$

and $\dfrac{dP}{dz} = -\dfrac{i}{q(z)}$ \hfill (17.69)

The solution $\varphi(x,y,z) = \exp\left\{-i\left[P(z) + \dfrac{k}{2q(z)}r^2\right]\right\}$

Is called the *fundamental Gaussian Beam* solution of the time-independent wave equation since its 'intensity' as a function of x *and* y is

$$UU^* = \varphi\varphi^* = \exp\left\{-i\left[P(z) + \dfrac{k}{2q(z)}r^2\right]\right\}\exp\left\{i\left[P^*(z) + \dfrac{k}{2q^*(z)}r^2\right]\right\}$$

$$(17.70)$$

where $P^*(z)$ *and* $q^*(z)$ is the conjugate complex of $P(z)$ *and* $q(z)$, respectively, so

$$UU^* = \varphi\varphi^* = \exp\{-i[P(z) - P^*(z)]\}\exp\left\{-\dfrac{ikr^2}{2}\left[\dfrac{1}{q(z)} - \dfrac{1}{q^*(z)}\right]\right\} \quad (17.71)$$

For convenience we introduce two real beam parameters $R(z)$ *and* $w(z)$ that are related to $q(z)$ by

$$\dfrac{1}{q} = \dfrac{1}{R} - \dfrac{i\lambda}{\pi w^2} \hfill (17.72)$$

Where both R and w depend on z. it is important to note that $\lambda = \lambda_0/n$ is the wavelength in the *medium*. From equations (17.71) and (17.72) above we can see that

$$UU^* \propto exp\,\dfrac{-ikr^2}{2}\left(\dfrac{-i\lambda}{\pi w^2} - \dfrac{i\lambda}{\pi w^2}\right) \propto exp\,\dfrac{-2r^2}{w^2} \hfill (17.73)$$

Thus, the beam intensity shows a Gaussian dependence on r, the physical significance of $w(z)$ is that it is the distance from the axis at the point z where the intensity of the beam has fallen to $1/e^2$ of its peak value on axis and its amplitude to $1/e$ of its axial value: $w(z)$ is called the *spot size*. With the parameters

$$U = \exp\left\{-i\left[kz + P(z) + \dfrac{kr^2}{2}\left(\dfrac{1}{Z} - \dfrac{i\lambda}{\pi w^2}\right)\right]\right\}$$

17.10.2 The Transformation of A Gaussian Beam By A Lens:

A lens can be used to focus a laser beam to a small spot, or systems of lenses may be used to expand the beam and recollimate it. An idea thin lens in such an

application will not change the transverse mode intensity pattern measured at the lens but it will alter the radius of curvature of the phase fronts of the beam.

We have seen that a Gaussian laser beam of whatever order is characterized by a complex beam parameter q(z) which change as the beam propagates in an isotopic, homogeneous material according to q(z) = q_0+z, where

$$q_0 = \frac{i\pi w_0^2}{\lambda} \quad and \quad \frac{1}{q(z)} = \frac{1}{R(z)} - \frac{i\lambda}{\pi w^2} \tag{17.74}$$

$R(z)$ is radius of curvature of (approximately) spherical phase front at z and is given by

$$R(z) = z\left[1 + \left(\frac{\pi w_0^2}{\lambda z}\right)^2\right] \tag{17.75}$$

This Gaussian beam becomes a true spherical wave as $w \longrightarrow \infty$. Now, a spherical wave changes its radius of curvature as it propagates according to $R(z) = R_0 + z$ where R_0 is its radius of curvature as z = 0. So the complex beam parameter of a Gaussian wave change in just the same way as it propagates as does the radius of curvature of a spherical wave.

When a Gaussian beam strikes a thin lens the spot size, which measures the transverse with of the beam intensity distribution, is unchanged at the lens. However, the radius of curvature of its wavefront is altered in just the same way as spherical wave. If R_1 and R_2 are the radii of curvature of the incoming and outgoing waves measured at the lens, as shown in Fig. (17.18), then as in the case of a true spherical wave

$$\frac{1}{R_2} = \frac{1}{R_1} - \frac{1}{f} \tag{17.76}$$

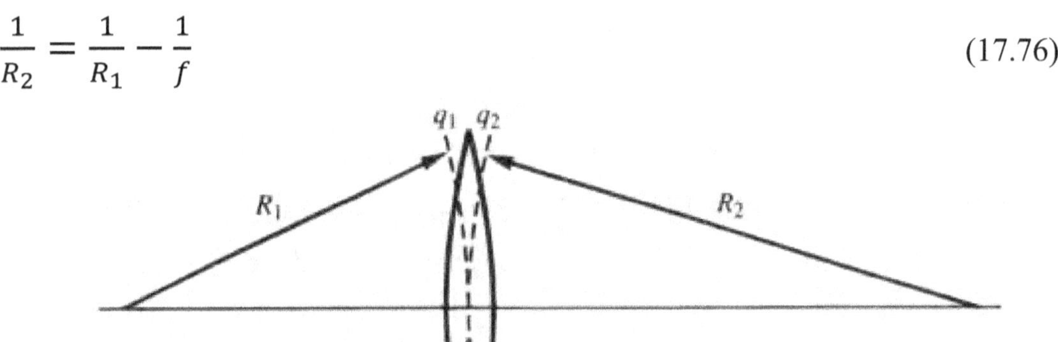

Figure 17.18: Transformation of a Gaussian beam by a lens. The input and output phase front curvatures and beam parameters are indicated.

So, for the change of the overall beam parameter, since w is unchanged at the lens, we have the following relationship between the beam parameters, *measured* at the lens

$$\frac{1}{q_2} = \frac{1}{q_1} - \frac{1}{f} \tag{17.77}$$

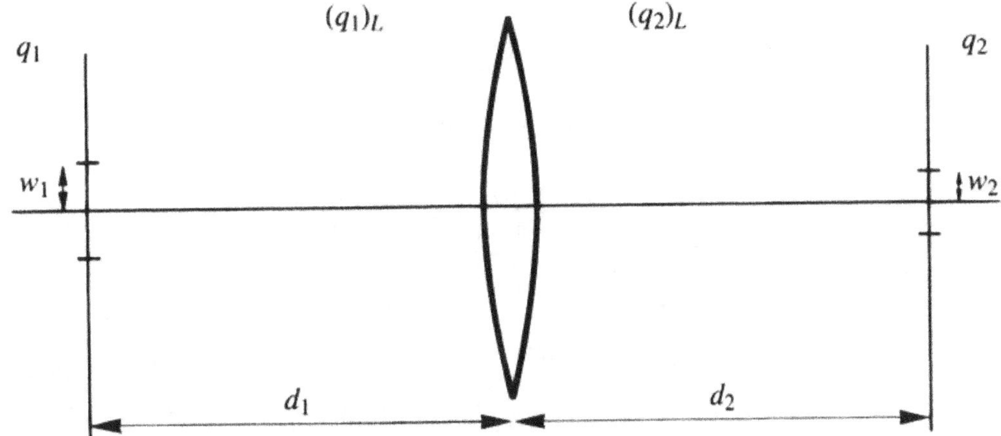

Figure 17.19: Optical system diagram for transformation of a Gaussian beam from an input to an output plane.

If instead q_1 and q_2 are measured at distances d_1 and d_2 from the lens as shown in Fig. (16.19) then at the lens

$$\frac{1}{(q_2)_L} = \frac{1}{(q_1)_L} - \frac{1}{f} \tag{17.78}$$

And since $(q_1)_L = q_1 + d_1$ and $(q_2)_L = q_2 + d_2$ we have

$$\frac{1}{q_2 - d_2} = \frac{1}{q_1 + d_1} - \frac{1}{f} \tag{17.79}$$

Which gives

$$q_2 = \frac{\left(1 - \frac{d_2}{f}\right)q_1 + (d_1 + d_2 - d_1 d_2/f)}{(-q_1/f) + (1 - d_1/f)} \tag{17.80}$$

If the lens is placed at the beam waist of the input beam then

$$\frac{1}{(q_1)_L} = \frac{-i\lambda}{\pi w_0^2} \tag{17.81}$$

Where w_0 is the post size of the input beam. The beam parameter immediately after the lens is

$$\frac{1}{(q_2)_L} = \frac{-i\lambda}{\pi w_0^2} - \frac{1}{f} \tag{17.82}$$

This can be rewritten as $\quad (q_2)_L = \dfrac{-\pi w_0^2 f}{i\lambda f + \pi w_0^2}$ \hfill (17.83)

At the distance d_2 after the lens

$$(q_2)_L = \frac{-\pi w_0^2 f}{i\lambda f + \pi w_0^2} + d_2 \tag{17.84}$$

This can be rearranged to give

$$1/q_2 = \frac{(d_2 - f) + \left(\frac{\lambda f}{\pi w_0^2}\right)^2 d_2 - i\left(\frac{\lambda f^2}{\pi w_0^2}\right)}{(d_2 - f)^2 + \frac{(\lambda f d_2)^2}{\pi w_0^2}} \tag{17.85}$$

The location of the new beam waist (which is where the beam will be focused to its new minimum spot size) is determine by the condition $\mathcal{R}(1/q_2) = 0$: namely

$$(d_2 - f) + \left(\frac{\lambda f}{\pi w_0^2}\right)^2 d_2 = 0 \tag{17.86}$$

which gives

$$d_2 = \frac{f}{1 + \left(\frac{\lambda f}{\pi w_0^2}\right)^2} \tag{17.87}$$

Almost always $\left(\dfrac{\lambda f}{\pi w_0^2}\right)^2 \ll 1$ so the new beam waist is very close to the very focal point of the lens.

Examination of the imaginary part of the right hand side of Eq. (17.85) reveals that the spot size of the focused beam is

$$w_2 = \frac{\frac{\lambda f}{\pi w_0}}{\sqrt{1 + \left(\frac{\lambda f}{\pi w_0^2}\right)^2}} \tag{17.88}$$

Which provided $\dfrac{\lambda f}{\pi w_0^2} \ll 1$, as is frequently the case, gives

$$w_2 \simeq f\theta \tag{17.89}$$

Where $\theta = \lambda/\pi w_0$ is the half angle of the divergence of the input beam.

It is straightforward to show that in general the minimum spot size of TEM_{00} Gaussian beam focused by a lens is

$$w_f = \frac{\lambda f}{\pi w_1}\left[\left(1 - \frac{f}{R_1}\right)^2 + \left(\frac{\lambda f}{\pi w_1^2}\right)^2\right]^{\frac{-1}{2}} \tag{17.90}$$

Where w_1 and R_1 are the laser-beam spot size and radius of curvature at the input face if the lens. If the lens is placed very close to or vary far from the beam waist of the beam being focused, Eq. (17.90) reduces to

$$w_f = f\theta_B \tag{17.91}$$

Where θ_B is the beam divergence at the input face of the lens.

Thus, if the focusing lens is placed a great distance from, or vary close to, the input beam waist then the size of the focused spot is always close to $f\theta$. If, in fact, the beam incident on the lens were a plane wave, then the finite size of the lens (radius r) would be the dominant factor in determining the size of the focused spot. We can take the 'spot size' of the plane wave as approximately the radius of the lens, and from Eq. (17.74), setting $w_0 = r$, the radius of the lens, we get

$$w_2 = \frac{\lambda f}{\pi r} \tag{17.92}$$

This focused spot cannot be smaller than a certain size since for any lens the value of r clearly has to satisfy the condition $r \leq f$

Thus, the minimum focal spot size that can result when a plane wave is focused by a lens is

$$(w_2)_{min.} \sim \frac{\lambda}{\pi} \tag{17.93}$$

We do not have equality sign in Eq. (17.93) because a lens for which r = f does not qualify as a thin lens, so Eq. (17.92) does not really hold exactly.

We can see from Eq. (17.74) that in order to focus a laser beam to a small spot be must either use a lens of very short focal length or a beam of small beam a divergence. We cannot however, reduce the focal length of the focusing lens indefinitely, as when the focal length does not satisfy $f_1 << w_0, w_1$, the lens ceases to satisfy our definition of it as a thin lens ($f_1 << r$). To obtain a laser beam of small divergence we must expand all recollimate the beam; there are to simple ways of doing this:

(i)   With a Galilean telescope as shown in Fig. (17.20). The expansion ratio for this arrangement is $-f_2/f_1$, where it should be noted that the focal length f_1 of the diverging input lens is negative. This type of arrangement has the advantage that the laser beam is not brought to a focus within the telescope, so the arrangement is very suitable for the expansion of high power laser beams. Very high power beams can cause air breakdown if brought to a focus, with considerably reduce the energy transmission through the system.

(ii)   With an astronomical telescope, as shown in fig. (17.20). The expansion ratio for this arrangement is f_2/f_1. The beam is brought to a focus within the telescope, which can be a disadvantage when expanding high intensity laser beams because air breakdown at the common focal point can occur. The telescope can be evacuated or filled to high pressure to help prevent as such breakdown occurring. An advantage of this system is that by placing a small circular aperture at the common focal point it is possible to obtain an output beam with a smoother radial intensity profile then the input beam. The aperture should be chosen to have a radius about the same size, or slightly larger than, the spot size of the focused Gaussian beam at the focal point. This process is called spatial filtering and is illustrated in Fig (17.22). In both the astronomical and Galilean telescope, spherical aberration is reduced by the use of bispherical lenses. This distributes the focusing power over the maximum number of surfaces.

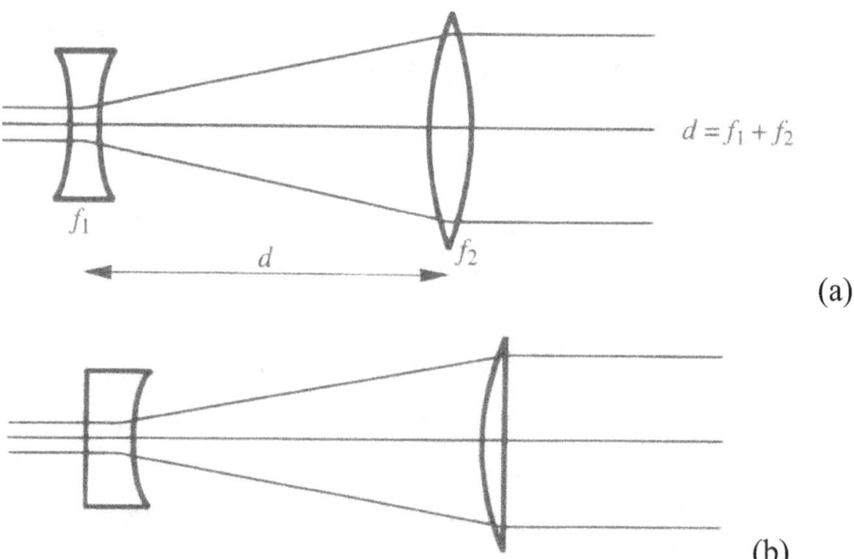

Figure 17.20: Galilean telescope laser beam expanders: (a) using bispherical lenses; (b) using plano-spherical lenses.

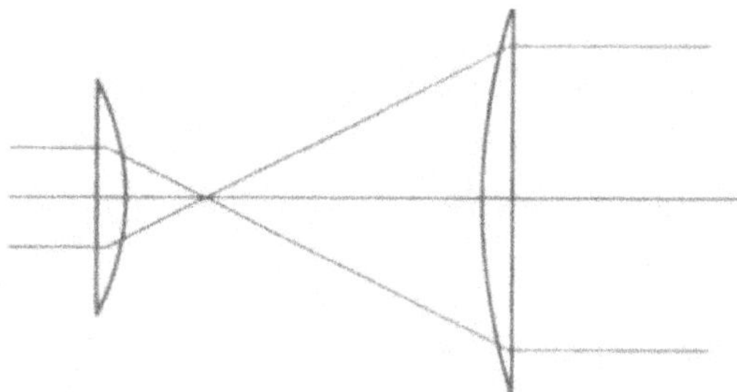

Figure 17.21: Astronomical telescope laser-beam expander in which the beam is brought to an internal focus.

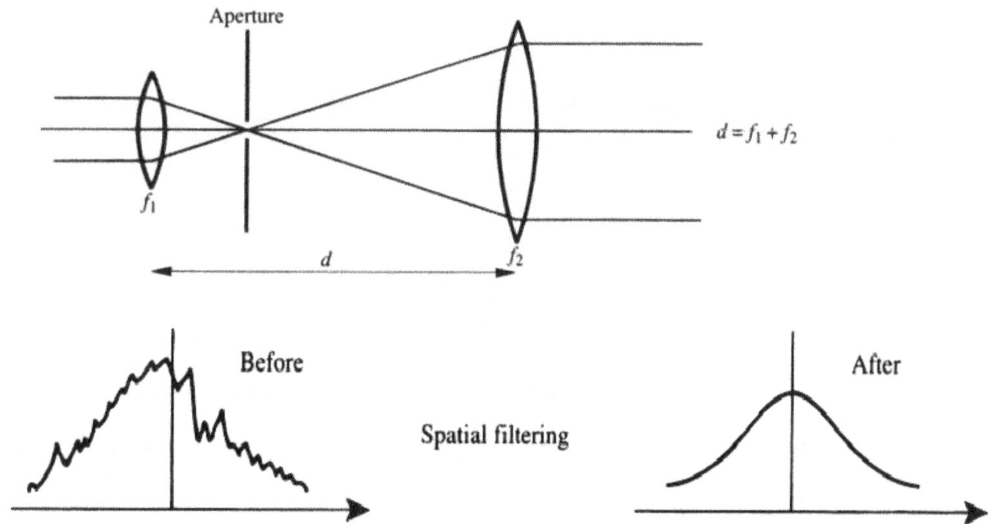

Figure 17.22: Beam expander incorporating a spatial filter for smoothing the intensity profile.

17.10.3 Transformation of A Gaussian Beams By The General Optical System

As we have seen, the complex beam parameter of a Gaussian beam is transformed by a thin lens in just the same way as the radius of curvature of a spherical wave. Now, the transformation of the radius of curvature of a spherical wave by an optical system with transfer matrix $\begin{pmatrix} A & B \\ C & D \end{pmatrix}$ obeys

$$R_2 = \frac{AR_1 + B}{CR_1 + D} \tag{17.94}$$

Thus by continuing to draw a parallel between the q of a Gaussian beam and the R of a spherical wave, we can postulate that the transformation of the complex beam parameter obeys a similar relation i.e.

$$q_2 = \frac{Aq_1 + B}{Cq_1 + D} \tag{17.95}$$

Where $\begin{pmatrix} A & B \\ C & D \end{pmatrix}$ is the transfer matrix for paraxial rays. Eq. (17.95) can be justified for a general optical system by considering that such a system could, in principle, be replaced by an arrangement of spaced thin lenses. Each thin lens, of length of uniform medium, obeys Eq. (17.95). We can illustrate the use of the transfer matrix in this way by following the propagation of a Gaussian beam in a lens waveguide.

17.10.4 Gaussian Beams In Lens Waveguides

In a biperiodic lens sequence containing equally spaced lenses of focal lengths f_1 and f_2 the transfer matrix for n unit cells of the sequence is,

$$\begin{pmatrix} A & B \\ C & D \end{pmatrix}^n = \frac{1}{sin\phi} \begin{pmatrix} A sin\, n\phi - sin(n-1)\phi & B sin\, n\phi \\ c\, sin\, n\phi & D\, sin\, n\phi - sin\,(n-1)\phi \end{pmatrix} \tag{17.96}$$

So from eq. (17.95) and writing

$$q_{n+1} = (1/\sin\phi)\, \frac{[A sin\, n\phi - sin(n-1)\phi]q_1 + BB sin\, n\phi}{C\, sin(n\phi)q_1 - D\, sin\, n\phi - sin\,(n-1)\phi} \tag{17.97}$$

The condition for stable confinement of the Gaussian beam by the lens sequence is the same as in the case of paraxial rays. This is condition that ϕ remains real, i.e. $|\cos\phi| \leq 1$, where

$$\cos\phi = \frac{1}{2}(A+D) = 1 - \frac{d}{f_1} - \frac{d}{f_2} + \frac{d^2}{2f_1 f_2} \tag{17.98}$$

Which gives

$$0 \leq \left(1 - \frac{d}{2f_1}\right)\left(1 - \frac{d}{2f_2}\right) \leq 1 \tag{17.99}$$

17.10.5 Gaussian Beams In Plane And Spherical Mirror Resonators

We have already mentioned that a beam-like solution of Maxwell's equations will be satisfactory transverse mode of a plane or spherical mirror resonator provided we place the resonator mirrors at point where there radii of curvature of the phase fronts of the beam. So for a Gaussian beam a double-concave mirror resonator match of phase front as shown in Fig. (17.23) and plano-spherical and concave-convex resonator as shown in Fig. (17.24) and (17.25).

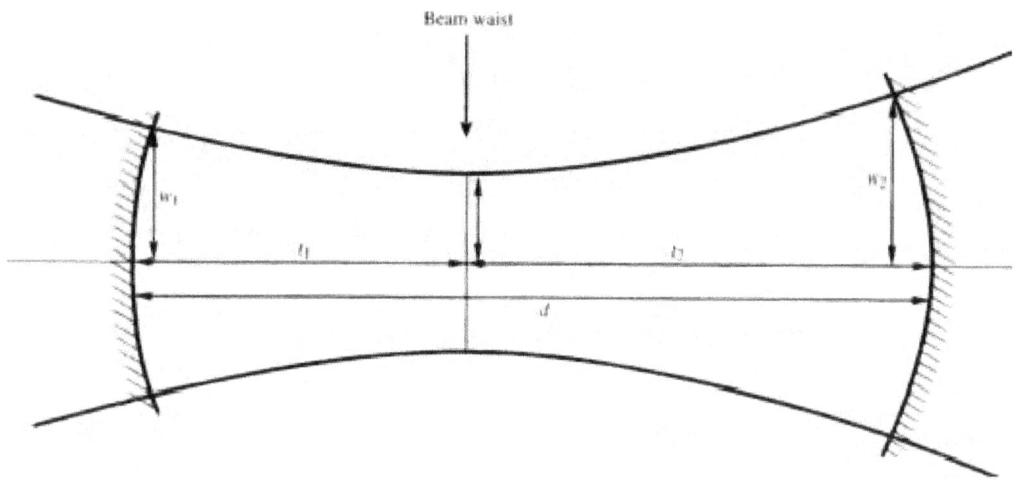

Figure 17.23: Resonator using two concave mirrors showing the location of the beam waist and the contour of the Gaussian beam

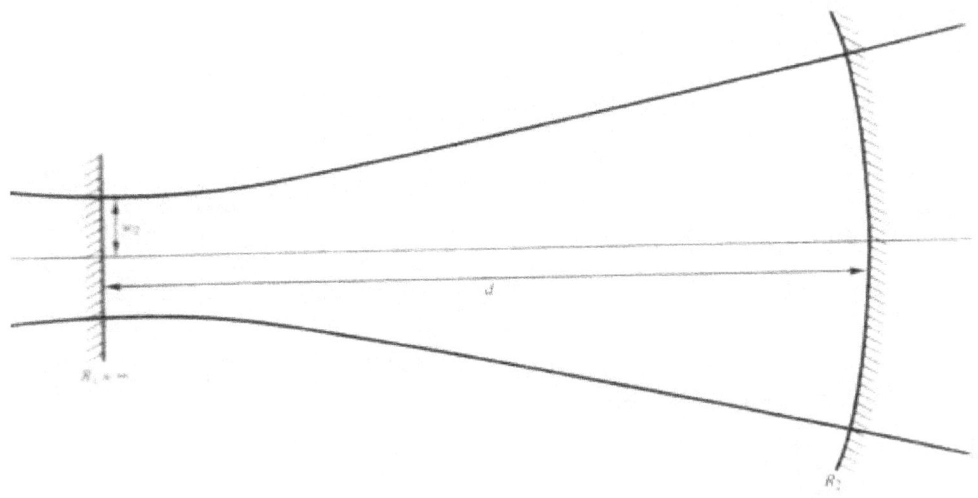

Figure 17.24: Laser resonator that uses a plane and a concave mirror.

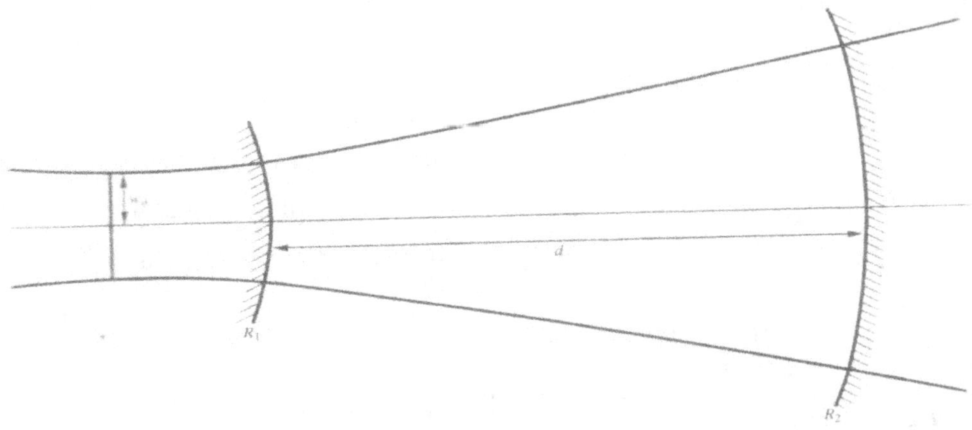

Figure 17.25: Convex- Concave Laser Resonator.

We can consider these resonator in terms of their equivalent biperiodic lens sequences as shown in Fig. (17.26). Propagation of a Gaussian beam from plane 1 to plane 3 in the biperiodic lens sequences in equivalent to one complete round trip inside the equivalent spherical mirror resonator. If the complex beam parameters at planes 1, 2, and 3 are $q_1, q_2,$ and $q_3,$ then

$$\frac{1}{q_1+d} - \frac{1}{f_1} = \frac{1}{q_2} \tag{17.100}$$

Which gives

$$q_2 = \frac{(q_1+d)f_1}{f_1-q_1-d} \tag{17.101}$$

and $$\frac{1}{q_2+d} - \frac{1}{f_2} = \frac{1}{q_3} \tag{17.102}$$

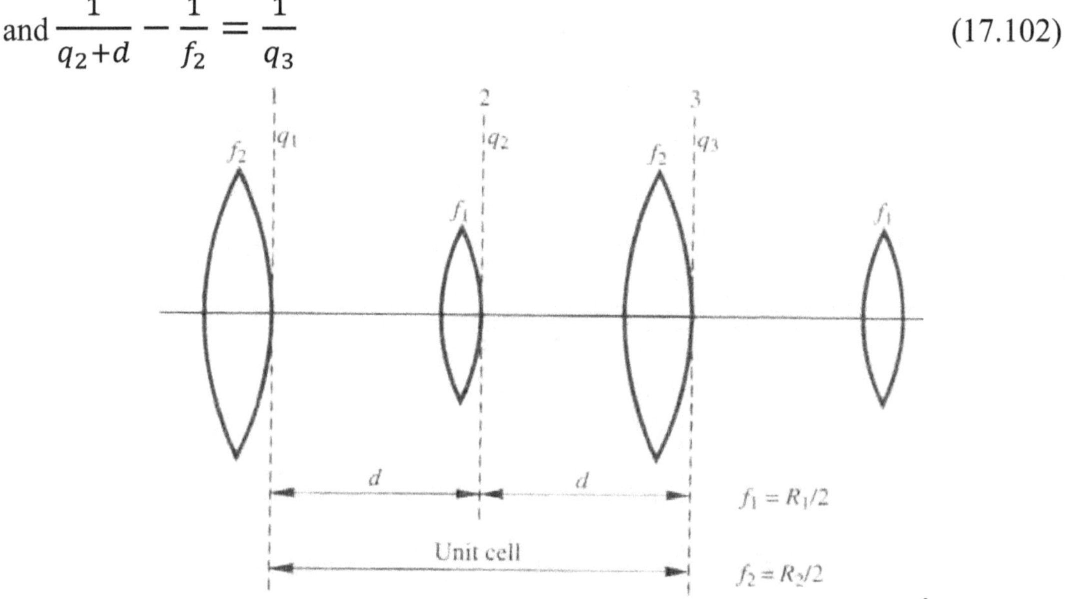

Figure 17.26: Biperiodic lens sequence equivalent to the resonator in figure 17.23.

If the Gaussian beam is to be a real transverse mode of the cavity, then we want it to repeat itself after a complete round trip- at least as far spot size and radius of curvature are concerned-that is we want

$$q_1 = q_1 = q \tag{17.103}$$

17.11 Self Learning Exercise

Q.1 What is the Long Radius Cavity?

Q.2 In a laser scheme emitting at a wavelength of $0.94\,\mu m$ are separated by $300Hz$ given that refractive index of the laser material is 3.3

(a) Calculate the length of the cavity

(b) Calculate the number of longitudinal modes emitted by the Scheme.

(c) Which of these modes will actually take part in the laser action?

Q.3 Give a brief description of cavity modes in a laser resonator with simple diagrams.

Q.4 How does Gaussian beam works in Lens Waveguides?

17.12 Summary

This unit is basically gives us knowledge about the analysis of optical system and working of Laser resonator. The mathematics of the resonator is also given in this unit. This unit consists a very important topic named as Optics of Gaussian Beams. This unit contains all about Gaussian Beams.

17.13 Glossary

Resonator: A device with population inversion, convert oscillator into amplifier, known as Resonator.

Laser Resonator: When resonator is used by the help of Laser beams, then it is known as Laser Resonators.

LOSER: Light Oscillation by Stimulated Emission of radiation.

Paraxial Rays: Those rays whose directions of propagation occur at sufficiently small angles θ to the symmetry axis of the system.

Gaussian Beams: We shall look from a wave standpoint at how narrow beams of light travel through optical systems and also we see that a special solution is exist for the electromagnetic wave equation that take the form of narrow beams called as *Gaussian Beams*.

17.14 Answer of Self Learning Exercise

Ans.1: A long radius cavity is a configuration of resonator whose reflectors have their radii of curvatures much longer than the distance that is separating them. The arrangement of a long radius cavity is shown below.

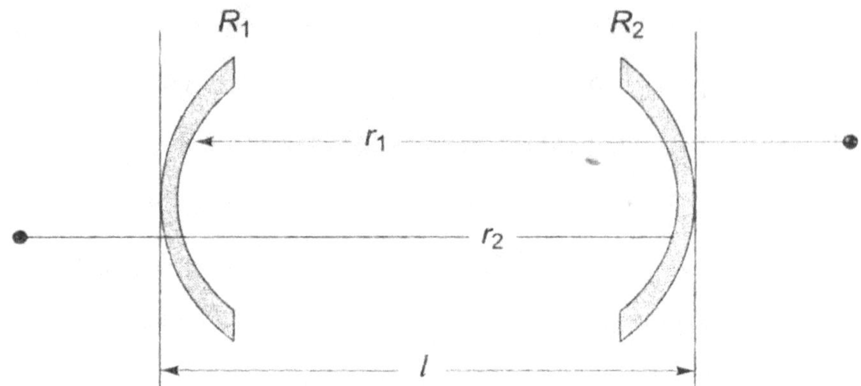

Figure: A Long Radius Cavity

In this resonator $r_1 = r_2 < l$.

This type configuration is a good compromise between the plane parallel and confocal configuration. Here, we find that a slight misalignment of reflectors will not result in severe problems because their curvatures would automatically focus the light back towards the first reflector. With this configuration a stable cavity is available due to the fact that the light rays reflected from R_2 to R_1 will keep bouncing back and forth indefinitely. Most of the commercial lasers adopt this type of arrangement of the reflectors and beams of light.

Ans. 2: (a) The cavity length l is given by the relation:

$l = c/2\,\Delta f\,\mu$,$c = 3 \times 10^8$,$\Delta f = 300 \times 10^9$,$\mu = 3.3$

So now $\quad l = \dfrac{3 \times 10^8}{2 \times 300 \times 10^9 \times 3.3} = 151.5\ \mu m$

(b) Number of longitudinal modes is given by the relation:$N_{lm} = 2\,\mu\,l/\lambda$

(lm Shown with N is only an acronym for longitudinal modes).

$$= 2 \times 3.3 \times 151.5/0.94 \approx 1064$$

(c) Out of these longitudinally modes only those modes which fall within the spontaneous emission spectrum will be taking part in the laser action.

Ans.3: It is known that a wave with a frequency f which is traversing a laser cavity gives a series of standing waves inside the resonator. The configuration and the dimensions of the cavity control these discrete resonant attributes.

The modes of the laser which are governed by the axial dimensions of the cavity are called longitudinal modes or axial modes. But if the laser modes are determined by the cross-sectional dimensions of the resonator then such modes are termed as transverse modes.

The arrangement of the two cavities indicating the longitudinal and transverse modes are shown in the diagrams (a) and (b), which are given below:

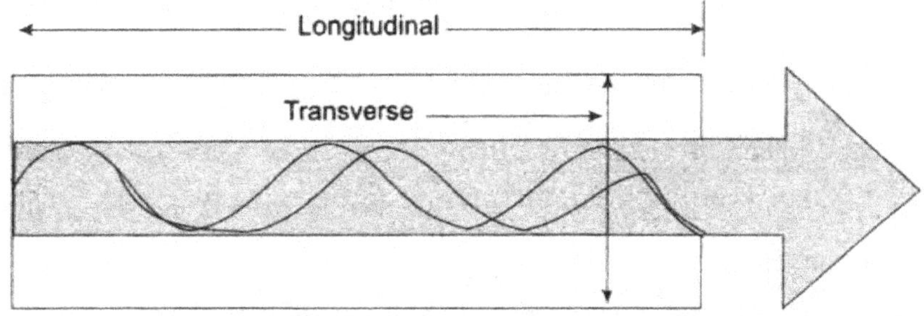

(a) Longitudinally travelling waves

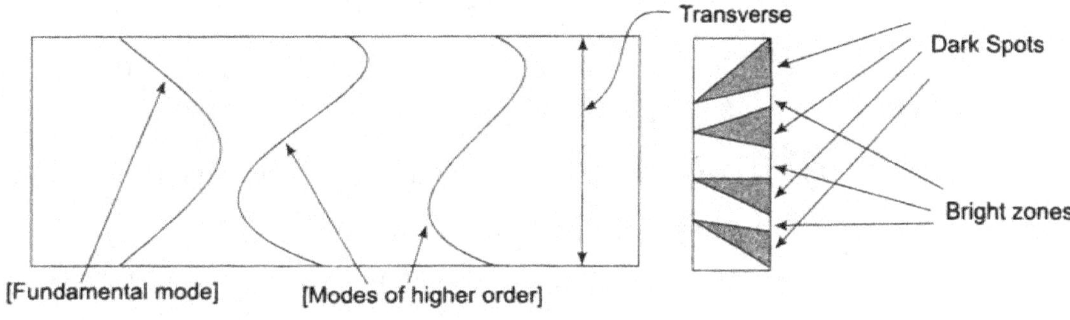

(b) Transverse modes

Figure for the modes of cavities.

Ans.4: In a biperiodic lens sequence containing equally spaced lenses of focal lengths f_1 and f_2 the transfer matrix for n unit cells of the sequence is, from eq. (15.3)

$$\begin{pmatrix} A & B \\ C & D \end{pmatrix}^n = \frac{1}{sin\phi}\begin{pmatrix} A\sin n\phi - \sin(n-1)\phi & B\sin n\phi \\ c\sin n\phi & D\sin n\phi - \sin(n-1)\phi \end{pmatrix}$$

So from eq. (16.88) and writing

$$q_{n+1} = (1/\sin\phi)\frac{[A\sin n\phi - \sin(n-1)\phi]q_1 + BB\sin n\phi}{C\sin(n\phi)q_1 - D\sin n\phi - \sin(n-1)\phi}$$

The condition for stable confinement of the Gaussian beam by the lens sequence is the same as in the case of paraxial rays. This is condition that ϕ remains real, i.e.

$|\cos\phi| \leq 1$, where $\cos\phi = \frac{1}{2}(A+D) = 1 - \frac{d}{f_1} - \frac{d}{f_2} + \frac{d^2}{2f_1f_2}$

Which gives

$$0 \le \left(1 - \frac{d}{2f_1}\right)\left(1 - \frac{d}{2f_2}\right) \le 1$$

17.15 Exercise

Q.1   What is a laser resonator? Define its various types.

Q.2   Give the brief description of Fabry-Perot resonator and its mathematical analysis.

Q.3   What do you know about the resonator? And give the laser resonator stability.

Q.4   Give a brief description on the Spherical resonator and its mathematic analysis.

Q.5   Give the description stable and unstable resonator.

Q.6   Develop a matrix approach for dealing with flat and spherical mirrors in paraxial ray analysis.

Q.7   Develop a matrix approach for dealing with flat and spherical mirrors in paraxial ray analysis.

Q.8   For a thin lens with $u = 50\ mm, f = 20\ mm$ draw a ray tracing diagram to illustrate the location of the image and its magnification.

Q.9   A Gaussian beam has a minimum spot size $w_0 = 10\mu m$ and a wavelength of 880 nm. Calculate: (a) The radius of curvature 1 m from the beam waist. (b) The spot size 1 m from the beam waist. Where does the beam have its minimum radius of curvature?

Q.10 Prove that $w_f = \frac{\lambda f}{\pi w_1}\left[\left(1 - \frac{f}{R_1}\right)^2 + \left(\frac{\lambda f}{\pi w_1^2}\right)^2\right]^{\frac{-1}{2}}$

where symbol have its own meaning.

References and Suggested Readings

1. Laser and Non-linear Optics by B.B. Laud.

2. Laser Physics by Peter W. Milonni and Joseph H. Eberly.

3. Laser physics and its application by L.V. Tarasov.

4. Atomic and Molecular Spectra: Laser by Raj Kumar.

UNIT-18
Coherence and
Some Specific Applications of Lasers

Structure of the Unit

18.0 Objectives

In this unit we study about the ability of different wave fronts to interfere with each other. This unit gives us very important knowledge about the uses of lasers and the advantages of Optical fiber. We classified the coherence in this unit and study knowledgeable topic wave propagation in optical fiber. The main objective of this unit is to study about the topics which contain any information related to the laser.

18.1 Introduction

In this unit, we study about the property of light and laser. The action and propagation of light in the Optical Fiber is the main topic of this unit. Coherence is the property of light in which we study the measure of the ability of two photons to interfere or intermix with each other. Coherence having two types:

(1) Spatial Coherence

(2) Temporal Coherence

This unit gives us a idea about the uses of the laser and some advantages of the Optical Fiber.

18.2 Optical Coherence

A wave which is appears to be a pure sine wave for an infinitely large period of time or in an infinitely extended space is said to be a perfectly coherent wave. In such a wave, there is a definite relationship between phase of the wave at a given time and at a certain time later or at a given point and at a certain distance away. No actual light source, however, emits a perfectly coherent wave. Light waves which are pure sine waves only for a limited period of time or in a limited space are partially coherent waves.

In other words, we can say that it is a measure of the ability of different wavefronts to interfere or intermingle with each other, when the wavefronts are combined, as in an interferometer. As such, coherence can also be termed as 'the measure of the ability of two photons to interfere or intermix with each other'.

Optical radiation belongs to the large spectrum of electromagnetic radiation and has a degree of coherence which is dependent on the mechanism used for generation of the very phenomenon 'Coherence'.

There are two types of optical coherence, which are define below:

18.3 Temporal Coherence

The oscillating electric field 'E' of a perfectly coherent light wave would have a constant amplitude of vibration at any point, while its phase would vary linearly with time. As a function of time, the field would appear as shown in figure (18.1). It is an ideal sinusoidal function of time.

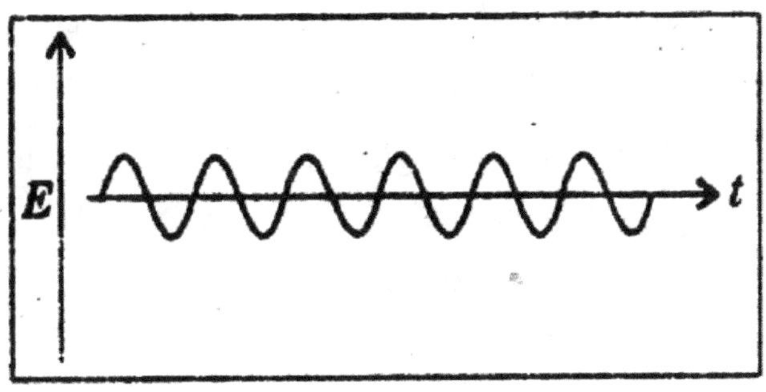

Figure:18.1

However, no light emitted by an actual source produces an ideal sinusoidal field for all values of time. This is because when an excited atom returns to the initial state, it emits light pulse of short duration such as of the order of 10^{-10} second for sodium atom. Thus, the field remains sinusoidal for time intervals of the order of 10^{-10} second after which the phase changes abruptly. Hence, the field due to an actual light source will be as shown in figure 18.2.

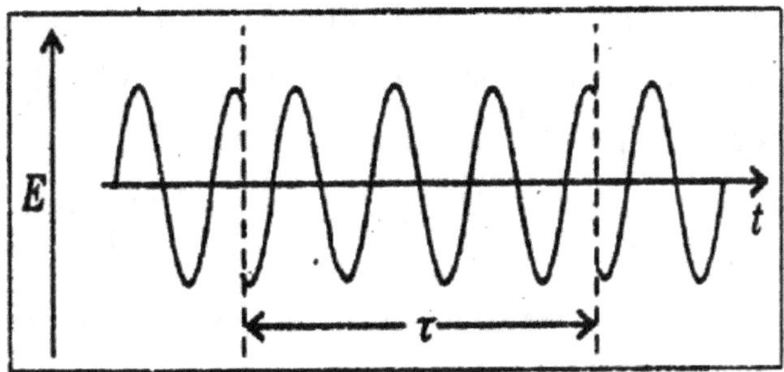

Figure: 18.2

The average time-interval for which the field remains sinusoidal (i.e., definite phase relationship exists) is known as "coherence time" or "temporal coherence" of the light beam and is denoted by. The distance L for which the field is sinusoidal is given by

$$L = \tau c$$

Where c is the velocity of light in vacuum. L is called the "coherence length" of the light beam.

Laser light is very high degree of temporal coherence, whereas conventional light has poor temporal coherence. In case of laser light, the path difference in two beams in Michelson's experiment may be a few meters or even more. For ordinary light, the difference may be a few centimeters. This path difference is also known as 'Coherence Length' of the source light and the coherence time can be calculated by dividing the coherence length by frequency.

The ability of light waves to create interference is measured in terms of the 'degree of coherence' of light waves. If the degree of coherence is higher, the probability of production of an interference pattern, with contrast, will also be higher.

According to ZERMIKE, the degree of coherence is equal to the visibility of fringes when the path difference between the beams is small and the amplitudes are equal and these are exactly the conditions required for the formation of fringes.

18.4 Spatial Coherence

The spatial coherence is the phase relationship between the radiation fields at different points in space. Let us consider light waves emitting from a source S (figure 18.3). Let A and B be two points lying on a line joining them with S.

The phase relationship between A and B depends on the distance AB and on the temporal coherence of the beam. If AB<<L (coherence length), there will be a definite phase relationship between A and B, i.e. there will be high coherence between A and B. On the other hand, if AB>>L, there will be no coherence between A and B.

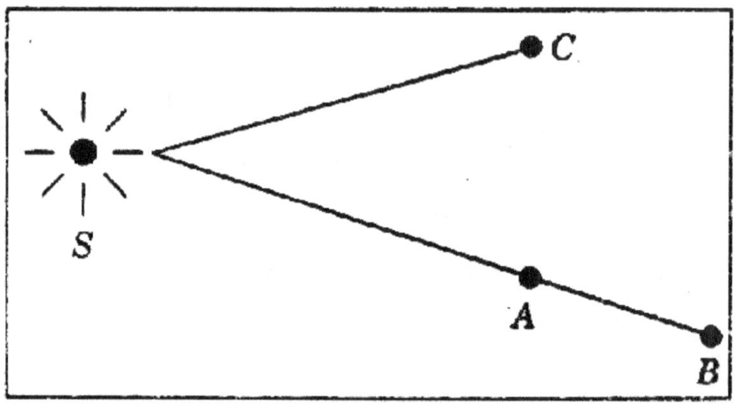

Figure 18.3

Spatial coherence is also known as 'Transverse Coherence' or 'Lateral Coherence'. Spatial or transverse coherence indicates how far apart from each other, can the two sources of light or two portions of the same source can be located in a direction transverse to the direction of observation so as to get the property of coherence over a number of observation points. This question depends on what is the distance of separation of two points, in the transverse direction in the area of observation.

If we use a laser beam, the dark and bright fringes, falling on the screen, will have the highest degree of contrast; because of spatial coherence of laser. If we use an ordinary beam of light, such as that form an incandescent bulb, we get the fringes with a very faint contrast between darkness and brightness.

18.5 Some Laser Applications

With the discovery of laser, man's control of light has been and will continue to extend to an unpredictably large and diverse number of applications. It would be practically impossible to list all the potential uses of the laser, but some of the important uses and applications of lasers are given below:

(i)Laser beam is used in the medical field. The laser beam is used in delicate surgery as cornea grafting. With laser beam the surgery is completed in much shorter time. This beam is also used in the treatment of kidney, stone, cancer, and tumor and in depositing and cutting the blood cells in brain operations. In most of these cases, the laser is like an optical knife which would be more accurate, less painful and faster than the scalpel.

(ii) Laser beam has a wide scope in the technical and engineering field. The laser beam is used for cutting steel sheets and melting and drilling hard material. It can create hole in diamond. Extremely thin wires used in cables can be drawn through the diamond hole. Metallic rods can be melted and joined hole by means of a laser beam (laser welding).

(iii)The application of the laser of the field of communications is receiving a large amount of technological attention. Since the light from the laser is coherent, it can theoretically carry messages in the same manner low frequency carriers.

(iv) Lasers play very important role during war-time. Lasers are used to detect and destroy the enemy missiles. Now, laser-refiles, laser-pistols and laser-bombs are also being made which can be aimed at the enemy in the night. In space, laser has been used to control rockets and satellites and in directional-communication.

(v) Laser is used for three-dimensional photography (holography).

(vi) Laser is useful in science and research. It has been used to perform Michelson-Morley experiment which is the building stone of Einstein-theory of relativity. It can be used to determine the temperature of plasma and the density of electron. Laser torch is used to see objects at long distances.

(vii) Since laser rays are very much parallel, so these are used in measuring long distances. The distance between earth and moon has been measured by laser rays.

(viii)Laser rays have proved to be useful in detecting nuclear explosions and earthquakes, in vaporizing solid fuel of rockets and in the study of the surfaces of distant planets and satellites.

(ix)Lasers are used in industry. The precision property of laser light has been immense help in industry, particularly in testing the quality of optical components e.g. prisms, lenses etc. The use of lasers in industry has considerably increased the accuracy in the measurement of the size of the physical quantities.

(x)Laser beam is used in the field of biology. The ability of laser beams to concentrate high power density of light at a focal point has opened a new era of micro Raman spectroscopic analysis. This enables one to extend such studies to biomedical samples available only in very small quantities.

(xi)Asers can be extremely useful tool in controlled fusion research (Laser-fusion).

(xii) Radio astronomers have found lasers extremely valuable for amplifying very faint radio signals from space. Radio telescopes have now an additional accessory-a laser (ruby). Thus, lasers are used in astronomy.

18.6 Wave Propagation in Optical Fibers

For the wave propagation of light waves in an optical fiber, we should have following mechanism and conditions:

(1) Mechanism: If light waves enter at one end of a fiber in proper conditions, most of it is propagated down the length of the fiber and comes out from the other end of the fiber. There may be some loss due to a small fraction leakage through the side-walls of the fiber. This type of a fiber is called light-guide or sometimes light- pipe. The reason of confining the light beam inside the fiber, is the total internal reflection and refraction of light waves. The light which enters at one end of a fiber at a single angle to the axis of the fiber follows a zig-zag path due to series of reflections down the length of the fiber.

(2) Conditions: Total internal reflection in the walls of the fiber can occur, if and only if, the following two conditions are satisfied:

(i)The glass at around the centre of the fiber should have higher refractive index (μ_1) than that of the material (cladding) surrounding the fibre(μ_2).

(ii)The light should be incident at an angle of θ (between the path of the ray and normal to the fiber wall) which will be greater than the critical angleθ_c.

$$\sin \theta_c = \frac{\mu_2}{\mu_1} \tag{18.1}$$

Reflection, refraction and total internal reflection of light waves are shown in given figure 18.4.

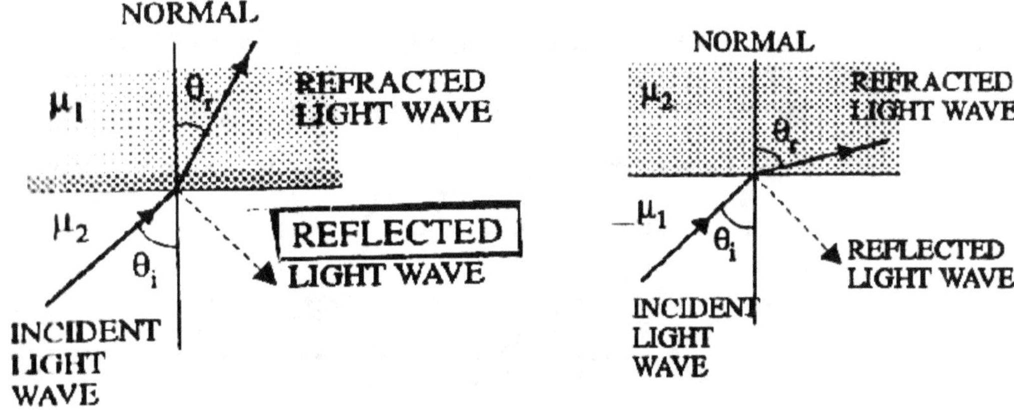

Figure 18.4(a) $\mu_1 > \mu_2$ **Figure 18.4 (b)** $\mu_1 < \mu_2$

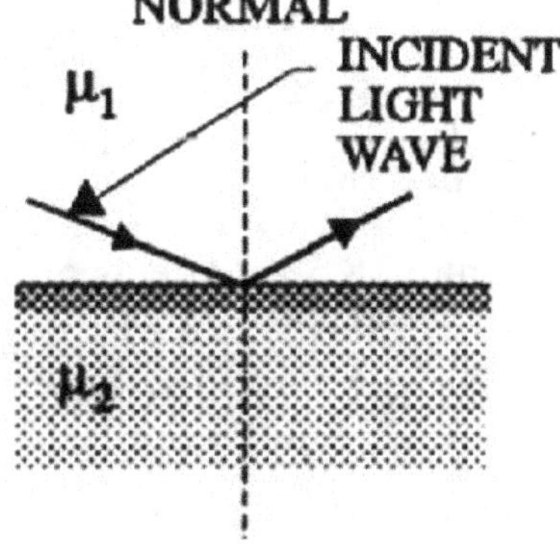

Figure 18.4(c) $\mu_1 > \mu_2$ *(Total Refelction)*

Let us now write down the conditions of reflection, refraction and total internal reflection.

(1) In reflection angle of incidence is equal to the angle of reflection.

(2) In refraction (i) $\mu_1 \sin \theta_i = \mu_2 \sin \theta_r$

(ii) The refracted wave should move towards the normal, if the light wave is incident from the optically lighter medium to an optically denser medium. And the refracted light wave should move away from the normal, if the light wave travels from the optically denser to optically lighter medium.

(3) The condition for total internal reflection is given in equation (18.1).

The angle of incidence at which total reflection first occurs is called the critical angle θ_c for the two mediums. Light waves incident at angles greater than θ_c will also be totally reflected.

18.6.1 Propagation of Light Wave Through an Optical Fiber

As we saw in above topic that the central core of an optical fiber consists of a glass core, with a certain refractive index μ_1 and totally enclosed by a glass cladding, having refractive index μ_2. $(\mu_1 > \mu_2)$

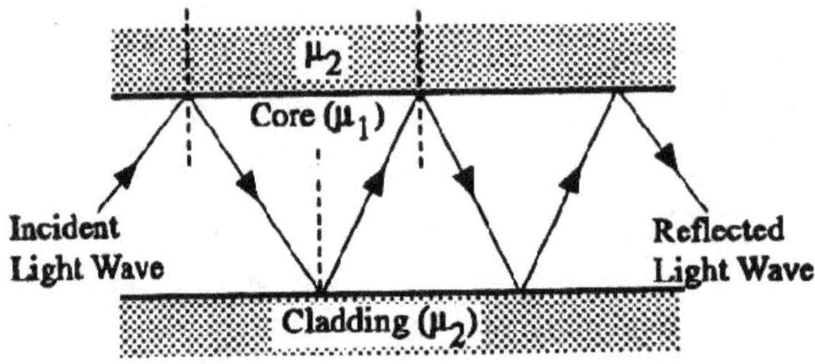

Figure 18.5 : *Light wave propagation along a glass fiber core*

The given figure 18.5 represents a longitudinal cross-section of a fiber. Any light wave, which travels along the core and meets the cladding at the critical angle of incidence, θ_c will be totally reflected. This reflected ray will be totally reflected. This reflected ray will then meet the opposite surface of the cladding, again at the critical angle θ_c and so again totally reflected. Therefore the light wave is propagated along the fiber core by a series of total internal reflections from the core-cladding interface. This is a sort of step index fiber, as there is clearly a sudden change of refractive index at the junction of the core and the cladding.

The path of only one light beam shown in the below given diagram 18.5, which is possible only from a very tiny point-source.

But practically it is not so. Light energy emanating from any practical point source, will have several paths with different angles of incidence at the boundary –layer. It may also contain different colours with different frequencies (and so wavelengths). Then it is called step – index multimode propagation as shown in below given figure 18.6. any other light wave which is meeting the core-cladding interface at or above the critical valueθ_c will also be totally reflected and hence will propagate along the core. However, any light wave, meeting the core-cladding interface at an angle belowθ_c will pass into and be absorbed by the cladding.

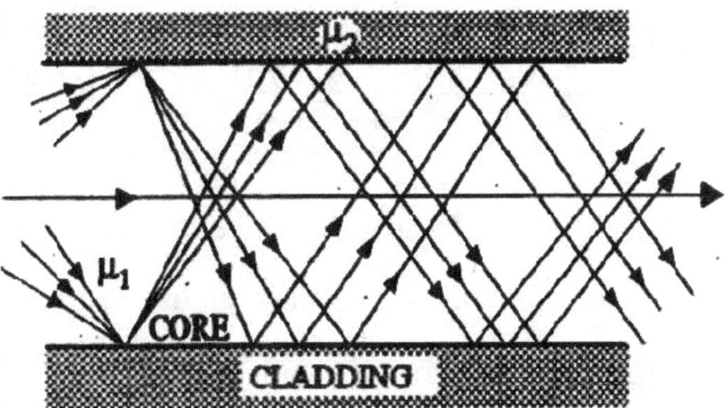

Figure 18.6: *Stepped index multimode propagation* $\mu_1 > \mu_2$

Thus, the various light waves, travelling along the core, will have propagation paths of different lengths. Hence they will take different times to reach a given destination. Thus a distortion is produced and is called transmitted time dispersion. This dispersion sets an upper limit on the rate at which the light can be modulated by an analogue or digital electrical signal. As a result of the distortion, the variations of successive pulses of light may overlap into each other, and thereby cause distortion of the information being carried. However, this defect can be minimized by making the core diameter of the same order as the wavelength of the light wave to be propagated. The resultant propagation is a single light wave, as shown in given figure 18.7. However this type of fiber is called a stepped index monomode fiber. This has very high capacity and large bandwidth.

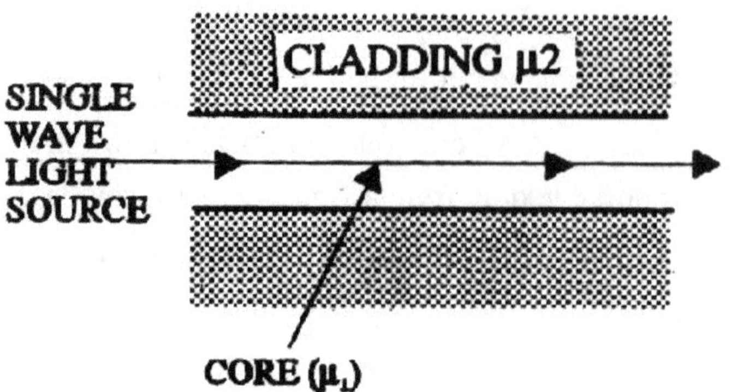

Figure 18.7 :*Stepped index monomode propagation* $\mu_1 > \mu_2$

Now, we discuss the propagation of multiwave light energy in graded index fiber, as shown in figure 18.8, with the individual waves being gradually refracted in the graded-index core, instead of being reflected by the cladding.

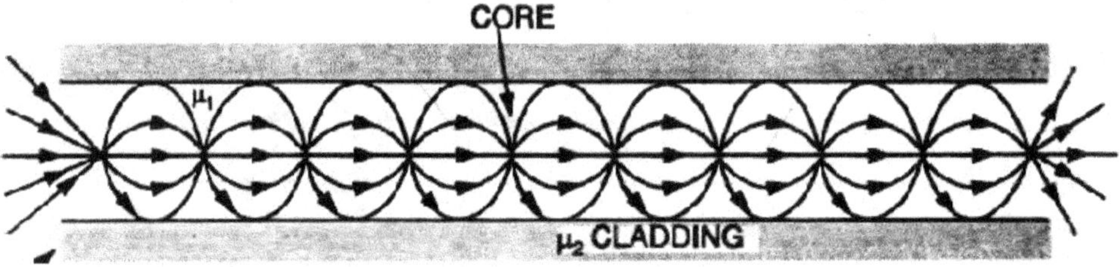

Figure 18.8 : *Graded index multimode propagation*

Hence waves travelling at different incident angles will travel different distances from the horizontal central axis, before being reflected back to recross the central axis.

It is obvious that the light waves with large angle of incidence travel more paths than those with smaller angles. But we know that the decrease of refractive index allows a higher velocity of propagation. Thus all waves will reach a given point along the fiber at virtually the same time. As a result the transit time dispersion is greatly reduced. This type of light wave propagation is referred to as graded index multimode propagation.

One important thing to mention here that in all three types of fibers (i.e. .n step index multimode ,step index monomode, graded index multimode), the thickness of cladding material around the fiber core should at least be several wavelengths. This arrangement will prevent light energy losses due to absorption and scattering.

18.7 Distance and Velocity Measurements

18.7.1 Distance Measurements

The large coherence length and high output intensity coupled with a low divergence, enables the laser to find applications in precision length measurements using interferometric techniques. The method essentially consists of dividing the beam from the laser by a beam splitter into two portions and then making and then making them interfere after traversing two different paths. One of the beams emerging from the beam splitter is reflected by a fixed reflector and the other usually by a retroreflector(it reflects an incident beam in a direction exactly opposite to that of an incident beam) mounted on the surface whose position is to be monitored. The two reflected beams interfere to produce either constructive or destructive interference. Thus, as the reflecting surface is moved, one would obtain alternatively constructive and destructive interference, which can be detected with the help of a photodetector. Since, the change from a constructive to a destructive interference corresponds to a change of a distance of half a wavelength, one can measure the distance traversed by the surface on which the reflector is mounted by counting the number of fringes which have crossed the photodetector. Accuracies up to $0.1\mu m$ can be obtained using such a technique.

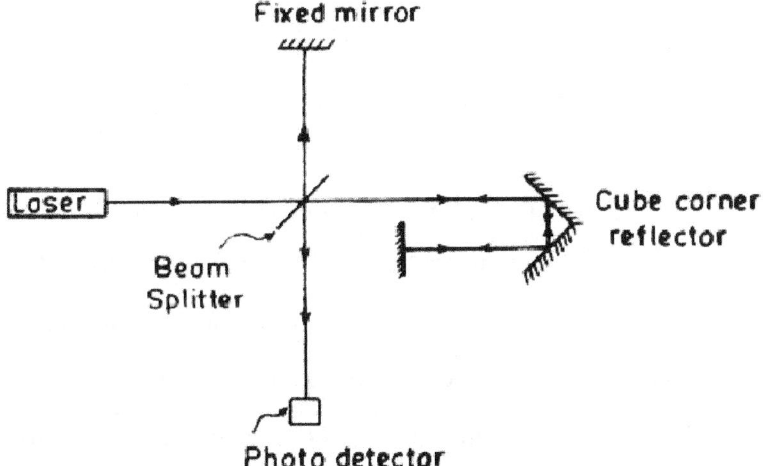

Figure18.9: *Laser interferometer arrangement for precision length measurements.*

This technique is being used for accurate positioning of aircraft components on a machine tool, for calibration and testing of machine tools, for comparison with standards, and many other precision measurements. The conventional cadmium light source can be used only overpath difference of about 20 cm. with the laser one can make very accurate measurements over very long distances because of the large coherence length. The most common type of laser used in such applications is the helium-neon laser, and since the distance measurements is being made in terms of wavelength, in these measurements, a high wavelength stability of the laser output must be maintained.

18.7.2 Velocity Measurements

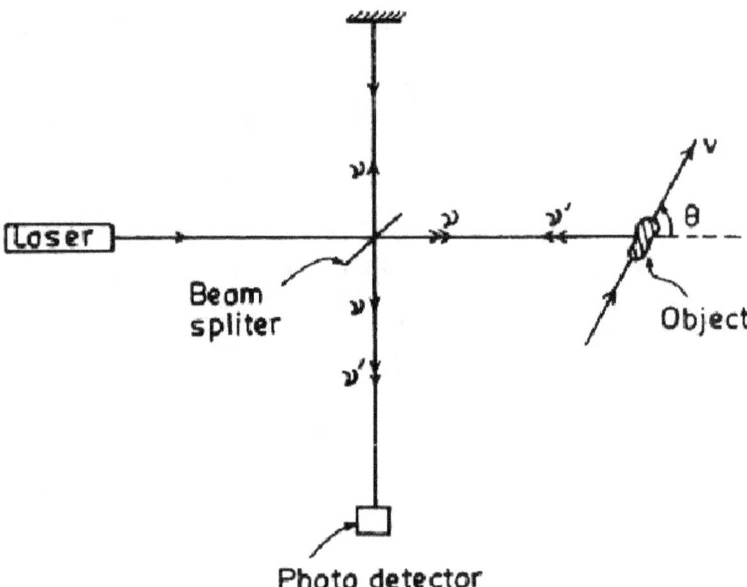

Figure 18.10: *Schematic of an arrangement for measuring the velocity of a moving object using Doppler shift.*

It is well known that when a light beam gets scattered by a moving object, the frequency of the scattered wave is different from that of the incident wave; the shift in the frequency depends on the velocity of the object. Indeed, if ν represents the light frequency and υ is the velocity of the moving object which is moving at an angle θ with respect to the incident light beam (fig. 18.10), then the change in frequency $\Delta\nu$ between the incident and the reflected beams is given by

$$\frac{\Delta v}{v} = \frac{2v}{c}\cos\theta$$

where c represents the velocity of light in free space. Thus the change in frequency

Δv is directly proportional to the velocity v of the moving object; this is the Doppler shift. Thus, by measuring g the change in frequency suffered by a beam when scattered by a moving object, one can determine the velocity of object. This method has been successfully used for velocity determination of many types of materials from about 10mm/min to about 150mm/min. further using the above principle, portable velocity-measuring meters have been fabricated which measure speeds in the range of 10-80 miles/hour; these have been used by traffic police. Laser Doppler velocity meters have also been used for measuring fluid flow rates.

18.8 Self Learning Exercise

Q.1 What is Numerical aperture? Give its relation with the acceptance angle.

Q.2 Compute the NA and the acceptance angle of an optical fiber from the following data.

$$\mu_1(core) = 1.55 \ and \ \mu_2(cladding) = 1.50$$

Q.3 Calculate the NA, acceptance angle, and the critical angle of the fiber having

$$\mu_1(refractive \ index \ of \ core) = 1.50$$
$$\mu_2(refractive \ index \ of \ cladding) = 1.45$$

Q.4 What are the applications of fibers ?

Q.5 What are the optical fibers ?

Q.6 Give some advantages of the optical fiber.

18.9 Summary

The summary of this unit is not one because it contains two different topics, (1) Propagation of light in optical fiber and optical fiber, (2) which is related to the property of laser and uses of laser.

We have studied the propagation of light in air and other medium, but in this chapter we have studied the propagation of light in optical fiber.

An important property of light is also given in this unit which is Coherence.

18.10 Glossary

Coherence: It is the property of light in which we study the measure of the ability of two photons to interfere or intermix with each other.

Optical Coherence: A wave which is appears to be a pure sine wave for an infinitely large period of time or in an infinitely extended space is said to be a perfectly coherent wave.

Spatial Coherence: The spatial coherence is the phase relationship between the radiation fields at different points in space. The spatial coherence is the phase relationship between the radiation fields at different points in space.

Light Guide: If light waves enter at one end of a fiber in proper conditions, most of it is propagated down the length of the fiber and comes out from the other end of the fiber. There may be some loss due to a small fraction leakage through the side-walls of the fiber. This type of a fiber is called light guide.

18.11 Answers to Self Learning Exercise

Ans.1: Numerical aperture is an important term associated with a fiber. Sometimes this term called as the figure of merit for optical fibers. Numerical aperture of an optical fiber is defined as

$$NA = \frac{\sqrt{\mu_1^2 - \mu_2^2}}{\mu_0}$$

If the fiber is surrounded by air $(\mu_0 = 1)$ then

$$NA = \sqrt{\mu_1^2 - \mu_2^2}$$

Generally μ_1 are only a few percentages greater than μ_2.

$$NA = \sqrt{(\mu_1 - \mu_2)(\mu_1 + \mu_2)} \simeq \sqrt{2\mu_1(\mu_1 - \mu_2)}$$

$$NA = \sqrt{2\mu_1^2 \frac{(\mu_1 - \mu_2)}{\mu_1}}$$

$$NA = \mu_1\sqrt{2\Delta}$$

where

$$(\mu_1 + \mu_2) \simeq 2\mu_1 \, and \, \Delta = \frac{(\mu_1 - \mu_2)}{\mu_1}$$

and also we know that the half acceptance angle

$$\theta_0 = \sin^{-1}\sqrt{\mu_1^2 - \mu_2^2}$$

Thus the light which travels within a cone defined by the acceptance angle is trapped and guided. This is the fundamental property of light propagation in a fiber. This cone is referred to as acceptance cone.

Hence

$$\theta_0 = \sin^{-1}(NA)$$

This is the required relation between numerical aperture and half acceptance angle.

Ans.2: We know that the Numerical aperture of an optical fiber is

$$NA = \mu_1\sqrt{2\Delta}$$

where

$$\Delta = \frac{(\mu_1 - \mu_2)}{\mu_1} = \frac{1.55 - 1.50}{1.55} = 0.03226$$

$$NA = 1.55\sqrt{2 \times 0.03226} = 0.394$$

Now for acceptance angle, we have the relation

$$\theta_0 = \sin^{-1}(0.394) = 23.2°$$

Ans. 3: We know that the relation between refractive index and Δ

$$\Delta = \frac{(\mu_1 - \mu_2)}{\mu_1} = \frac{1.5 - 1.45}{1.5} = 0.033$$

$$NA = 1.55\sqrt{2 \times 0.033} = 0.387$$

Now for acceptance angle, we know that the relation

$$\theta_0 = \sin^{-1}(0.387) = 22.78°$$

Now the critical angle is given by

$$\theta_c = \sin^{-1}\frac{\mu_2}{\mu_1}$$

$$\theta_c = \sin^{-1}\frac{1.45}{1.5} = 75.2°$$

Ans.4: There is some application of fiber, which is given below:

(1) When the transmission medium has a very large bandwidth, a single mode fiber is used. Example: Undersea cable system.

(2) When the system bandwidth requirements are between 200 MHz and 2GHz km, a graded index multimode fiber would be the best choice. Example: in intra-city trunks between telephone central offices.

(3) When the system bandwidth requirements are lower, a step-index multimode fiber would be better. Example: Data links.

Ans.5: Optical fibers are the light equivalent of microwave waveguides with the additional advantages of a very wide bandwidth. Physically an optical fiber is a very thin and flexible medium, having a cylindrical shape consisting of three sections: (i) the core, (ii) the cladding and (iii) the jacket. Out of these, the core is the innermost section and is made of glass or plastic. This is the actual fiber and has the remarkable property of conducting an optical beam. It is surrounded by its own cladding, a glass or plastic coating, which has optical properties which are different from those of the core. The outer section is called the jacket made of plastic or polymer and other materials and is provided for protection against moisture, absorption, crushing and other-environment dangers.

The core acts like a continuous layer of two parallel mirrors. A signal is first encoded into a beam, which is then passed in between the two boundaries and propagated as a result of multiple internal reflections.

Ans.6: An optical fiber has very clear advantages over wire or radio system. The following are the main advantages of optical fibers:

(1) *Attenuation* in a fiber is markedly lower than that of coaxial cable or twisted pair and is constant over a very wide range. So transmission within wide range of distance is possible without repeaters etc.

(2) *Smaller size and lighter weight.* Optical fibers are considerably thinner than coaxial cable or bundles twisted-pair cable. So they occupy much less space.

(3) *Electromagnetic isolation* .Electromagnetic waves generated from electrical disturbances and electrical noises do not interfere with light signals.

As a result, the system is not vulnerable to interference, impulse noise or cross-talk.

(4) *No physical electrical connection* is requires between the sender and the receiver.

(5) The fibre is *much more reliable,* because it can better withstand environment conditions, such as pollution, radiation and salt produces no corrosion. Moreover, it is nominally affected by the nuclear radiation. Its life is longer in comparison to copper wire.

(6) Almost there is *no cross − talk* in optical fibers and hence transmission is more *secure* and private, as it is very difficult to tap into a fibre.

(7) *Greater bandwidth.* Bandwidth of the optical fibre is higher than that of an equivalent wire transmission line.

(8) As fibers are *very good dielectrics,* isolation coating is not required.

(9) Due to the *non − inductive* and *non − conductive* nature of a fiber, there is no radiation and interference on other circuits and systems.

(10) *Greater Repeater Spacing.* Fewer repeaters indicate lower cost and fewer sources of error. It has been observed that a fibre transmission system can achieve a data of 5 *Gbps* over a distance of 111 Km, without repeaters, whereas coaxial and twisted-pair systems generally have repeaters every few kilometres.

18.12 Exercise

Q.1　What is coherence or optical coherence?

Q.2　What is the Optical Fiber?

Q.3　Give some advantages of the Optical Fiber.

Q.4　Give the brief classification of the Coherence.

Q.5　Give some uses of Lasers.

Q.6　What is the behaviour of Light, when it propagates in the Optical Fiber medium?

Q.7　What is acceptance angle? Give its relation with the Numerical Aperture.

Q.8 Give the application of Laser in Medical Sciences and Holography.

Q.9 Calculate Δ for a step index fiber having core & cladding refractive indices 1.48 and 1.46 respectively.

Q.10 Calculate the NA, Δ and acceptance angle of a fiber having the following characteristics:

$$\mu_1 (refractive\ index\ of\ core) = 1.44$$
$$\mu_2 (refractive\ index\ of\ cladding) = 1.40$$

References and Suggested Readings

1. Laser and Non-linear Optics by B.B. Laud.

2. Laser Physics by Peter W. Milonni and Joseph H. Eberly.

3. Laser physics and its application by L.V. Tarasov.

4. Atomic and Molecular Spectra: Laser by Raj Kumar.

UNIT-19
Laser Gyroscope, Applications of Lasers

19.0 Objectives

There is an ever-increasing demand for accurate, low-cost and highly reliable guidance, control, and navigation systems for air, land, sea and space vehicles. The heart of these systems is gyroscope. Gyroscope is a device used to maintain orientation in space during motion and determine the angular rate of its carrying vehicle with respect to a reference frame. At present time, the best gyroscopes are still mechanical in nature using a large rotating mass, but now there are made a lot of efforts to develop high performance non-mechanical gyroscope like laser gyroscope. In this unit, we will discuss about the laser gyroscope and its working principle, and applications of lasers in defense, industry and medicine.

19.1 Introduction

The Laser gyroscope works on a physical principle discovered by the French Physicist George Sagnac in 1913. Unlike the conventional spinning gyroscope with its gimbals, bearing and torque motors, the laser gyroscope uses a ring of laser light together with rigid mirrors and electronic devices. Thus, the laser gyroscope has more attractive features than conventional gyroscope including much greater reliability, lower maintenance requirements, light weight and compact in size. The LASER light has four main characteristics including (i) coherency (ii) high monochromaticity (iii) narrow angular spread and (iv) high intensity. Due to these characteristics, the lasers have many specialized applications is the field of defense, industry and medicine.

19.2 Sagnac Effect

In 1913, George Sagnac performed a ring interferometry experiment to point out the effect of rotation on the propagation of light traveling along a closed path. The first ring interferometry experiment aimed at observing the correlation of angular velocity and phase –shift performed by the French Physicist George Sagnac and his result came to be known as the Sagnac effect. This experimental arrangement is called a Sagnac interferometer as shown in figure 19.1.

In this experiment, a beam of light in slipt by a half-silvered mirror into two beams and sent in two opposite directions around a closed path on a revolving platform (turntable). If the apparatus is stationary, the two beams will travel equal distance around the closed loop and arrive at the detector simultaneously in phase. However, if the entire device is rotating, the beam traveling around the loop in the direction of rotation will travel a slightly greater distance than the beam traveling counter to the direction of rotation, because during the period of travel the mirrors and detector will all move slightly toward the counter rotating beam and away from the co-rotating beam. Consequently, the beam will reach the detector at slightly different times and slightly out of phase, producing optical fringes. Therefore, the interference pattern obtained at each angular velocity of the plate-form features a different phase-shift particular to that angular velocity. Thus, the position of the interference fringes is dependent on the angular velocity of the set up.

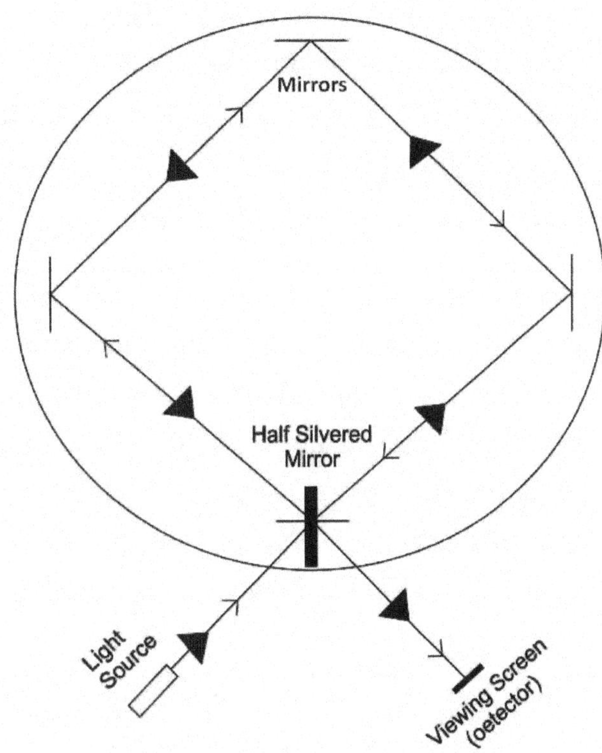

Figure 19.1 Sagnac interferometer

19.3   Laser Gyroscope or Ring Laser Gyroscope

Laser gyroscopes are based on Sagnac effect. The modern Sagnac gyroscope either uses a coil of fiber-optic cable to make the path length around a ring interferometer very long or use a ring laser cavity. Therefore, there are two types of gyroscopes.

1. Fiber-Optic Gyroscope (FOG)
2. Ring Laser Gyroscope (RLG)

In this unit, we will study about a laser gyroscope, which have two counter-rotating laser beam travel around a ring or closed path. Such a laser gyroscope is referred to commonly as a ring laser gyroscope as shown in figure 19.2.

The first experimental ring laser gyroscope was demonstrated in the USA by Macek and Davis in 1963. A ring laser gyroscope uses interference of laser light within a bulk optic ring to detect changes in orientation and spin. RLG can be used as the stable element in an inertial reference system. Physically an RLG consists of a three or four sided block of low expansion coefficient material, which defines a closed optical cavity. The light path is defined by mirrors mounted at the corners.

The light travels through holes bored in the block containing a low pressure gas, usually a He-Ne mixture which lases when the anode and cathode are excited. Thus, RLG is itself actually a laser, therefore there is no need of any external light source in this gyroscope.

19.4 Working of Laser Gyroscope

According to the Sagnac effect, if two identical light waves circulate in opposite directions along a closed path undergoing a rotation (ω), then the light beam traveling in the same direction as the rotation takes longer time to travel around the path than the other beam, i.e. the resonant frequencies of the two standing waves produce in cavity are different. Therefore, a beat frequency (difference between the two frequencies) is obtained, which is directly proportional to the rotation rate. This gives change in the interference pattern of the beams.

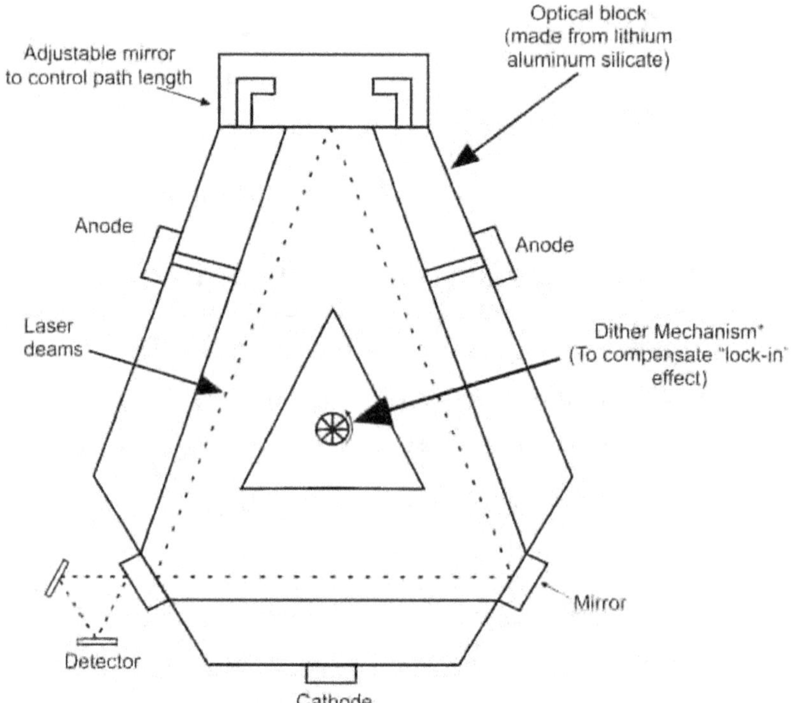

Fig. 19.2 Ring Laser Gyroscope

Hence, the angular rate is measured by counting the interference fringes. The phase shift produced is given by

$$\Delta\phi = \frac{2\pi}{\lambda} \times (path\ difference) \tag{19.1}$$

The path difference between the counter-rotating laser beams traveling around closed path is calculated as follows:-

Usually several mirrors are used, so that the light beams follow a circular path. Let R be the radius of the platform and ω be its angular velocity, then the circumferential tangent speed will be $v = \omega R$. Assuming entire device is rotated clockwise, then the time taken by light traveling in the co-rotating direction in one complete rotation is

$$t_1 = \frac{2\pi R}{c-v} = \frac{2\pi R}{c-\omega R} \tag{19.2}$$

and that in the counter-rotating direction is

$$t_2 = \frac{2\pi R}{c+v} = \frac{2\pi R}{c+\omega R} \tag{19.3}$$

The difference between the travel times is

$$\Delta t = t_1 - t_2 = 2\pi R \left(\frac{1}{c-R\omega} - \frac{1}{c+R\omega} \right)$$

$$\Delta t = \frac{2\pi R \times 2R\omega}{c^2 - R^2\omega^2} = \frac{4\pi R^2 \omega}{c^2 - R^2\omega^2}$$

Since $R\omega$ is very small in comparison to c ($R\omega \ll c$), when c is the speed of the light, then

$$\Delta t \approx \frac{4A\omega}{c^2} \tag{19.4}$$

where $A = \pi R^2$ is the area enclosed by the loop.

The path difference is $\Delta x = c\Delta t = \dfrac{4A\omega}{c}$ $\tag{19.5}$

The number of fringes shifted $\Delta n = \dfrac{c\Delta t}{\lambda} = \dfrac{4A\omega}{c\lambda}$ $\tag{19.6}$

This is Sagnac's formula. The amount of displacement is proportional to angular velocity of rotating platform or turntable.

Thus, the phase shifted is (using eq. 19.5)

$$\Delta\phi = \frac{2\pi}{\lambda} \left(\frac{4A\omega}{c} \right)$$

$$\Delta\phi = \frac{2\pi\omega}{\lambda} \left(\frac{4A\omega}{c} \right)$$

$$\Delta\phi = \frac{8\pi A\omega}{\lambda c} \tag{19.7}$$

It there is N turns in the circular part, then total phase difference $\emptyset = N\Delta\emptyset$

$$\phi = \left(\frac{8\pi A\omega}{\lambda c}\right) N$$

$$\phi = \left(\frac{8\pi AN}{\lambda c}\right) \omega \tag{19.8}$$

where $\left(\frac{8\pi AN}{\lambda c}\right)$ = scale factor for the gyro $\tag{19.9}$

A = Area enclosed by the beam path

N = No. of times beam has gone around the path

ω = Rotation of system

As we know that the laser gyroscope measures rotation rate by sensing frequency differences. When the rate of rotation is very small and thus the frequency difference between two beams is also small. There is a tendency for the two frequencies to couple together. This effect is known as "Lock-in". It limits the accuracy of the laser gyroscope at important low rotation rates. The problem of lock-in is compensated by introducing a randomized "dither" by mechanically rotating the gyroscope.

The laser gyroscope is insensitive to variations in the earth's magnetic and gravitation fields due to the use of solid-state components and mass-less light. These are especially attractive for high performance aircraft, remotely piloted vehicles and missiles.

19.5 Applications of Lasers

Lasers are used in variety of applications. These are based on the four important properties of laser beam viz. coherency, high intensity, monochromaticity and narrow angular spread (high directionality). Many of the interferometric techniques in optics (example-Holography) use laser sources on account of their high coherent nature. High intensity of laser beam makes it convenient for material processing and surgery. The high monochromaticity of laser beam has been used to separate isotopes. The high directionality of the laser beam makes it useful for ranging *i.e.* to measure distance of an object and the speed with it moves.

19.5.1 Applications of lasers in Defense

Lasers have been used for various military technologies.

a) Laser finder: Generally, laser beam is being used for effective and automatic control and guidance of rockets and satellites. When used in radars, it can be employed to destroy aeroplanes and missiles. Focused laser beams would become the legendary death ray of fictions that would burn everything standing in the way of beam. With improved firing accuracy of guns and enhanced destructive power ammunition, it is imperative to have accurate information about range to improve the *First Hit Probability*. Lasers play an active role in it and it is no wonder the weapon platforms like battle tanks, all over the world, are being with laser range finders. Therefore, lasers are ideal tools, available for range finders. Finders interfaced with computers are available. The *First Hit Probability* has thus improved tremendously.

b) Laser Radar: The technique used in laser radar is similar to electronic radar. In the conventional radar, the electromagnetic pulse is transmitted and the reflected echo is recorded. Similarly, in laser radar, light energy is sent out in pulses and the reflected light echo is collected. However, here, since light is diffused by targets, there is a requirement of high peak power; that must have suitable pulse repetition rate, which is governed by the distance that is required to be measured. So, a laser radar scheme uses a laser scanner to observe the field of view. The scanner can be a system with mechanical devices formed with rotating mirrors. Scanning of the beam is done through the field of view in a pattern somewhat similar to the technique used in televisions.

19.5.2 Applications of lasers in Industry

The Lasers are used in many material processing applications such as laser surface modification (Cladding, Alloying and surface hardening), welding, drilling, cutting, scribing, thermal treatment, marking, etc.

The laser beam is usually a few millimeters in diameter. For material processing applications, the laser beam has to be focused using lens or a combination of lenses. If λ is the wavelength of light, a is the radius of the beam and f is the focal length of the lens, the incoming beam will get focused into a region of radius b given by

$$b \approx \frac{\lambda f}{a}$$

If P is the power of the incoming beam, them the intensity I obtained at the focused region is given by

$$I \approx \frac{P}{\pi b^2} = \frac{Pa^2}{\pi \lambda^2 f^2}$$

Thus on focusing a $1W$ laser beam (with λ = 1.06 μm and having a beam radius of 1 cm) by a lens of focal length 2 cm, the intensity at the focused spot is given by $\approx$ 7.08 x 10^6 W/cm^2. Because of such large intensities, enormous heat is generated which can melt metals. Therefore, Lasers have been employed to modify the surface properties of a material.

(a) Laser Surface Modification (LSM)

Laser surface modification (LSM) refers to any process in which the heat produced by the interaction of the laser beam and surface of the material is used to bring about a desirable alteration in material properties. It is used to increase wear resistance, corrosion resistance or strength. The techniques usually employed are laser chemical vapour deposition (LCVD), cladding, alloying, surface hardening and surface melting. The set-up used for LSM is relatively simple. The beam is directed perpendicular or nearly perpendicular to the surface to be treated. The spot size of the laser beam is usually large and the beam or the surface is moved to facilitate the scanning of the entire surface.

Laser chemical vapor deposition (LCVD) is a process whereby the laser is used to heat a substrate so that a vaporous material can be deposited on it. It is one of the several techniques used in thin film technology. The substrate is heated for better adhesion. Laser is an inappropriate heat source for depositing on large areas the substrate. It is ideal and appropriate for localized heating and for complex substrate geometries which are encountered in integrated circuits. In some cases, laser could also be used to dissociate molecules or excite atoms in the gas phase or on a surface to form a thin film, without actually heating the substrate.

Cladding and Alloying: Cladding and alloying are similar processes. In both these techniques, an appropriate powder is laid down on the surface and is melted by the heat from the laser beam. During cooling the melt forms a chemical bond with the base metal surface. This process is also known as hard facing. In alloying, the base metal is melted to a substantial depth. The alloying material is mixed into the base metal to produce a new alloy. These two processes are shown in Fig. 19.3. Cladding and alloying may be used for improving wear resistance, corrosion resistance and impact strength.

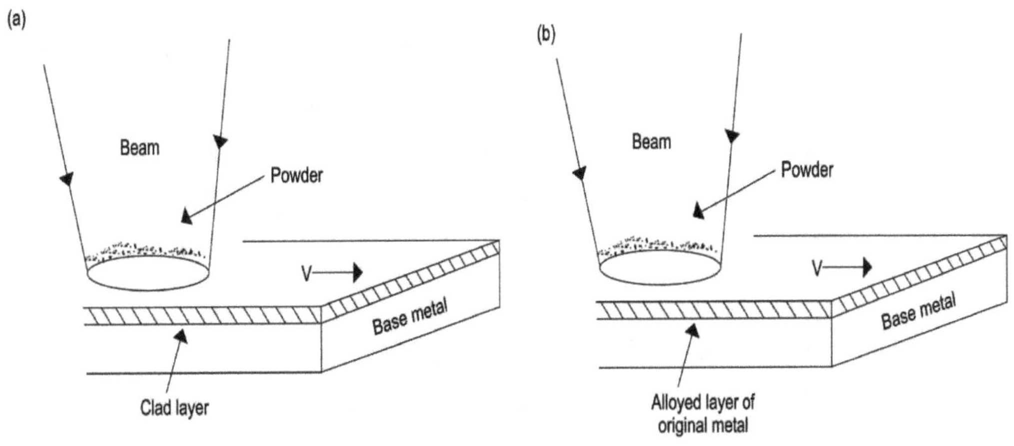

Fig. 19.3 Cladding and Alloying

Surface hardening: It is well known that many materials such as cast iron and steel become hard after heat treatment. The laser is an ideal tool for selective and localized heating. Surface hardening with the laser improves wear resistance, impact strength and fatigue strength. The structural changes that occur on the surface due to rapid heating and cooling introduce residual stresses. These residual compressive stresses are responsible for the improved properties.

(b) Soldering and Welding

Lasers are used for both welding and soldering. In soldering, the laser acts as a heart source to reflow the solder which has previously been placed on the metallic part or parts. Laser soldering is preferred because the heat input can be localized and minimal heating is required. Hot plate or hydrogen flame techniques heat either the entire substrate or a large portion of it. This can damage sensitive electronic components.

Two regimes of operation occur in laser welding depending on the incident power and power density. These are referred to as thermal conduction welding and keyhole welding. In thermal conduction welding, the laser welding is predominantly controlled by thermal conduction. This means that the power is absorbed at the surface of the metal and diffuses into the metal. Melting occurs for a depth determined by the thermal characteristics of the material and the welding parameters. Obviously, there must be sufficient heat to raise the material's temperature above the melting point. In thermal conduction welding, a great deal of energy is lost due to reflection. Although, the reflectance of metal decreases as the temperature increases. Molten metal is still highly reflective.

In key-hole welding, the welding process is made more effective. The vapor pressure of the heated metal overcomes the surface tension of the molten pool and opens up a cavity into the metal which allows the laser beam to penetrate deeper as shown in fig. 19.4. The molten metal flows around the key hole and fills the hole, after the laser beam has passed over. Typical laser weld configurations are shown in the figure. Usually the welding operation is carried out in an atmosphere of flowing insert gas such as Helium, argon and nitrogen as shown in fig 19.5. This ensures the protection of the focusing optics and control of weld plasma. The plasma is produced by the heating of the work piece which emits small particles, electrons and ionized atoms. The advantages of laser welding are(a) flexibility (b) no need of a filter material (c) weld in most atmospheres (d) weld many dissimilar metals (e) deep penetration (f) high speed (g) no special joint preparation (h) minimal part distortion and (i) small heat affected zone.

(c) Hole drilling

There are two methods of piercing holes with lasers. They are percussive drilling and gas assist process. In percussive drilling, the energy from the laser pulse causes material to vaporize rapidly enough so that the molten and the solid material are expelled from the hole due to the rapid high pressure build-up. In the Gas assist process, air or an insert gas is blown into the hole so that the molten material and solid particles are expelled. Holes ranging from 0.025 mm in diameter up to 1.5 mm in diameter are routinely drilled by laser up to depths of 25 mm.

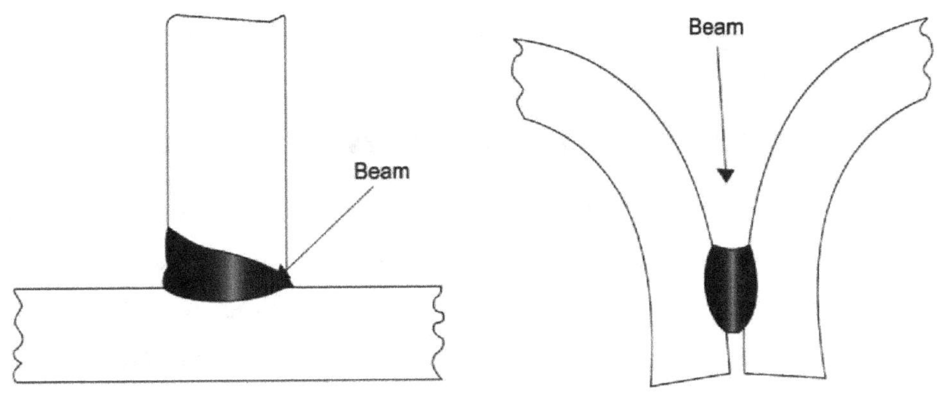

Fig. 19.4 Welding

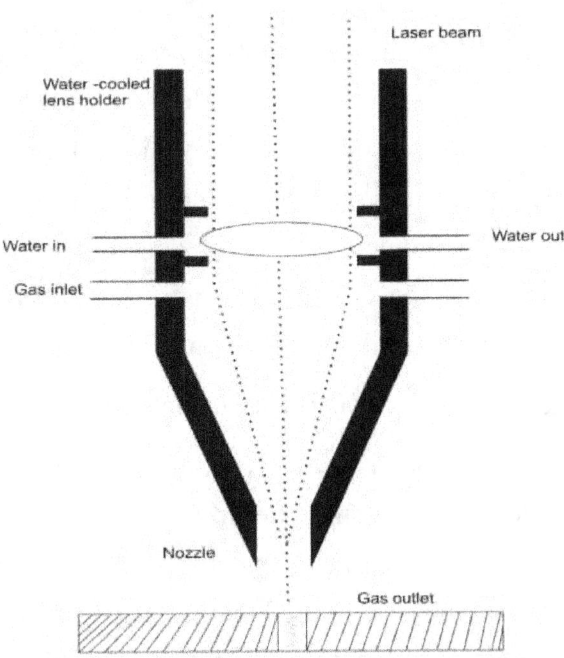

Fig. 19.5 Gas flow welding apparatus

(d) Cutting

Cutting of metals with lasers is a gas assisted process. The purpose of using the gas is to blow away the molten material and prevent it from damaging the focusing optics. Since the molten material is blow away, the amount of vaporization required is minimized. This improves both the cutting rate and the quality of the cut. Laser cutting is also employed in garment industry.

(e) Marking

One of the major uses of lasers is in marking (sometimes referred to as engraving). CO_2 lasers are used to produce entire patterns, bar codes, serial numbers, logos etc. with one or two laser pulses.

19.5.3 Applications of lasers in medicine

Lasers are widely used in medicine both for therapeutics as well as for diagnosis. When a human tissue is exposed to laser radiation its temperature raises. The extent of damage depends on the time for which the tissue is at elevated temperatures. The nature of laser tissue interaction process may be divided into several regions, determined primarily by the intensity of the laser beam and its interaction time with the tissue. Lasers are used for tissue cutting and relatively high beam intensities with exposure times of milliseconds to seconds. This results in rapid deposition of heat and subsequent vaporization or decomposition. These are processes that are characteristic of thermal or thermo-acoustic ablation. For nanosecond pulses at relatively high photon energies, photons can directly break specific chemical bonds resulting in strong absorption and particularly clean cut. Lasers have been extensively used in ophthalmology, oncology, dermatology, cardiology, gynecology, dentistry and acupuncture. In many cases, the laser is precisely targeted with the help of an optical fiber.

The advantages of using laser include

(a) greater precision in targeting the diseased tissue
(b) less bleeding and swelling
(c) greater control over the input energy by mean of varying the laser pulse duration as well the Number of pulses per second
(d) Less pain and dispensing with general anesthesia.

(a) Ophthalmology

The eye is roughly spherical in shape and consists of an outer transparent wall called cornea as shown in fig.19.6. This is followed by iris which controls the amount of light entering the eye and the eye lens. The space between the cornea and the lens is filled with aqueous humor. The rear portion of the eye consists of retina. Light falling on the eye is focused onto the retina. The photosensitive pigment contained in the retinal cells convert the light energy to electrical pulses. These are carried by the optic nerve to the brain for processing. Sometimes the retina many get detached from the underlying tissue causing blurred vision or blindness. Lasers are primarily used for treating the retinal detachment. It is used for photocoagulation of the retinal blood cells *i.e.*, the blood vessels are heated to the point where the blood coagulates and forms a viscous mass. Photocoagulation is useful for repairing retinal tears and holes that develop prior to retinal detachment.

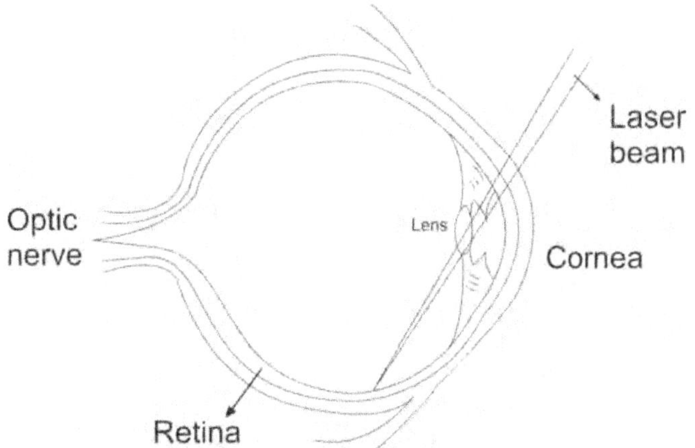

Figure 19.6 Laser eye surgery

Lasers can be used for correction of focusing defects of the eyes. In the method referred to as LASIK (Laser In-Situ Keratomileules), the cornea of the eye can be crafted to adjust its curvature. This leads to proper focusing of light on to the retina. This treatment is suitable for people who use glasses with high lens powers.

(b) Oncology

Laser is an ideal tool for burning away the cancerous tissues. The treatment has the advantage that the laser beam can be focused to a very small region. This helps in retaining the healthy tissues which are otherwise destroyed along with cancerous cells. Some cancerous cells selectively absorb a phototoxic dye. These dyes become toxic when exposed to light and hence destroys the cancer cells on exposure to laser. This is known as photodynamic therapy.

(c) Dermatology

Lasers can be used to remove tattoos and skin blemishes. It is also used to remove port-wine birth marks, which appear as a result of concentration of blood vessels in the patch of the skin.

(d) Cardiology

Cholesterol clots and plaques in coronary arteries obstruct the smooth flow of blood and cause malfunctioning of the heart. Lasers have been used to vapourise cholesterol clots and plaques which are deposited in the coronary arteries. The presence of clots and plaques in coronary arteries alters the speed of blood flow. Hence a measurement of the speed of blood could provide an early diagnosis for taking remedial measures. Laser Doppler velocimetry is employed to measure the speed of blood flow.

(e) Gynecology

Blocks in the fallopian tubes are one of the major reasons for the infertility in women. These blocks can be removed with laser. Cysts and tumors in uterus have been cauterized using laser.

(f) Urology

Lasers have been successfully used to remove kidney stones from the bladder.

(g) Dentistry

Lasers have been used to drill out the decayed portion of the tooth or sterilize the decayed tooth cavity. It has been used to fuse the enamel when there is a slight porosity or cracks. Laser brushing of teeth is found to alter the surface texture of the teeth and make it resistant to decay. In dental restorative operations, laser is used for welding gold and other dental alloys with various dental bridge work elements.

(h) Acupuncture

Pain due to disorders of many internal organs of the body travels away from the source of disorder and is experienced at sites remote from the ailing organs. This is because the pain manifests itself due to the irritation of a nerve trunk, its root or its terminal ends. There are a number of areas in the human skin where pain from the organ appears and this is known as *referred pain or reflex pain.* About 700 sites on the skin surface are identified which corresponds to various inner organs. Acupuncture is traditionally performed with sharp pointed needles. Depending on the diagnosis of the medical experts, these needles are inserted into the relevant areas of the skin to overcome pain and cure the disease. In recent years, acupuncture has been tried with laser pulses instead of sharp pointed needles.

19.6 Illustrative Examples

Example 1   In a laser gyroscope, fringe shifted of 5 are seen due to rotation of platform of area $1m^2$, when a laser light of wavelength 600 *nm* is used in it. Calculate the angular speeds of rotating platform.

Sol.   Number of Fringes shifted due to rotation of platform

$$\Delta n = \frac{4A\omega}{c\lambda}$$

$$\omega = \frac{c\lambda\Delta n}{4A}$$

Given $\lambda = 600$ nm $= 6\times10^{-7}$m, A $= 1m^2$, $\Delta n = 5$ and c $= 3\times10^8$m/s

$$\omega = \frac{3\times10^8 \times 6\times10^{-7}\times5}{4\times1}$$

$$\omega = 225 \text{ revolutions/second}$$

Ans.

Example 2   A laser light of wavelength 500 nm is circulating in laser gyroscope along a closed path of area $2m^2$ and rotating with the angular speed of 20 revolutions/second in 5 turns. Calculate the phase shift produced in circulated light by the rotation of turntable.

Sol.   The total phase difference is

$$\phi = \left(\frac{8\pi A\omega}{\lambda c}\right) N$$

Given $\lambda = 500\times10^{-9}$m $= 5\times10^{-7}$m, A $= 2m^2$,

$\qquad$ N $= 5$, $\omega = 20$ rev/s   and   C $= 3\times10^8$m/s

$$\phi = \frac{8\times2\times20\times5}{5\times10^{-7}\times3\times10^8}\pi$$

$$\phi = \frac{1600}{150}\pi = 10.7\ \pi \text{ radius}$$

19.7 Self Learning Exercise

Q.1 What is Gyroscope?

Q.2 Write the Sagnac's formula.

Q.3 Laser gyroscope is based on the……………………… .

Q.4 Write the types of modern gyroscopes.

Q.5 What is L.S.M.?

Q.6 What is LCVD process?

Q.7 Define the word LASIK.

Q.8 What do you mean by laser Radar?

19.8 Summary

This unit starts with the introduction of laser gyroscope. By giving the concept of Sagnac's ring interferometry experiment, the laser gyroscope has been explained in detail. The applications of laser in defence, industry and medicine have also been studied in this unit. Some examples on the above concept are given in the end of the unit.

19.9 Answers to Self Learning Exercise

Ans.1: Gyroscope is a device used to maintain orientation in space during motion and determine the angular rate of its carrying vehicle with respect to a reference frame.

Ans.2: $\Delta n = \dfrac{4A\omega}{c\lambda}$

Ans.3: Sagnac effect

Ans.4: Two types (i) Fiber-optic gyroscope and (ii) Ring laser gyroscope

Ans.5: Laser surface modification

Ans.6: Laser chemical vapour deposition

Ans.7: Laser In-situ keratomileules

Ans.8: In laser radar, light energy is sent out in pulses and the reflected light echo is collected.

19.10 Exercise

Section-A (Very Short Answer Type Questions)

Q.1   Which laser source is used in laser gyroscope?

Q.2   What is beat frequency?

Q.3   What are the advantages with laser gyroscope?

Q.4   Which laser is used for marking purpose?

Q.5   Who discovered the principal on which laser gyroscope is working?

Section-B (Short Answer Type Questions)

Q.6   What is the aim of ring interferometry experiment?

Q.7   Why laser gyroscope is referred a ring laser gyroscope.

Q.8   Which technique is used to measure the speed of blood flow in human body?

Q.9   Why laser soldering is preferred?

Q.10   What is photodynamic therapy?

Section-C (Long Answer Type Questions)

Q.11   Describe the construction and working of laser gyroscope?

Q.12   What do you mean by gyroscope and Sagnac effect. Explain the working of Laser gyroscope.

Q.13   Discuss the various applications of lasers in medicine?

Q.14   How laser can be used in industry applications.

Q.15   Explain the various application of lasers in defense?

19.11 Answers to Exercise

Ans.1: He-Ne laser

Ans.2: Difference between the two frequencies in laser gyroscope experiment.

Ans.3: Light weight, compact in size, lower maintenance and much greater reliability.

Ans.4: Co_2 Laser

Ans.5: French Physicist George Sagnac.

References and Suggested Readings

1. O. Svelto and David C.Hanna, Principals of Laser, Plenum Press, New York, 1976.

2. A.K. Ghatak and K.Thyagarajan, Optical Electronics, Cambridge Press, New York, 1989.

3. W.T. Silfvast, Laser Fundamentals, Cambridge Press, New York, 1996.

4. K.Thyagarajan and A.K. Ghatak, Lasers-Theory & applications, Macmillan, 1981

5. B.A. Lengyel, Introduction to Laser Physics, John Wiley & Sons, New York, 1967.

6. Peter W. Milonni, Joseph H. Eberly, Laser Physics, John Wiley & Sons, New York, 1970